Evolutionary Dynamics
of a Natural Population

Evolutionary Dynamics of a Natural Population

The Large Cactus Finch of the Galápagos

B. Rosemary Grant and Peter R. Grant

The University of Chicago Press
Chicago and London

B. Rosemary Grant is a research scholar in biology at Princeton University. *Peter R. Grant* is professor and director of the Program in Ecology, Evolution, and Behavior at Princeton University.

The University of Chicago Press, Chicago 60637
The University of Chicago Press, Ltd., London
© 1989 by The University of Chicago
All rights reserved. Published 1989
Printed in the United States of America
98 97 96 95 94 93 92 91 90 89   54321

Library of Congress Cataloging-in-Publication Data

Grant, B. Rosemary.
Evolutionary dynamics of a natural population : the large cactus finch of the Galapagos / B. Rosemary Grant and Peter R. Grant.
p.   cm.
Bibliography: p.
Includes indexes.
ISBN 0-226-30590-2 (alk. paper). — ISBN 0-226-30591-0 (pbk.   :
alk. paper)
1. Large cactus ground finch—Galapagos Islands—Evolution.
2. Animal populations—Galapagos Islands.   3. Birds—Galapagos
Islands—Evolution.   I. Grant, Peter R., 1936–   .   II. Title.
QL696.P2438G73   1989
598.8'83—dc19   89-30024
                                                       CIP

This book is printed on acid-free paper.

Contents

List of Illustrations		ix
List of Tables		xiii
Preface		xvii

1 The Variation Problem   1

Introduction 1   Darwin's Finches 1   Theoretical
Framework 10   Scope and Organization of the
Study 11   Duration of the Study 14   Organization
of the Book 15   Summary 15

2 The Island Setting   17

The Physical Environment 17   Climate 19
Vegetation 24   Plant Phenology 27
Animals 34   Age of the Community 38
Summary 41

3 Demography: Survival   42

Introduction 42   Sources and Study Areas 43
Birds of Known Age 43   Birds of Unknown Age 46
The Initial Group of Adults 49   Individual
Cohorts 52   A Synthetic Survivorship Curve 56 Envi-
ronmental Determinants of Survival 57   Age Structure
of the Breeding Population 61
Comparative Demography 63   Summary 67

4 Demography: Reproduction   69

Introduction 69   Timing 70   The Rate of
Reproduction 76   The Amount of Reproduction 80
Effects of Age and Experience 87   Individual
Variation 92   Lifetime Reproductive Success 96
Summary 98

5   Survival and Reproduction   100

Introduction 100   Survival and Reproduction 100
Comparative Demography 106   Summary 113

6   Song and Territories   115

Introduction 115   Song Structure 116   Transmission of
Song 118   Territories and the Role of Song 125 Song
and Breeding 129   Song Type, Mate Choice, and the
Distribution of Territories 132   The Role of Song in
Species, Individual, and Kin Recognition 138   Origin
and Maintenance of the Two Song
Types 143   Summary 145

7   Mate Choice   148

Introduction 148   The Mating System: Mainly
Monogamy 149   Courtship, Pairing, and Parental
Care 150   The Choosing Sex 150   Cues Used in
Mate Choice 151   Fitness Consequences of Mate
Choice 156   Courtship Behavior and Experience 158
Extra-pair Copulations 159   Mate Changes 163
Synthesis 169   Summary 172

8   Phenotypic and Genetic Variation   174

Introduction 174   Morphological Variation 174
Genetic Variation 177   Heritability 178   When Are
Offspring Fully Grown? 178   Estimates of
Heritability 181   Potential Distortions to the
Estimates 182   Genotype-Environment
Correlations 184   The Expression of Genetic
Variation during Ontogeny 186   Genetic Correlations
among Traits 188   Discontinuous Variation: A Beak
Color Polymorphism 189   Summary 195

9   Hybridization and Selection   197

Introduction 197   Introgression 197   Effects of
Interbreeding on Morphological Variance 203
Selection 207   Dry-Season Survival 210
Reproduction 223   Directional and Stabilizing
Selection 227   The Maintenance of Variation 230
Summary 233

10   Population Subdivision and Sympatric Speciation   235

Introduction 235   The Initial Observations 237
Subsequent Observations 240   How the Population

Became Subdivided 242   Sympatric Speciation? 246
Population Subdivision 248   Summary 252

11   The Place of *G. conirostris* in Its Community   253

Introduction 253 Evolutionary History 253
Community Ecology 260   Some Speculations on
Evolution and Competition 273   The Populations on
Española and Gardner 276   Summary 277

12   Summary, Synthesis, and Some Implications for
Conservation   279

Introduction 279   Synthesis 280   Implications for
Conservation 285

Appendixes
  I Scientific Names of the Bird Species 293
  II Annual Summaries of Reproductive
     Characteristics 295
  III Demographic Characteristics of *A* Males and *B*
      Males 298
  IV Effective Population Size and Neighborhood
      Size 303
  V Nestling Beak Color Morphs Produced by *A* Males
     and *B* Males 306

References   307
Author Index   331
Subject Index   337

Illustrations

Color Plates (*following page 44*)

1   *Geospiza conirostris*, the large cactus finch, and beak color polymorphism shown by nestlings
2   Cactus bush in flower and a view of the island from Darwin Bay
3   The appearance of the vegetation in a normal wet season and in a dry season
4   Developing foliage, flowers, and fruits of *Bursera graveolens*
5   *Cordia lutea* flowers and fruits, and flowers of *Ipomoea habeliana*
6   The contrast between a very wet year and a dry year
7   Growth of grasses and sedges early in the wet season of 1987, and extensive growth of *Mentzelia aspera* later in the same year
8   Rampant growth of an *Ipomoea habeliana* vine in July 1983 compared with the remnants of dead stems twelve months later
9   Differences in the effects of rain on the vegetation in two El Niño years
10   The four species of Darwin's Finches on Genovesa
11   *Cordia lutea* leaves stripped by sphingid caterpillars and regenerated two weeks later
12   Caterpillars on *Cordia lutea* and *Croton scouleri* flowers, and a spider

Figures

1.1   The large cactus finch, *Geospiza conirostris*, and its diet 2
1.2   The Galápagos archipelago 3
1.3   Large variation in *conirostris* beak size contrasted with small variation in the beak size of bullfinches (*P. pyrrhula*) from the Azores island of São Miguel 4

1.4 Six different male *conirostris* to illustrate the variety of beak sizes and shapes 8
1.5 Six different female *conirostris* 9
1.6 A model for the maintenance of genetic variation in a population 10
1.7 Measurements 13
2.1 Smooth and rocky lava 18
2.2 Isla Genovesa 19
2.3 The similarity of the habitat on Genovesa as shown by views of the north and the south 20
2.4 Arcturus lake 21
2.5 Daily maximum shade temperatures expressed as weekly averages 22
2.6 Annual rainfall 23
2.7 Monthly rainfall in 1982 and 1983 24
2.8 Similarity of the vegetation on the north and the south sides of Genovesa 26
2.9 *Lantana peduncularis* and *Waltheria ovata* 28
2.10 *Eragrostis cilianensis* 29
2.11 A clump of *Opuntia helleri* cactus bushes 29
2.12 The southeastern arm of Genovesa 30
2.13 The flowers and fruits of *Croton scouleri* 30
2.14 Flowering phenology in 1978 and 1983 31
2.15 A *Bursera graveolens* tree smothered by *Ipomoea habeliana* vines 35
2.16 *Opuntia helleri* cactus smothered by *Ipomoea habeliana* vines 36
2.17 Cactus on the northeast coast that collapsed in the El Niño conditions of 1983 37
2.18 The six principal land bird species on Genovesa 38
2.19 The four principal land bird species other than finches 39
2.20 Weekly counts of caterpillars on 12 plant species in relation to weekly amounts of rain in 1987 40
2.21 Pollen profiles constructed from Arcturus lake sediment cores 40
3.1 The two study areas on Genovesa 44
3.2 Plumage categories 47
3.3 Plumage categories of males in relation to age 48
3.4 Survivorship of four cohorts of males and females 53
3.5 A synthetic survivorship curve for *conirostris*, with survivorship of the temperate zone white-crowned sparrow (*Zonotrichia leucophrys*) shown for comparison 57
3.6 The number of breeding males on study area 1 58
3.7 Nests of *conirostris* 60
3.8 Changes in the age distributions of breeding males 62
3.9 Maximum longevity as a function of size among 15 species of temperate zone finches and sparrows and 5 tropical species 64

4.1   Initiation of breeding in relation to rainfall 71
4.2   Clutch initiation in relation to caterpillar abundance in 1987 72
4.3   *Opuntia helleri* flowers 74
4.4   Sections of an *Opuntia helleri* cactus flower 75
4.5   The decline in caterpillar abundance at the end of three breeding seasons 76
4.6   The mean number of clutches per pair as a function of the amount of rain in nine breeding seasons 81
4.7   Annual variation in the frequencies of different clutch sizes 84
4.8   Seasonal variation in the frequencies of different clutch sizes in 1983 85
4.9   Recruitment from 86 females 96
4.10   A genealogical tree of a successful male parent, No. 6108 97
4.11   Lifetime production of eggs and fledglings by 86 females 98
5.1   The product of survival and reproduction in relation to age $(l_x m_x)$ 104
5.2   Reproductive value in relation to age (V_x) 105
6.1   The subtypes of song *A* and of song *B* 116
6.2   The songs of fathers and sons 119
6.3   Two examples of bilingual birds 120
6.4   The location of fledglings on and off territory in relation to age in days after fledging 121
6.5   A rare example of nonvertical transmission of song 124
6.6   Territories in study area 1 134
6.7   The songs of *conirostris*, *magnirostris*, and *difficilis* on Genovesa 140
6.8   Frequencies of dispersal distances from natal to breeding territories of 33 females and 30 males 142
6.9   Songs of two species of cactus finches compared 144
7.1   Frequencies of dispersal distances by females who changed mates 165
8.1   Composite growth curves for four morphological traits 179
8.2   Offspring values for tarsus length regressed on midparent values 183
8.3   Estimates with 95 percent confidence limits of the slope of the regression of nestling measurements on the midparent values 187
8.4   Frequencies of yellow morphs among nestlings of 97 families 192
9.1   Beak depth frequency distributions of adult male and female *difficilis*, *conirostris*, and *magnirostris* 198
9.2   Probable hybrids on a principal components plot of morphological trait variation 199
9.3   Songs of *conirostris*, *magnirostris*, and a *magnirostris* × *conirostris* hybrid on Genovesa 200
9.4   Songs of *conirostris* on Genovesa and Española, and of *magnirostris* on Genovesa and Santa Cruz 202

9.5   Sexual dimorphism of *conirostris* 204
9.6   *G. conirostris* with an immigrant *fortis* and with an immigrant *scandens* 205
9.7   Components of fitness 208
9.8   *G. conirostris* stripping bark from a *Croton* bush, tearing open an *Opuntia* pad, and cracking small seeds on the ground 211
9.9   Cactus exploited by *conirostris* 213
9.10 *G. conirostris* cracking an *Opuntia* seed 214
9.11 *G. conirostris* exploiting *Opuntia helleri* cactus for food 215
9.12 Termites exposed by the removal of bark from dead *Bursera graveolens* branches 216
9.13 Forces on the beak during biting in two extreme forms of *conirostris* 217
9.14 Decline in the production of *Opuntia* flowers and fruits in the dry period after 1983 219
9.15 Fruits on an *Opuntia* pad 220
9.16 Numbers of finches captured in mist nets in March of several years 231
10.1 Beak characteristics of song *A* males and song *B* males in 1978 238
10.2 Mean beak lengths of song *A* males and song *B* males 242
10.3 Fluctuations in the proportions of yellow morphs among nestlings of *A* and *B* males on study area 1 and study area 2 249
11.1 Española with its satellite island, Gardner 256
11.2 Average beak sizes of *conirostris* on Genovesa and Española, and of three other ground finch species on Daphne Major 258
11.3 *G. conirostris* from Genovesa and Española compared with *magnirostris* on Genovesa 259
11.4 *G. magnirostris* cracking the hard stone of *Cordia lutea* 261
11.5 *G. difficilis* and *Certhidea olivacea* feeding on nectar in the flowers of *Waltheria ovata* 262
11.6 Deciduous forest habitat on Marchena 263
11.7 *G. scandens* on Marchena compared with *conirostris* on Genovesa 266
11.8 *Camarhynchus psittacula* male on Marchena 267
11.9 *Platyspiza crassirostris*, the vegetarian finch, stripping the bark off a twig of *Croton scouleri* on Marchena 268
11.10 Expected population density of finches in relation to beak depth on four islands 269
12.1 Frequency distributions of *conirostris* and *magnirostris* beak depths before and after the long dry period of 1984 and 1985 289
A1   Number of years lived by *A* males and *B* males 298
A2   Frequencies of subtypes of song *A* in different parts of the island in 1987 305

Tables

1.1 Average Coefficients of Variation for Bill Dimensions of Male Ground Finches 4

1.2 Coefficients of Variation for Bill Depths of Male Ground Finches 5

1.3 Coefficients of Variation for Bill Dimensions of Male Sparrows from North and Central America 5

1.4 Coefficients of Variation for Bill Dimensions of Male Island Finches 6

2.1 Monthly Totals of Rainfall 23

2.2 Plants of Genovesa 25

2.3 Mean Percentage Cover of Plant Species in Random Quadrats 32

2.4 Numbers of Seeds and Fruits of the Principal Plant Species 33

3.1 Survival and Recruitment of Birds Banded as Nestlings 43

3.2 Annual Survival of Birds of Known Age 45

3.3 Frequencies of Transitions between Plumage States 49

3.4 Survival of Birds Present as Adults in 1978 50

3.5 Partial Life Table of the 1980 Cohort 55

4.1 Nestling Diets 75

4.2 Incidence of Brood Overlapping 78

4.3 Average Intervals in Days between Successive Broods 78

4.4 Production of Eggs, Nestlings, and Fledglings 82

4.5 Nesting Success 83

4.6 Reproductive Success in Relation to Potential 83

4.7 Reproductive Performance of Experienced and Inexperienced Birds 90

4.8 Correlations between Mean Clutch Size and the Timing of Egg Laying in Successive Years 93

4.9 Repeatabilities of Clutch Size and the Timing of Egg Laying 94

5.1 Survival and Fertility Table for Females of the 1976 and 1980 Cohorts 103

5.2   Reproductive Characteristics of *Geospiza* and Mainland Species 107

5.3   Indices of Lifetime Reproductive Success 109

5.4   Variation in Annual Survival 112

6.1   Measurements of Song Types and Subtypes 117

6.2   Territories of Sons Whose Paternal Song Type Is Known 122

6.3   Responses of Males to Playback of Tape-Recorded Song 129

6.4   Frequencies of Males of Different Song Types 132

6.5   Shared Boundaries between Males of Homotypic and Heterotypic Song Types 136

6.6   Mating Success of Males in Relation to Song Neighborhood 136

7.1   Breeding Status of Males 149

7.2   Mating Success of Males in Relation to Plumage 153

7.3   Mating Success of Males in 1986 155

7.4   Selection Analysis for the Order in Which Males Obtained New Mates 156

7.5   Selection Analysis for the Number of Fledglings Produced by Mated Males 157

7.6   Number of Mates of Nine Females That Bred in at Least Five Years 164

8.1   Morphological Characteristics of Males and Females 175

8.2   Phenotypic Correlations between Morphological Traits 176

8.3   Morphological Parameters for Parents and Their Offspring 180

8.4   Heritabilities of Morphological Traits 182

8.5   Results of a Principal Components Analysis 183

8.6   Genetic Correlations between Morphological Traits 189

8.7   Frequencies of Pink and Yellow Morph Nestlings 191

8.8   Nestling Beak Color Morphs of Parents and Their Offspring 193

9.1   Effects of Suspected Hybrids on Phenotypic Variation 206

9.2   Morphological Comparison of *G. conirostris* with Its Two Sympatric Congeners 206

9.3   Selection on Young Birds Feeding in a Dry-Season Manner, 1984–86 212

9.4   Incidence of *Opuntia* Stigma Damage by Cactus Finches 221

9.5   Selection over the Dry Period, 1983–85 221

9.6   Diets in the Dry Seasons of 1983–85 222

9.7   Sexual Selection in 1984 224

9.8   Components of Adult Female Fitness in Relation to Bill Morphology 225

9.9   Changes in Bill Variation between the Beginning and End of the Study 228

9.10 Selection during the Dry Period, 1984–85, among *G. magnirostris* Males 232

10.1   Morphological Characteristics of Song *A* Males and Song *B* Males 237

10.2   Frequencies of Males That Sang Song *A*, Song *B*, and Both Songs 243

11.1   Minimum Forces of Selection Involved in Some Evolutionary Transitions 258

11.2   Dry-Season Diets of Four Species of Darwin's Finches on Genovesa 260

11.3   Dry-Season Diets of Seven Species of Darwin's Finches on Marchena 264

11.4   Plant Species Whose Seeds Are Eaten by *G. conirostris* on Genovesa and Española 270

11.5   Proportions of Time Spent Foraging for Different Food Items in the Dry Season of 1979 on Genovesa and Española 271

A1   Mean Clutch Sizes 295

A2   Mean Number of Clutches per Season 296

A3   Nest Failures 296

A4   Mean Number of Fledglings per Brood 297

A5   Mean Annual Production of Fledglings per Pair 297

A6   Numerical Changes of *A* and *B* Males on Study Area 1 299

A7   Breeding Success of the Two Song Groups of Males 300

A8   Frequencies of Nestling Beak Color Morphs among Offspring Produced by *A* Males and *B* Males on Areas 1 and 2 306

Preface

Genetic variation in quantitative characters is the raw material for much of evolution. A substantial body of theoretical work deals with the maintenance and significance of such genetic variation. Field studies of the subject have been largely neglected, yet such studies that employ a theoretical framework can be immensely valuable. Not only can they further our understanding of the evolutionary changes that lead to speciation, they can provide information necessary for the conservation of small isolated populations of endangered species in which genetic variation is often depleted.

In 1973 we began a study of the medium ground finch, *Geospiza fortis*, on the small island of Isla Daphne Major, Galápagos. This population has two advantages for the study of variation: it occurs on a small island where it is possible to uniquely band every individual, and it is extremely variable in body size and bill dimensions. It is, however, close to larger neighboring islands, and from them it receives a trickle of immigrants, a few of which stay to breed with the residents and thereby contribute to the genetic variation. We wanted to study at the same time a small and variable population on a well-isolated island where any change in variation would be due to factors intrinsic to the island and not influenced by gene flow from other populations of the same species. The large cactus finch, *Geospiza conirostris*, on Isla Genovesa satisfied these two conditions. First, the population is small and unusually variable in bill dimensions. Second, Isla Genovesa is well isolated, being a low flat island in the extreme northeast of the Galápagos archipelago. A combination of the prevailing southeast trade winds, which tend to deflect westward any migrant leaving the more southerly islands, and the low topography, which limits the visibility of Genovesa, makes it one of the most effectively isolated islands in the archipelago. We studied the population for eleven years, and this book is the result.

We continued for more than a decade because a long-term study is needed to appreciate and interpret the dynamics of a population living in a climatically variable environment. It took us several years to obtain sufficient measurements for estimating the heritability of morphological traits, to document natural selection and hybridization, and to establish some of the factors involved in mate choice. Abundant rain fell in years 6 and 10. Droughts occurred in years 8 and 11. Without them our understanding of the population would have been much less complete.

There are many routes to the same objective. A diversity of approaches to the same problem can work synergistically, so that the integration of the results from different approaches leads to a fuller understanding of both the problem and its solution. In this study we simultaneously examined the ecology, behavior, and genetics of the population, mainly by direct observation. Experiments were used sparingly, partly because the population is entirely natural, and partly because experiments are strongly circumscribed by National Park regulations. Experiments can be extremely valuable, especially when combined with other approaches, but they do have the disadvantage of sometimes causing unforeseen, even permanent, changes, distorting the results of future research work. For example, exchanging eggs between clutches would have helped us to interpret the results of heritability analysis, but the offspring would have imprinted on their foster fathers' song instead of their fathers' song, and this could have altered the future mating structure of the population. If experiments of this sort are to be conducted, they should be carried out on one of the many populations in the world that have already been perturbed.

Studying this population has been a privilege. The array of thirteen species of Darwin's Finches is unique in being the only group (of this size) of closely related species of birds still intact in the archipelago where it evolved. In all other related groups, like the Hawaiian honeycreepers for example, some species have become extinct through human interference of one sort or another. Even within the Galápagos archipelago, the populations of organisms on Isla Genovesa are among the few that have been untouched by human habitation, feral animals, and introduced plants. They deserve to be preserved.

Our research has been carried out with the permission and support of the Dirección General de Desarrollo Forestal, Quito, Ecuador, the Servicio Parque National Galápagos (SPNG), the Charles Darwin Foundation, and the Charles Darwin Research Station. We must single out for special acknowledgment of their considerable help, Miguel Cifuentes and Humberto Ochoa, Intendentes of SPNG; Craig MacFarland, Hendrick Hoeck, Friedemann Köster, and Günther Reck, directors of the Charles Darwin Research Station since 1978; Don Ramos for his efficient and always cheerful help with our supplies; Sylvia Harcourt for administrative assistance at often crucial times; and finally, the many boat captains and their crews who have taken us to and from the islands, but particularly Pancho Castaneda and Bernardo Guttierez. The research has been funded by the National Science Foundation (USA). As part of the terms for receiving support from the National Science Foundation we declare that "any opinions, findings, and conclusions or recommendations expressed in this publication are those of the authors and do not necessarily reflect the views of the National Science Foundation."

In fieldwork we have been ably assisted by David Anderson, Chris Chappell, James Gibbs, Ann Heise, Bill Johnson, Kristen Nelson, Scott Stoleson, James Waltman, Michael Wells, Tom Will, and our daughters Nicola and Thalia. To all of them we express our gratitude for their invaluable help.

We began work on this book when one of us (P.R.G.) was supported by a fellowship from the John Simon Guggenheim Memorial Foundation. We are grateful for this support, and for extra assistance from the Foundation to defray the costs of publishing the color plates. The book has been improved in a multitude of ways by stimulating discussions and suggestions from many people: Robert Curry, Tjitte de Vries, Lisle Gibbs, Henry Horn, Ola Jennersten, Åke and Ulla Norberg, Stuart Pimm, Trevor Price, Gunilla Rosenqvist, Dan Rubenstein, Dolph Schluter, Staffan and Astrid Ulfstrand, Robert Vrijenhoek, and an anonymous reviewer. Ola Jennersten kindly supplied us with many of the photographs. Laurene Ratcliffe, Robert Bowman, Peter Marler, Susan Peters, and Andrea Cohen helped us with the sonagrams, and T. McQuistion identified coccidea parasites. We especially thank Joan Crespi, Barbara Delanoy, and Joan Nielsen for their patience and considerable help in typing the manuscript. Finally, throughout the study we have had the enthusiastic and stimulating friendship and help of our daughters Nicola and Thalia.

1

The Variation Problem

Introduction

A major problem in evolutionary biology is to understand why populations vary to the extent they do in morphological, ecological, behavioral, and physiological traits. This has an important bearing on all other evolutionary issues, since evolutionary change takes place as a result of some variants in a population contributing more offspring to the next generation than others. Understanding variation helps us to understand change.

In this book we describe an eleven-year investigation into the causes and significance of morphological variation in a population of the large cactus finch, *Geospiza conirostris* (Plate 1), on Isla Genovesa in the Galápagos archipelago. Cactus finches are 20–30 g birds that feed on arthropods and seeds, and to a large extent on various parts of prickly pear cactus (*Opuntia helleri;* Fig. 1.1). Our principal concern is with their unusually large variation in beak size and shape, since variation in these traits is relatively easy to understand ecologically. We present background information on the environment and demographic features of the population before discussing the range of genetic, ecological, and behavioral factors responsible for maintaining the particular level of variation. We then place the species in its community setting in order to understand its evolution. We conclude by discussing the implications of this study for understanding the genetic structure of small populations and the problems of conserving them.

Darwin's Finches

Darwin's Finches on the Galápagos Islands (Fig. 1.2) and Cocos Island are related to sparrows and buntings, members of a subfamily (Emberizinae) of the finch family Fringillidae. They comprise a group of six species of ground finches, six species of tree finches, a warbler finch, and the Cocos finch (Grant 1986a). Some populations are extremely variable, especially

1

in beak dimensions (Fig. 1.3). This can be seen by comparing coefficients
of variation for the six species in the ground finch group (Tables 1.1 and
1.2) with coefficients for species of finches and sparrows living elsewhere
(Tables 1.3 and 1.4).

Typical coefficients for species in continental regions are 2–5 percent
(Table 1.3); that is, the standard deviation is about 2–5 percent of the
mean, regardless of whether the mean is small or large. Some populations
of the ground finches (genus *Geospiza*) have coefficients in this range, as

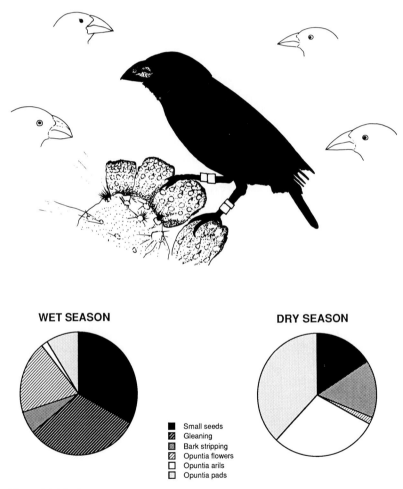

WET SEASON   DRY SEASON

- ■ Small seeds
- ▨ Gleaning
- ▨ Bark stripping
- ▨ Opuntia flowers
- □ Opuntia arils
- □ Opuntia pads

Figure 1.1 The large cactus finch, *Geospiza conirostris*, and its diet. Variation in beak
size and shape is illustrated by outline drawings of photographed birds.

do the tree finches (Grant 1986a), but many others have larger coefficients (Table 1.2). Table 1.1 shows the averages of the coefficients for the various populations of each ground finch species. Average coefficients for *fuliginosa, difficilis,* and *scandens* are similar to the coefficients for continental species, whereas *magnirostris, fortis,* and *conirostris* have distinctly larger coefficients. Many of the differences between populations are statistically significant (*F* tests; Grant et al. 1985) and therefore cannot be attributed to chance.

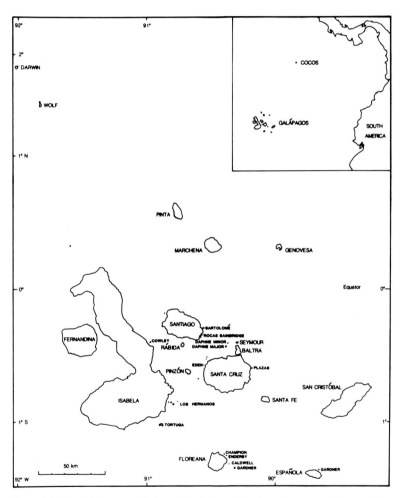

Figure 1.2 The Galápagos archipelago. Populations of *conirostris* occur on Genovesa in the northeast, and on Española and its satellite Gardner in the south. From Grant and Grant (1982).

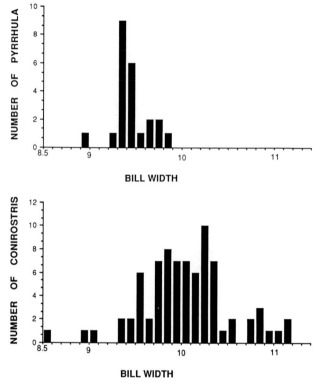

Figure 1.3 Large variation in *conirostris* beak size contrasted with small variation in the beak size of bullfinches (*P. pyrrhula*) from the Azores island of São Miguel; male specimens only.

Table 1.1 Average Coefficients of Variation for Bill Dimensions of Male Ground Finches (*Geospiza* spp.), and Individual Coefficients for *G. conirostris*

| Species | Coefficients of Variation | | | No. Populations |
	Length[1]	Depth[2]	Width[1]	
G. difficilis	4.66	3.70	5.73	6
G. scandens	5.27	4.29	5.78	9
G. fuliginosa	5.36	4.54	5.24	16
G. magnirostris	5.68	6.54	5.98	6
G. fortis	6.11	7.04	7.73	12
G. conirostris	6.06	7.23	7.41	3
Genovesa	5.92	7.38[3]	7.56	
Española	7.19	6.91[3]	7.86	
Gardner	6.75	7.39[3]	6.84	

Source: Grant et al. (1985)[1], Grant (1986a)[2], and Lack (1945, 1947)[2,3].

Note: See Appendix I for common names of the finches.

Table 1.2 Coefficients of Variation for Bill Depths of Males from Various Populations of Ground Finches (*Geospiza* spp.)

Species	Coefficients of Variation			No. Populations
	Minimum	Maximum	Average	
G. difficilis	3.40	4.16	3.70	6
G. scandens	3.20	4.96	4.29	9
G. fuliginosa	2.60	5.73	4.54	14
G. magnirostris	4.34	9.17	6.54	7
G. fortis	3.38	10.30	7.04	12
G. conirostris	6.91	7.39	7.23	3

Source: calculated from data in Lack (1947) and taken from Grant (1986a).

Note: Coefficients for other dimensions are given in Grant et al. (1985). Individual coefficients for the three populations of *conirostris* are given in Table 1.1. See Appendix I for common names of the finches.

The most variable populations are *fortis* on Isla Santa Cruz and *magnirostris* on Isla Darwin. The *fortis* population has a coefficient for bill depth of 10.30 (Table 1.2). This is more than twice as large as the largest coefficient for this dimension of any population on the mainland. The population of *magnirostris* on Isla Darwin appears to be even more variable, with a coefficient for bill depth of 13.13 (Bowman 1961). However there are uncertainties about whether the sample of 16 males and 8 females belongs to one species or two (Grant et al. 1985), and for this reason we have not used it in the calculations of the averages in Table 1.1

Table 1.3 Coefficients of Variation for Bill Dimensions of Male Sparrows from North and Central America

Species	Coefficients of Variation			Samples
	Length	Depth	Width	
Song sparrow[1]	4.25	2.83	3.59	147
Fox sparrow[2]	4.90	4.40	3.70	(7)
Bridled sparrow[3]	2.33	4.58	5.19	44
Cinnamon-tailed sparrow[3]	4.08	3.33	6.83	47
Savannah sparrow[4]	4.64	4.62	—	46
Dark-eyed junco[5]	4.03	4.23	—	125
Mexican junco[5]	4.27	5.05	—	105
White-crowned sparrow[6]	3.92	—	—	42
Seaside sparrow[7]	3.49	—	—	98
Le Conte's sparrow[7]	4.64	—	—	87
Sharp-tailed sparrow[7]	3.33	—	—	67

Source: Smith and Zack (1979)[1], Zink (1986)[2], Wolf (1977)[3], Stobo and McLaren (1975)[4], Miller (1941)[5], Banks (1964)[6], and Murray (1969)[7].

Note: Average coefficients are given for seven populations (denoted by parentheses) of the fox sparrow. See Appendix I for scientific names of the species.

or in Table 1.2. If only one species is involved, it is the most variable population of birds known anywhere in the world.

There has been speculation about whether bird populations on islands are likely to be generally more variable or less variable than their mainland counterparts (Van Valen 1965; Grant 1967). Individual cases of high and low variation can be identified in the tables, but no strong trend in either direction emerges from a comparison of the coefficients of mainland (Table 1.3) and island (Table 1.4) populations. Darwin's ground finches are exceptionally variable in comparison with them all. They vary more than other species of finches on islands in general (e.g., Fig. 1.3), and other

Table 1.4 Coefficients of Variation for Bill Dimensions of Male Finches on Various Islands around the World

Islands and Species	Coefficients of Variation			
	Length	Depth	Width	Samples
Hawaii[1]				
Laysan finch	5.44	4.16	4.51	31
Kona grosbeak*	3.37	2.05	2.00	16
Greater koa-finch*	3.21	3.14	2.93	20
O'u	6.41	4.16	5.26	34
Palila	4.36	3.84	3.50	24
Tristan da Cunha group[2]				
Rowettia goughensis	7.90	5.63	3.43	31
Nesospiza acunhae	5.19	5.73	3.58	16
Azores[3]				
Bullfinch	4.09	2.87	2.07	23
Canaries[3]				
Blue chaffinch	2.79	3.70	2.67	38–93(2)
Canaries, Madeira, and Azores[3]				
Chaffinch	4.04	4.29	3.48	15–180(11)
California, Channel[4]				
House finch	5.02	4.52	4.14	27–283(4)
Tres Marías[5]				
Cardinal	4.18	—	3.59	44
Lesser goldfinch	2.16	—	2.19	27
Aleutians[6]				
Gray-crowned rosy finch	3.11	2.12	—	32
Sable[7]				
Ipswich sparrow	4.79	4.84	—	51
Guadalupe[8]				
Dark-eyed junco	4.01	4.77	—	95
Vancouver[9]				
White-crowned sparrow	4.29	—	—	61

Source: unpublished[1], Abbott (1978)[2], Grant (1979)[3] with Bullfinch coefficients corrected, Power (1983)[4], Grant (1965)[5], Johnson (1977)[6], Stobo and McLaren (1975)[7], Miller (1941)[8], and Banks (1964)[9].

Note: Average coefficients are given for some species, together with the range of sample sizes, with the number of populations in parentheses. An asterisk signifies the species is extinct. See Appendix I for scientific names of the species.

species of finches on tropical islands in particular. An interesting comparison is made with the Hawaiian honeycreeper finches, which are cardueline finches in the subfamily Fringillinae according to the classification of Sibley et al. (1988). These finches have undergone a more spectacular adaptive radiation and diversification than Darwin's Finches, and they include species that are approximate ecological analogues of the seed- and fruit-eating members of Darwin's Finches. Their coefficients of variation are similar to those of continental species and hence lower than the most variable Darwin's Finch populations.

THE CHOICE OF SPECIES AND LOCATION FOR STUDY

The larger the scale of a phenomenon the easier it is to dissect for analysis. On this basis giant polytene chromosomes in the salivary glands of *Drosophila* flies were chosen many years ago for intensive cytogenetic research. Along the same lines, the obvious candidate for our study was *magnirostris* on Isla Darwin. The island is very remote (Fig. 1.2), however, and it is apparently impossible to climb. All museum specimens were collected nearly one hundred years ago on a small talus slope. Dolph Schluter (pers. comm.) visited this slope in 1980 and saw two large finches, scarcely an encouragement for a population study. The second obvious choice was *fortis* on Isla Santa Cruz. This suffers from a different disadvantage. The habitat has been much disturbed by human activity, which may have contributed to the variation, obscured the causes, or both.

Apart from the remote islands of Darwin and Wolf, only two of the major islands have not been disturbed by human modification of the habitat or by the introduction of mammals, fire ants, and exotic plants. These are Fernandina and Genovesa. Both islands support variable finch populations, but Fernandina is extremely large and not easy to work on because the terrain is difficult to traverse. We chose Genovesa.

Genovesa supports variable populations of two ground finch species: *magnirostris,* the large ground finch, and *conirostris,* the large cactus (ground) finch. *G. magnirostris* has a more powerful bill and jaw musculature than *conirostris* (Bowman 1961), and is more efficient at removing the bands we placed on their legs to mark them individually. Once this discovery had been made we concentrated on *conirostris*. It is one of the most variable species of Darwin's Finches (Figs. 1.4 and 1.5; Table 1.1). For example, the coefficient of variation for bill depth is exceeded by only ten others, mainly those of *fortis,* among the fifty-two individual coefficients that have been calculated for this dimension. The Genovesa population is as variable as the other two conspecific populations far away in the southern part of the archipelago (Fig. 1.2).

Long-term studies have been carried out on two other species of ground finches on Isla Daphne Major (Grant 1986a). They are *G. fortis,* the

Figure 1.4 Six different male *conirostris* to illustrate the variety of beak sizes and shapes. Photos by P. R. Grant.

Figure 1.5 Six different female *conirostris*. Photos by P. R. Grant.

medium ground finch, and *G. scandens,* the cactus finch. We will have several occasions to compare *conirostris* with these two species. The largest difference between the islands is their location. Daphne is much closer to another large island than is Genovesa (see Fig. 1.2) and is consequently more accessible to wandering birds. This is relevant to the maintenance of population variation.

Theoretical Framework

Variation in continuously varying characters like beak size is governed by genetic and environmental factors. For example, adult birds with certain genetic constitutions are more likely to be large than are others with different combinations of genes. Environmental factors determine how the genetic potential of individuals becomes expressed in the course of development. We would like to understand the full range and interplay of genetic and environmental effects on phenotypic variation.

The large amount of phenotypic variation in the *conirostris* population may reflect a large amount of underlying genetic variation. Whether it does or not, it is important to know the factors which determine and maintain a particular level of genetic variation. Figure 1.6 summarizes the major influences as we understand them. Hidden genetic variation is translated

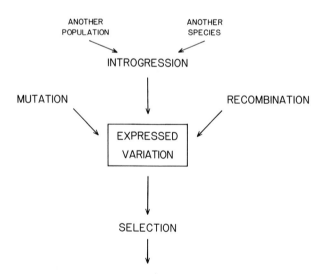

Figure 1.6 A model for the maintenance of genetic variation in a population. Genetic variation is inherited from the previous generation. Recombination exposes hidden variation which is stored in negatively correlated and linked combinations of alleles. Variation is enhanced by mutation and introgression and reduced by most forms of selection, as indicated by the arrows. Adapted from Grant (1986a).

into expressed genetic variation each generation through the recombination of alleles and their expression in phenotypic variation through development. Against this background of recurring recombination, two counteracting processes operate. New alleles enter the population, and new and old ones are lost. The amount of genetic variation in the population at a given time is determined by the resolution of these opposing processes. They may, or may not, be in a state of balance.

Alleles enter a population in two ways: (1) by mutation, and (2) by introgression of genes, either through interbreeding with immigrant conspecifics from another population, or by hybridization with either a sympatric or an immigrant allopatric species. Alleles are lost through most forms of selection and through genetic (random) drift. Thus the *conirostris* population could be unusually variable because the rate of genetic input through mutation and/or introgression is unusually high, or the rate of genetic depletion through selection or drift is unusually low (Grant and Price 1981). It could also be unusually variable for strictly environmental reasons; the set of environmental factors to which the birds are exposed during growth and development may be unusually variable, in time and space, and may have lasting effects on adult phenotypes.

Scope and Organization of the Study

The theoretical framework sets the scope of the study and helps us to identify major factors in need of measurement. The genetic structure needs to be known. To what extent is phenotypic variation governed by the underlying genetic variation? What are the rates of mutation and of introgression from different sources? What is the breeding structure of the population? What are the selective forces acting on the population, and why do they act? What are their strengths, directions, and frequencies? Do they vary in time and space? Do they act equally or unequally on different age and sex classes of individuals in the population?

Mutation, like recombination, is not easily studied in the field. The rate of mutation at loci that contribute to the determination of polygenic, quantitative traits is important to the maintenance of variation (Lande 1976), but to an argued extent (Turelli 1984, 1985; Gimmelfarb 1986), in the absence of introgression. In the presence of introgression it is likely to be relatively unimportant (Grant and Price 1981). Therefore we need to know if introgression is occurring, and if so, at what rate. It is a relatively straightforward matter to determine this. The identities of breeding birds and their offspring are needed, the first to determine if individual *conirostris* breed with any birds other than locally born conspecifics, the second to find out if their offspring survive to breed with *conirostris*.

It is also a relatively straightforward matter to determine the degree to which phenotypic variation in the population reflects underlying genetic variation. The degree of resemblance between offspring and their parents in a trait can be taken, with appropriate safeguards against bias (e.g., van Noordwijk et al. 1980; Boag and van Noordwijk 1987), as a measure of the proportion of the total variation attributable to the additive effects of genes; in other words, the heritability of the trait (Falconer 1981). Measurements of breeding adults and of their offspring when fully grown are needed for this determination.

The operation of natural and sexual selection on phenotypic variation can be detected and the effects measured relatively easily, providing a substantial sample of birds is marked in such a way that individuals can be recognized and their fates determined by nearly continuous monitoring.

To obtain the required measurements we have done the following. Each year since 1978 we have put up mist nets to capture birds in an area around Bahía Darwin on the southern side of Genovesa; this constitutes about 10 percent of the total area of the island occupied by breeding birds. A numbered metal band and three, colored, plastic (polyvinyl chloride) bands have been placed on the legs of each bird we captured. The color and positions of the ten plastic bands were coded to correspond to the numbers. Plumage and beak color were recorded, the birds were weighed, and measurements of wing length, tarsus length, beak length (upper mandible), total beak depth, and beak width (upper and lower mandibles separately) were taken (Fig. 1.7). The birds were then released, unharmed, and observed. They occasionally remove one or more of the colored bands, so upon recapture they were rebanded, if necessary, as well as remeasured.

Offspring were banded as nestlings but measured as adults. An attempt was made to find all nests in the study area in every breeding season. The fates of the nests were then followed from the day the first egg was laid or soon after until the day the last nestling fledged. The owners of the nest were identified. We took care to avoid visiting nests both at the time of hatching, in case this caused desertion, and when owls and mockingbirds (potential predators) were in the vicinity. The nestlings were weighed and banded on day 8, counting the day of hatching as day 0. This is the last day they can be handled without the risk of inducing premature fledging. In 1978 only, nestlings were also measured every other day in all dimensions except lower mandible width to establish the pattern of nestling growth. Offspring were frequently observed after fledging. A small proportion of them were captured in mist nets at various ages, and they were all measured at this time. This enabled us to determine when growth ceases. Altogether we banded approximately 2500 adults and young.

It is much more difficult to interpret selection, if it occurs, than to detect it. For one thing we need to know as much as possible about the structural and behavioral properties of birds, and about the environmental forces

Figure 1.7 Measurements. Upper left: bill length. Lower left: bill width. Upper right: bill depth. Lower right: tarsus length. The leg carries a black plastic band and a numbered metal band. Photos by P. R. Grant.

impinging upon them, in order to identify the reasons for the relative success of some individuals and the relative failure of others. That is a tall order. If we are fortunate enough to identify some possible causal links, we cannot proceed further to test them by experimental methods, because experimental manipulations of either the birds or their food would disturb, thereafter, the natural processes we are trying to understand. Some discoveries in the course of this study could not have been made, or would have been impossible to interpret, had they been preceded by experimental manipulation of the birds or their food supply. Galápagos National Park regulations severely limit the scope of experiments in order to maintain the biota as close to the natural state as possible.

Given these difficulties and constraints, we placed our emphasis on documenting three things: the breeding success of known individuals, as described above, their survival, and their feeding habits. In a parallel study on Isla Daphne Major, it had been shown previously that individuals of different beak sizes in the variable population of medium ground finches (*fortis*) fed on different foods, to some extent, and exhibited different efficiencies in dealing with common food items (Grant et al. 1976). Therefore we followed the protocol established in that study of recording the diets of as many individual *conirostris* of known identity as possible: initially by observing individuals for a maximum of 300 seconds, later by simply recording the first identified food item seen to be consumed by a banded individual on morning surveys.

To place the observations of feeding in the context of food supply, we estimated seed, fruit, and flower abundance in a 23,000 m^2 study plot established by Abbott et al. (1977) in 1973. At each sampling time 50×1 m^2 quadrats were chosen randomly. Seeds, fruits, and flowers on plants in each quadrat were counted, as were seeds and fruits on the ground in two 25-cm^2 subquadrats. The percent cover of plants within the quadrat was also noted. The samplings were done when our other activities permitted in 1978 and 1983–85. In all years the presence of leaves, buds, flowers, fruits, and caterpillars was recorded once a week on the twelve most abundant species of plants, and at the same time numbers of buds, flowers, and fruits were counted on ten marked bushes of *Opuntia helleri*. In 1978, late 1983, and 1987 caterpillars were counted once a week on one hundred leaves and adjacent stems and flowers on ten bushes or trees of the twelve most common plant species. The data are sufficient for making broad comparisons between times of contrasting conditions, but they cannot be used as an estimation of the complete food supply because they do not include many types of arthropods exploited by finches, especially spiders when the finches are breeding and various insect larvae and termites when they are not.

Duration of the Study

Field studies such as ours should continue until the answers have been obtained to all the major questions. Two things became apparent after the first field season. First, this would have to be a long-term study, preferably spanning two generations at least. There was, surprisingly, almost no adult mortality in the first year, and almost no juvenile survival. Second, we would need to study the population under contrasting climatic conditions. We had the bad luck to start after a very dry year (1977) when some interesting changes in the structure of the population probably took place, changes that we could only infer from some unusual observations made in 1978. On the other hand, we had the good fortune to witness extremely wet conditions (1983 and 1987) and extremely dry conditions (1985 and 1988) after having become acquainted with "normal" processes in the population. We therefore ended the detailed study in 1988, a little more than ten years after the study began. By this time the effects of the 1985 drought had passed to the next generation.

Organization of the Book

The content and sequence of the chapters reflect the major components of our study, that is, ecological, behavioral, and morphological aspects of the evolution of a single species. We start with general features of the environment and the population, and follow with particulars of population structure and morphological variation.

Chapter 2 provides a description of the island environment. It shows that climatic conditions fluctuate enormously from year to year. These fluctuations have strong effects on finch demography, which forms the substance of the next three chapters. Annual survival differs among cohorts and age groups (Chapter 3), and annual reproduction varies even more (Chapter 4). Birds breed repeatedly for several months in wet years and do not breed at all in droughts. In Chapter 5 we integrate survival and reproduction, then place the features of *conirostris* in perspective by comparing them with passerine birds, especially finches, in other regions.

Much of the rest of the book explores how environmental fluctuations affect the lives of finches. Chapter 6 deals with territory and song. This leads logically to a consideration of the cues birds use in choosing mates (Chapter 7) and the bearing these have on population structure. From the ecology of populations and the behavior of individuals, we proceed to morphological variation and the degree to which it is governed by underlying genetic variation (Chapter 8). We demonstrate the relationship between beak shape and feeding habits, and then examine how variation in beak dimensions is influenced by natural selection and by hybridization (Chapter 9). The potential microevolutionary consequences of selection and population structure, and in particular the potential for sympatric speciation, are considered in Chapter 10. The next chapter (11) uses all that we have learned about population variation to explain why the cactus finch is a cactus finch; to place the species in a community setting in order to understand its evolution. The final chapter (12) first summarizes the major findings of the study, and then discusses the implications for the genetic structure of small populations and the problems in conserving them.

Summary

Some populations of Darwin's Finches are remarkably variable in beak size and are thus suitable for addressing the general evolutionary question of why natural populations are as variable as they are. In this book we present the results of an eleven-year investigation into the causes of variation in a population of the large cactus finch, *Geospiza conirostris,* on Isla Genovesa, Galápagos. We chose this population for three reasons: it is one of the most variable populations of Darwin's Finches; it is strongly isolated geographically; and its habitat has never been disturbed by human activity.

Variation in continuously varying traits is governed by genetic and environmental factors. Genetic variation is enhanced by new alleles introduced through mutation and introgression; it is diminished by most forms of selection and by drift. The level at which genetic variation is maintained is set by a balance between these opposing processes.

This conceptual framework helps to identify the major factors in need of measurement. Mutation cannot be measured in a field study like ours, but this omission is not important if some degree of introgression is occurring. We measured birds and marked them individually, recorded their diets, measured their food supply, counted the offspring they produced, and registered when they died. These data enabled us to: (1) estimate the amount of genetic variation underlying the observed variation in bill dimensions, (2) establish patterns of differential success in terms of reproduction and survival, (3) determine the incidence of interbreeding with other populations, (4) document the occurrence of natural and sexual selection, and (5) interpret selection in terms of fluctuating environmental conditions.

2

The Island Setting

The Physical Environment

Genovesa is one of the most isolated islands in the Galápagos archipelago (see Fig. 1.2). The nearest island, Marchena, is approximately 50 km to the west, and the next nearest islands, Pinta, Santiago, and Baltra, are all about 80–90 km away. The most likely source of organisms reaching the island by passive transport is none of these: it is San Cristóbal to the southeast, because for most of the year the ocean surface current and the prevailing, occasionally strong, winds come from this direction (Houvenaghel 1984). The distance between Genovesa and San Cristóbal is a little more than 125 km.

The second distinctive feature of the island is its low elevation (maximum 76 m) in relation to its size, given as 17.35 sq km by Wiggins and Porter (1971) and 14.1 sq km by Black (1974); no larger island is lower, but several smaller islands are higher. Colonization could be affected by the inability of organisms to see the island until relatively close to it (Plate 2).

All Galápagos islands are volcanic in origin, and the oldest may have been formed 4–5 million years ago (Cox 1983). Genovesa is relatively young. Potassium-argon dating has placed its origin as no more than 1 million years ago, and the magnetic structure of its rocks suggests a date of about 750,000 years (Cox 1983). The island is heavily faulted. Walking across the island one encounters large areas of flat, generally smooth, platelike lava over much of the northern part (Fig. 2.1a), and more boulders in a faulted and dissected topography in the southern part (Fig. 2.1b). The faults are aligned with the contours of two caldera-like formations, one in the center of the island, like a hole in a biscuit, the other constituting a ''bite'' out of the southern half (Figs. 2.2 and 2.3). The center one is filled with Arcturus lake (Fig. 2.4; Beebe 1926), which is rich in minerals, and the southern one is Bahía Darwin.

Figure 2.1 Upper: smooth lava. Lower: rocky lava. Photos by P. R. Grant.

Climate

Despite lying almost on the equator, the island has a markedly seasonal climate. Seasonality arises from a regular alternation in the influence of two water masses; a cool one coming from the coast of Peru and flowing northwestward, and a warm one from the north that flows southeastward (Colinvaux 1984; Houvenaghel 1984).

In a typical year the warm water influence starts at the beginning of a calendar year and lasts for about four or five months. Air temperatures rise, cumulus clouds develop in an otherwise clear sky, and precipitation occurs sporadically as heavy showers. This is the wet season. Later the warm waters recede to the north, air temperatures decline, and rain no longer falls heavily. At this time of year, from about May to December, cloud cover may be extensive and precipitation takes the form of a light misty rain, known as garúa. It scarcely amounts to anything recordable in a rain gauge.

A biologically most important feature of the climate is that no two years are the same. Variation among years in temperature regime is detectable

ISLA GENOVESA

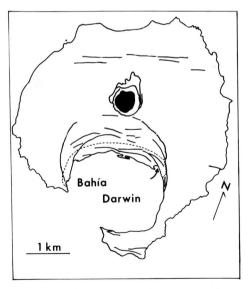

Figure 2.2 Isla Genovesa. Cactus is present everywhere except on the southeastern peninsula. Study area 1 is shown by a broken line, and the arrow points to the beach and adjacent seed-sampling area. Arcturus lake is shown in black, and several faults are indicated by unbroken lines.

Figure 2.3 The similarity of the habitat on Genovesa is shown by views of the north (upper) and the south (lower). See also Fig. 2.8. Photos by P. R. Grant (upper) and O. Jennersten (lower).

Figure 2.4 Arcturus lake. Photos by P. R. Grant.

but moderate (Fig. 2.5). Annual variation in rainfall is extreme (Fig. 2.6; Table 2.1) and unpredictable (Grant and Boag 1980).

In some years rain is both exceptionally heavy and prolonged. Unusually heavy rainfall is associated with a combined meteorological and oceanographic phenomenon known as the El Niño–Southern Oscillation, or ENSO (Philander 1983; Cane and Zebiak 1985). El Niño is the name given to unusually warm surface water of low salinity which usually appears in December or January along the coasts of Peru and Ecuador and then extends westward to the Galápagos. In other years, when the pendulum of atmospheric conditions swings across the Pacific Ocean to its opposite extreme, sea and air temperatures in the Galápagos are low, and little or no rain falls in the usually wet season. These are years of drought.

El Niño events occur irregularly at intervals of two to eleven years, with an approximate frequency of once every three to five years in the eastern Pacific (Graham and White 1988) and perhaps at slightly longer intervals in the archipelago. This last estimate, as well as other comments on typical patterns of temperature and rainfall variation, is based on climatic observations registered at various sites in the archipelago since 1950 (Grant and Boag 1980; P. R. Grant 1985). Rainfall maxima occurred in 1953, 1957, 1965, 1972, 1975, 1983, and 1987. Records on Genovesa span one decade only, and even then they are incomplete. Nevertheless they encompass the most extreme fluctuations in annual rainfall known for the Galápagos; the wettest conditions on record, in 1983 (Fig. 2.7), and the driest, two years later (Table 2.1). A second El Niño event occurred in 1987.

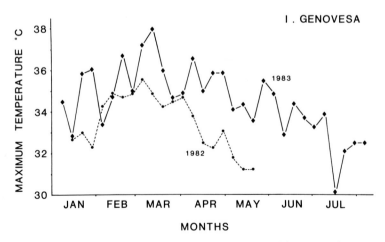

Figure 2.5 Daily maximum shade temperatures expressed as weekly averages for two years of contrasting conditions. 1983 was an El Niño year; 1982 was fairly dry (see Fig. 2.7). From Grant and Grant (1985).

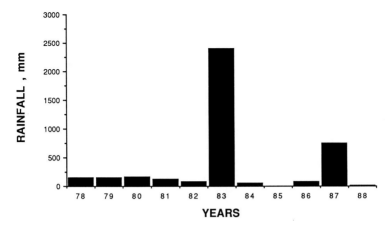

Figure 2.6 Annual rainfall. Rain falling in late 1982 has been combined with 1983 rain (see also Table 2.1). Records for 1985 are not missing; no rain fell that year.

The El Niño event of 1982–83 that gave rise to the extraordinary amount of rainfall has been described as the strongest oceanographic warming trend off the coast of South America this century (Cane 1983). It is probable that more rain fell on the Galápagos than at any other time this century (P. R. Grant and B. R. Grant 1985, 1987). Heavy rains began in the northern part of the archipelago in early November 1982, so Genovesa was probably one of the first islands to receive them. No one was present on the island then, but we were there on the last day, 21 July 1983, when 153 mm of rain fell in less than 24 hours. This amount, exceptional even

Table 2.1 Monthly Totals of Rainfall in Millimeters

Year	(Dec)	Jan	Feb	Mar	Apr	May	Jun	Jul	Total
1978	—	>8	43	24	78	0	—	—	>153
1979	—	?	?	?	?	?	—	—	>150
1980	—	25	116	0	23	0	—	—	164
1981	—	1	15	104	1	0	—	—	121
1982	—	0	35	47	0	0	—	—	82
1983	>150	375	65	210	240	398	465	505	>2408
1984	2	0	11	46	0	0	—	—	59
1985	—	0	0	0	0	0	—	—	0
1986	—	20	22	34	0	3	—	—	79
1987	—	165	200	96	244	44	0	0	749
1988	—	3	10	0	0	0	—	—	13

Note: December is placed in parentheses to indicate that it refers to the previous year. Blanks indicate no records. The rain gauge was situated at the top of the beach in Bahía Darwin. Rainfall was not recorded in 1979, but a known 4–5 month breeding season is consistent with a minimum total of 150 mm.

by El Niño standards, is roughly equal to the total rainfall in the wet season of the other *wettest* years!

Vegetation

The island is floristically simple. Sixty-three species of plants have been recorded, but most are rare. Two were probably brought to the island by visitors in the El Niño year of 1983 only to die and disappear within a year. Another, of similar origin and fate, was recorded in 1980. Ignoring these casuals, the total stands at sixty.

Spatial heterogeneity of the environment can have an important influence on the structure of animal populations. For *conirostris* the environment is relatively homogeneous and simple. A single type of vegetation covers the island quite uniformly (Hamann 1981). A representative sample of quantitative data from repeated sampling of the study area (Abbott et al. 1977; P. R. Grant and B. R. Grant 1980, 1985, 1987) gives a reliable picture of the vegetation over the whole island (Table 2.2), except that *Chamaesyce viminea* is distinctly commoner across much of the island than in the study area. The vegetation type is low, dry-deciduous forest (Fig. 2.8). Only six species of plants exceed 2 m in height. These are *Bursera graveolens, Cordia lutea, Croton scouleri, Opuntia helleri* (rarely above 2 m), *Rhizophora mangle,* and *Erythrina velutina* (a single individual). Four to five meters is a typical maximum canopy height for the dominant tree species, *Bursera graveolens.*

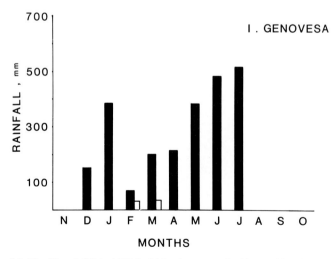

Figure 2.7 Monthly rainfall in 1983 (solid bars) contrasted with monthly rainfall in 1982 (open bars). From Grant and Grant (1985).

Table 2.2 Plants of Genovesa

| | 1978 | | | 1983 |
	Jan	Mar	Nov	Jul
Cordia lutea	36	40	38	42
Croton scouleri	70	56	58	68
Bursera graveolens	70	66	68	42
Opuntia helleri	26	20	20	16
Lantana peduncularis	4	4	2	26
Waltheria ovata	14	6	8	14
Heliotropium angiospermum	X	X	X	22
Sida salviifolia	—	—	—	4
Cacabus miersii	—	2	—	4
Ipomoea linearifolia	X	2	X	X
Ipomoea habeliana	2	X	2	2
Merremia aegyptica	—	—	—	X
Cryptocarpus pyriformis	X	4	X	X
Chamaesyce amplexicaulis	2	X	X	X
Chamaesyce recurva	X	X	X	2
Chamaesyce viminea	6	2	2	2
Mentzelia aspera	—	X	—	—
Abutilon depauperatum	—	2	—	X
Portulaca howelli	X	X	X	X
Cyperus anderssoni	X	X	X	—
Cyperus confertus	X	X	X	18
Eragrostis ciliaris	—	X	—	—
Eragrostis cilianensis	—	6	—	2
Trichoneura lindleyana	4	X	—	8

Note: The first six species are trees or shrubs, the remainder are annual or perennial herbs, sedges, and grasses. The frequency of occurrence of species (%) in 50 random quadrats of 1 m^2 each in the vegetation study plot on the south side of the island (see Fig. 2.2) is given in the upper part of the table. X = presence in the vegetation study area but not in the quadrats. Blanks signify the species was not recorded in the study area at a particular time. From Grant and Grant (1980, 1985), with corrections and additions. Superscripts refer to species recorded by Wiggins and Porter (1971) or contributors to the official list at the Charles Darwin Research Station but not by us[1], and species probably brought to the islands by human visitors and recorded in only 1980[2] and 1983.[3]

Additional species not present in the vegetation study area: *Rhizophora mangle, Sesuvium edmonstonei, Trianthema portulacastrum, Amaranthus sclerantoides, Amaranthus squamulatus, Mollugo flavescens, Mollugo snodgrassii, Boerhaavia caribaea, Tiquilia fusca, Heliotropium curassavicum, Blainvillea dichotoma, Pectis tenuifolia, Cuscuta acuta, Cuscuta gymnocarpa, Evolvulus simplex, Physalis pubescens, Borreria ericaefolia, Brachycereus nesioticus, Chamaesyce abdita, Chamaesyce nummularia[1], Desmodium decumbens, Desmodium procumbens, Desmodium canum, Ipomoea pes-caprae, Erythrina velutina, Rhynchosia minima, Stylosanthes sympodiales, Tribulus cistoides, Antephora hermaphrodita[1], Aristida repens,[1] Aristida subspicata, Cenchrus platyacanthus, Distichlis spicata, Panicum fasciculatum, Sporobolus virginicus, Digitaria horizontalis[3], Eleusine indica[3], Cassia sp.[2],* and *Polypodium insularum.*

Figure 2.8 Similarity of the vegetation on the north (upper) and the south (lower) sides of Genovesa. At both sites the principal elements are *Bursera* trees, *Opuntia* cactus, and a ground cover of *Chamaesyce viminea*. Photos by P. R. Grant.

The remaining species are shrubs or herbs. Three of them, together with young members of the tree species, dominate the shrub layer. These are *Waltheria ovata, Lantana peduncularis* (Fig. 2.9), and *Chamaesyce viminea.* The seasonally abundant herbs include several species of grasses and other seed-bearing plants of importance to finches (Fig. 2.10). Altogether less than two dozen species of plants are common enough to have much influence on the finches. This has been a factor of great convenience. An appreciation of the small number of common species can be gained from Table 2.2.

The breeding and feeding activity of *conirostris* is centered on cactus, as implied by its vernacular name (Fig. 2.11). Cactus is patchily distributed on a local scale, such as units of 100 m^2, but on a larger scale it is distributed approximately uniformly throughout the island. The only part of the island lacking cactus, and breeding cactus finches, is the southeastern peninsula (Fig. 2.12).

Plant Phenology

In this seasonally arid environment, where moisture retention is low because the soil is shallow and poorly formed, and where evaporation is high owing to intense sunlight, convection air currents, and winds, all but eight species of plants either die or lose their leaves in the dry season. The evergreen perennials are *Rhizophora mangle* (mangrove), *Sesuvium edmonstonei, Waltheria ovata, Cryptocarpus pyriformis, Chamaesyce viminea, C. amplexicaulis,* and the two cacti, *Opuntia helleri* and *Brachycereus nesioticus.* The loss of leaves from *Croton scouleri* during the dry season is a long-drawn-out process. *Cordia lutea* is slow to produce leaves at the beginning of the wet season, and retains them in the early part of the dry season; a few individuals have leaves even in the latter half of the dry season. These two species tend to blur an otherwise sharp distinction: the island is green in the wet season, because all species have leaves, and grey in the dry season, because all but eight have lost them (Plate 3).

The breeding season of plants is not tied to the calendar, but to the pattern of rainfall. Flowering and the production of fruits and seeds coincide with or rapidly follow the formation of leaves a short time after the first heavy rainfall at the beginning of the wet season, whenever that occurs, or after a heavy rainfall at the end of the wet season following a dry period. For example, *Croton scouleri* (Fig. 2.13) responded to late and heavy rain in April 1978 by flowering within 48 hours; *Lantana peduncularis* (Fig. 2.9) did so two days later and was followed by *Bursera graveolens* (Plate 4) on the next day. Three species that are important to finches are exceptions to this pattern in that they produce many flowers two or more months before the start of the wet season. These are *Waltheria ovata* (Fig. 2.9), *Opuntia helleri* (Plate 2), and *Cordia lutea* (Plate 5).

Figure 2.9 *Lantana peduncularis* (above and below left) and *Waltheria ovata* (below right). Both are eaten by caterpillars, which in turn are eaten by all finch species. Seeds of *Lantana* are moderately hard to crack; they are eaten by *conirostris*. *Waltheria* nectar is an important source of food in the dry season for *difficilis* and *Certhidea olivacea*. Photos by P. R. Grant (above) and O. Jennersten (below).

Figure 2.10 *Eragrostis cilianensis,* one of several plants producing abundant small seeds, which are consumed principally by *difficilis*. *G. conirostris* occasionally eat the seeds and the caterpillars that feed on the leaves. Photo by P. R. Grant.

Figure 2.11 A clump of *Opuntia helleri* cactus bushes, source of both food and nest sites for *conirostris*. Photo by P. R. Grant.

Figure 2.12 The southeastern arm of Genovesa has low shrubby vegetation, almost no cactus, and a large population of petrels (*Oceanodroma* spp). Photo by P. R. Grant.

Figure 2.13 The flowers (left) and fruits (right) of *Croton scouleri*. All ground finch species feed on the seeds. Photos by P. R. Grant.

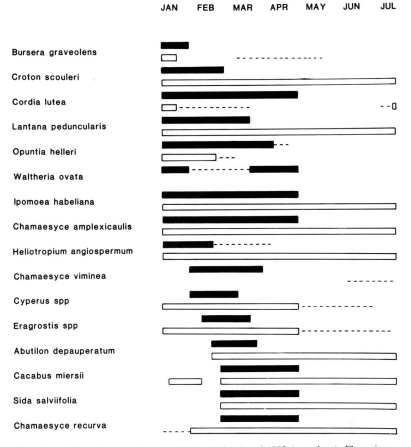

Figure 2.14 Flowering phenology in 1978 (solid bars) and 1983 (open bars). Flowering began in December 1982 before observations were made in 1983. Broken lines indicate a continuation of flowering by just one or a very few plants. From P. R. Grant and B. R. Grant (1987).

Superimposed on the basic patterns of seasonal production is an annual variation caused by the variable and unpredictable rainfall regime, which can be appreciated by considering the extremes, starting with very wet years.

The exceptionally heavy and extensive rains associated with El Niño events have three main effects on plants. The first is to lengthen the period of production (Fig. 2.14). To take an extreme example, *Croton scouleri* flowers once or twice during a typical wet season, but in 1982–83 individual trees flowered as many as five times, possibly more, and seeds produced in December 1982 developed into 2 m trees that were, themselves, flowering five months later. *Bursera graveolens* appeared to be an exception to the general trend by not flowering over a longer period than normal, but it did start to reproduce unusually early like all of the other species. The two species that normally reproduce after the rains cease, *Cordia lutea* and *Waltheria ovata,* had to wait a very long time. Their wet season production was devoted to vegetative growth, and it was considerable.

The second effect of unusual rains is the production of thicker, more dense vegetation (Plate 6). Table 2.3 illustrates this by comparing vegetation cover in 1983 and in 1978 (a typical year). The dense vegetation is the result of vegetative growth, as in the case of *Cordia lutea;* of germination of seeds and the production of new plants, most noticeably by *Heliotropium angiospermum, Lantana peduncularis,* and the *Cyperus* sedges (Plate 7; see also Table 2.2); or of both processes, as exemplified by several species but especially by *Croton scouleri.* The leaves of *Croton scouleri* and *Cordia lutea* were exceptionally large in 1983 and again in 1987. Annuals were unusually tall. For example, *Abutilon depauperatum* and *Sida salviifolia* normally grow to a maximum height of 0.5 m, but in 1983 the maximum height of both species was approximately 2 m.

The third effect is a greatly enhanced production of seeds from most species of plants. Production in 1983 was not measured, but the high value can be inferred from a comparison of the standing crop of seeds and fruits on plants in 1978 and 1983 (Table 2.4). It is unfortunate that sampling times differed, both absolutely and relative to the onset and cessation of the

Table 2.3 Mean Percentage Cover of Plant Species in Fifty Random Quadrats of 1 m² Each in the Vegetation Study Area on the South Side of the Island

	1978 March	1983 July	1985 July
Cordia lutea	5.5	29.2	18.6
Croton scouleri	13.2	33.1	17.5
Bursera graveolens	11.4	15.4	14.4
Opuntia helleri	10.1	11.0	12.2
Lantana peduncularis	1.2	11.4	0.0
Waltheria ovata	0.5	5.1	1.7
Heliotropium angiospermum	0.0	8.2	0.0
Sida salviifolia	0.0	2.1	0.0
Cacabus miersii	0.1	5.3	0.0
Ipomoea linearifolia	0.1	0.0	0.0
Ipomoea habeliana	0.0	1.9	0.1
Cyperus spp.	0.0	3.9	0.0
Eragrostis and *Trichoneura* spp.	0.2	2.4	0.0

Note: 1978 was a normal year, 1983 was exceptionally wet, and in 1985 there was a drought. Compare the stability of *Bursera* and *Opuntia* with the fluctuations of *Cordia, Croton,* and others.

Table 2.4 Numbers of Seeds and Fruits of the Principal Plant Species in the Fifty Quadrats, Expressed as Mean Number per m² ± SE

Seed Species	Seasons		Years		
	1978 March	1978 November	1983 July	1984 July	1985 July
Large					
Cordia lutea	5.9 ± 3.9	1.8 ± 0.6	0	0.5 ± 0.1	1.2 ± 0.4
Medium					
Bursera graveolens	7.7 ± 2.8	0	0	1.2 ± 0.5	0
Opuntia helleri	1.7 ± 0.8	1.6 ± 0.6	0.5 ± 0.1	0	0
Lantana peduncularis	38.7 ± 29.0	0	9.0 ± 2.8	0	0
Ipomoea linearifolia	0.5 ± 0.3	2.2 ± 1.6	0	0	0
Small					
Heliotropium angiospermum	0	0	37.8 ± 13.3	0	0
Chamaesyce recurva	0	0	20.0 ± 5.3	0	0
Croton scouleri	9.4 ± 2.7	0	15.7 ± 5.3	0	0
Sida salviifolia	0	0	Many	0	0
Cacabus miersii	0	0	Many	0	0
Cyperus spp.	0	0	464 ± 184.4	0	0.2 ± 0.2
Grass species	129.3 ± 124.1	51.7 ± 32.4	Many	0.8 ± 0.4	0

Note: Seeds and fruits on the ground were not counted in 1983 (only). Compare the abundance in the very wet year of 1983 with the paucity in the subsequent dry years and the dry season (November) of a normal year (1978). Seeds of three species were known to be abundant in 1983, but were not counted. They are indicated by the word, Many. Waltheria ovata is omitted from the table because it scarcely ever produces seeds. Size categories of seeds are based on measurements of depth (the second linear dimension, in mm) and hardness measured in kgf (see Abbott et al. 1977).

flowering season. The sampling in 1983 was incomplete and the comparison takes no account of seeds on the ground. Despite these limitations the data in Table 2.4 show that some species of plants carried unusually large numbers of seeds in 1983. Plants like *Croton scouleri* must have produced enormous crops of seeds in 1983; not only was the standing crop high in July, it was the fifth to be produced that year!

There is one clear exception to this consistent pattern of prolific production in an El Niño year. *Opuntia helleri,* the cactus which *conirostris* depend upon for food, fared poorly. By the middle of 1983 fertile fruits were relatively scarce. One reason for this is that many bushes were overgrown with vines. *Ipomoea habeliana* (Plate 8) and *Merremia aegyptica* vines grew rampantly in the more open areas in 1983, first over bare ground, then over low vegetation, and finally even up to the top of the tallest *Bursera* trees (Fig. 2.15). Many *Opuntia* bushes were smothered by *Ipomoea* and collapsed (Fig. 2.16). In addition, excessive uptake of water rendered them top heavy and susceptible to being blown down in the strong winds that accompanied storms in 1983. By the middle of 1984 large areas of cactus, especially on parts of the north coast, had been destroyed (Fig. 2.17).

The less extreme but still pronounced El Niño event of 1987 produced similar responses in the vegetation (Plate 9). The major difference between the years was an explosive growth of a small herb, *Mentzelia aspera* (Plate 7), in 1987 but not in 1983. By April patches of ground of up to $500-1000$ m^2 in the central part of the island were covered with a dense, near-monoculture of this plant. Beebe (1924) recorded it as being common on the island in 1923, which was not an El Niño year. Until 1987 we thought he might have misidentified the plant. Another plant he recorded as being common, *Gossypium* (cotton), has never been seen since.

The effect of the opposite climatic extreme, a drought, can be stated simply: little or no production. In the middle of the year when the rains failed to materialize altogether (1985), there were seeds on the plants of only two species (Table 2.4), and one of these (*Cordia lutea*) usually produces flowers and fruits in the dry season anyway. A few perennials died, and the vegetation reverted to something close to its pre–El Niño character. Most species did nothing in the drought and became active only with the resumption of rains in 1986.

Animals

The small community of Darwin's Finches on Genovesa comprises four species (Plate 10). They are also illustrated in Figure 2.18 together with the only other common land birds, a mockingbird and a dove. A seventh land bird, the yellow warbler, is rarer; the eighth and final species of land bird,

Figure 2.15 A *Bursera malacophylla* tree smothered by *Ipomoea habeliana* vines in the El Niño year of 1987 (upper), still covered in 1988 (middle), and more exposed in the following dry year (lower). Photos by P. R. Grant.

Figure 2.16 *Opuntia helleri* cactus smothered by *Ipomoea habeliana* vines in 1983 (above), and five years later (below). Photos by P. R. Grant.

the short-eared owl, preys on all the rest (Fig. 2.19). Mockingbirds prey on finch eggs and nestlings (P. R. Grant and N. Grant 1979).

There are more species of water birds than of land birds. They are largely independent of the finches, but interact with them in three ways. First, they provide alternative food for owls; this is especially true of two species of petrels, the wedge-rumped and band-rumped storm petrels. Second, the lava heron and the black-crowned night heron eat finches, but very rarely. Third, finches occasionally eat the regurgitated fragments of fish brought to the island by masked and red-footed boobies.

All other animals of importance to finches are invertebrates which constitute a large fraction of their diet (Fig. 1.1). There are no terrestrial mammals or reptiles. In the wet season an eruption of moth caterpillars accompanies the outburst of leaves and flowers soon after the first heavy rainfall (Fig. 2.20; Plate 11). All common plant species have their own species of caterpillars, except *Opuntia helleri* and *Chamaesyce viminea,* and individually they range in size from a fraction of a gram to 14 g in the case of a sphingid (hawkmoth; Plate 12). There is an abundance of other arthropods at this time on leaf surfaces, branches, and the ground, the most conspicuous types being spiders and orthopterans (grasshoppers and

Figure 2.17. Cactus on the northeast coast that collapsed in the El Niño conditions of 1983. Photo by P. R. Grant.

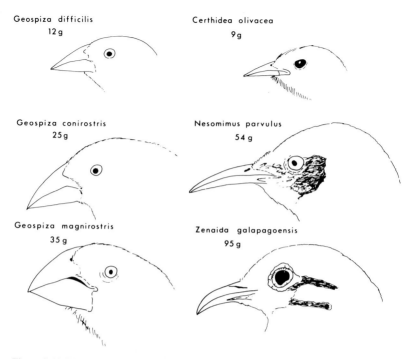

Figure 2.18 The six principal land bird species on Genovesa. Scale approximately life size. From ''The Breeding and Feeding Characteristics of Darwin's Finches on Isla Genovesa, Galápagos,'' *Ecological Monographs*, 1980, *50*, 381–410. Copyright © 1980 by the Ecological Society of America. Reprinted by permission.

crickets). In the dry season, by contrast, arthropod prey are concealed in chambers in dead branches of trees or beneath the bark (e.g., termites, cerambycid beetles), in the soil and very shallow leaf litter beneath *Cordia* bushes and trees (silverfish), or in the rotting pads of *Opuntia* cactus (fly and beetle larvae and pupae). The seasonal difference in the locations of potential prey is important to the finches, because different skills are required to find and secure the food.

Age of the Community

The Galápagos climate has been relatively stable for the last three thousand years; prior to that it was drier for at least another three thousand years. This conclusion has been reached from a paleoecological analysis of particles and plant products (pollen grains and spores) in cores taken from the sediment of El Junco, a lake in the highlands of Isla San Cristóbal (Colinvaux 1972). Applying the same methods of analysis to cores taken from Arcturus lake on Genovesa, Goodman (1972) has identified a major

Figure 2.19 The four principal land bird species other than finches. Top: mockingbird. Middle left: short-eared owl. Middle right: yellow warbler. Bottom: dove. Photos by K. T. Grant (mockingbird), P. R. Grant (owl), and O. Jennersten (warbler and dove).

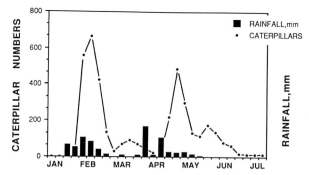

Figure 2.20 Weekly counts of caterpillars on 12 plant species in relation to weekly amounts of rain in 1987. Rainfall has the same scale as caterpillar numbers. See text for details.

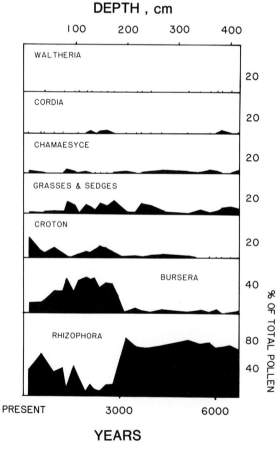

Figure 2.21 Pollen profiles constructed from Arcturus lake sediment cores and redrawn from Goodman (1972). Note the change in proportions about three thousand years ago, principally in mangrove (*Rhizophora*) pollen. *Opuntia* pollen is rarely registered in cores anywhere in the Galápagos, not being released into the atmosphere (P. A. Colinvaux 1987, pers. comm.).

change in the plant composition that took place three thousand years ago. After that time the composition was recognizably modern, but before then *Bursera, Croton,* and *Cordia* appear to have been relatively rare for a long time, and mangroves were much more common than now (Fig. 2.21). Unfortunately the relative abundance of *Opuntia* cactus is not known. If it too was relatively rare before three thousand years ago, then the *conirostris* population would have been much smaller than now, if indeed it was present at all.

Summary

Genovesa is a low, flat, and fairly small volcanic island, well isolated in the northeast of the Galápagos archipelago. It receives sporadic and occasionally heavy rain during the first four months of the year (wet season), and scarcely any during the remainder of the year (dry season). In some years little or no rain falls, whereas in others the rainfall is heavy and the wet season lasts for up to eight months. Annual rainfall is both extremely variable and unpredictable.

The amount and duration of plant and animal production is strongly governed by the unpredictable rains. Low, dry-deciduous forest covers the island relatively uniformly and comprises less than two dozen common species of plants. These produce no seeds, or almost no seeds, in a drought and an abundance of seeds in exceptionally wet years. Together with nectar and a variety of arthropods, they form the food supply of the four finch species present. Other land birds are a warbler, a dove, a mockingbird, and one predator, the short-eared owl. There are no terrestrial mammals or reptiles on the island. The community of plants and animals is likely to have been approximately stable for the past three thousand years. Before that time the climate was drier, the principal tree and shrub species were rarer, and cactus finches may have been much rarer or absent from the island.

3

Demography: Survival

Introduction

Finches spend most of the daytime foraging and feeding. The only time of the year when this is not true is prior to and during the incubation phase of a breeding cycle. Thus finding enough food to survive is a major, time-consuming necessity. The second most time-consuming activity is reproduction, where success is measured ultimately in terms of the number of offspring an individual contributes to the next generation. These two activities are obviously linked; birds must not only survive to reproduce, they must survive while reproducing. The two are also possibly linked in the not-so-obvious way that a given amount and a particular schedule of reproductive activity could affect subsequent survival, and hence prospects for subsequent breeding.

Some birds succeed much better than others. The major problem for us is to determine why. What are the ingredients of success? Is it just a matter of chance who survives to breed and who does not, who reproduces once and who reproduces many times? Or do some birds succeed because they are better equipped than others to exploit the environment and avoid its hazards?

We approach this problem in two steps. First, we characterize demographic patterns of survival and reproduction, then we examine the properties of individuals to see why some succeed better than others. The first is a population-level description and analysis; the second is an individual-level scrutiny of behavior and morphology.

This chapter and the next two provide the population-level characterization of survival and reproduction. The question of whether being large or small makes a difference to an individual's chances of surviving and breeding successfully will be considered in Chapters 7 and 9. In this chapter we examine how survival varies with age and year of birth, and how it differs between wet and dry seasons, and between wet and dry years. We conclude the chapter by comparing the *conirostris* population on

Genovesa with other populations of Darwin's Finches on the Galápagos and with other species of passerine birds, mainly finches, elsewhere.

Sources and Study Areas

Only 8 of the 120 banded nestlings that fledged in 1978 were seen in subsequent years (Table 3.1). To increase the number of banded nestlings, in order to increase our sample of adults of known age, we added a second study area in 1980, contiguous with the first and containing a similar density of cactus (Fig. 3.1). Our estimates of survival of birds of known age come from birds banded as nestlings in these two areas and subsequently seen anywhere on the island. Each year we searched extensively outside the two areas, mainly in the neighborhood, but despite these efforts to detect long-distance dispersers, we rarely found banded birds elsewhere. For example, we made the greatest search during the breeding season of 1987 in the immediate vicinity of the two study areas on the northern borders, in the center of the island, around the lake, and along west, north and east coasts. Yet we found only three individuals that had been banded as nestlings out of a total of more than 100 birds seen, and all three were within 300 m of the study areas.

Birds of Known Age

Birds born in the first half of the study give us a picture of survival over the whole eleven years of the study. The vast majority of those that fledge disappear and presumably die in their first year. In the years 1978–83 we

Table 3.1 Survival and Recruitment of Birds Banded as Nestlings and Known to Have Fledged

| | | Proportions | |
| | | Survived to Next Calendar Year[1] | Attempted to Breed in Study Areas |
Year	Cohort Size		
1978	120	0.07 (8)	0.04 (5)
1980	181	0.15 (28)	0.09 (16)
1981	242	0.12 (26)	0.02 (5)
1982	281	0.11 (30)	0.05 (14)
1983	420	0.16 (69)	0.02 (7)
1986	156	0.08 (12)	0.03 (5)
1987	359	0.12 (42)	<0.01 (1)
Total	1759	$\bar{x} = 0.11$	$\bar{x} = 0.04$

Note: Numbers of birds in parentheses.

[1] For most individuals this is equivalent to survival until age 1 year, but for some it is 6–12 months.

banded approximately 1250 nestlings that survived to fledging (Tables 3.1 and 3.2), and exactly 100 of them, or 8 percent, were seen more than one year later (Table 3.2). For the next few years annual survival was much higher and approximately constant at 50 percent. The oldest bird lived for nearly ten years.

Similar survival was observed at the end of the study following a hiatus in reproduction in 1984 and 1985 (next chapter). Eight percent of the 1986 nestlings survived for one year, and at least 2 percent survived two years. A minimum of 12 percent of the 1987 nestlings survived into early 1988 (Table 3.1).

Among the 47 birds that were banded as nestlings in the years 1978–83 and that survived to breed in the study areas, males survived as well or better than females at all ages (Table 3.2). Mean expectations of further life for one-year-olds were calculated from life tables to be 2.8 years for males but only 2.1 years for females. Age-specific survival was not constant but decreased with age in each sex over the first four years: that is, the

ISLA GENOVESA

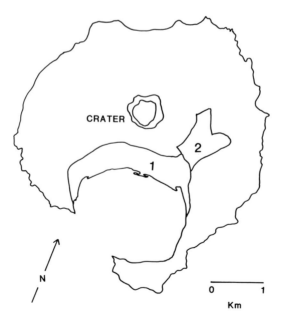

Figure 3.1 The two study areas on Genovesa.

Plate 1 *Geospiza conirostris*, the large cactus finch. Upper left: Male at the top of a *Bursera* branch. Photo by D. Schluter. Upper right: Beak color polymorphism shown by two nestlings, pink morph on the left and yellow morph on the right. Photo by P. R. Grant. Bottom: Banded male on a cactus pad. Photo by P. R. Grant. Middle right: Banded female on a *Bursera* branch. Photo by R. L. Curry.

Plate 2 Cactus bush in flower (upper) and a view of the island from Darwin Bay (lower). Photos by P. R. Grant.

Plate 3 The appearance of the vegetation in a normal wet season (upper) and in a dry season (lower). Photos by P. R. Grant (1978).

Plate 4 Developing foliage, flowers, and fruits of *Bursera graveolens*. Finches eat the red arils, then crack the stones to obtain the kernels. Photos by P. R. Grant.

Plate 5 *Cordia lutea* flowers (upper left) and fruits
(upper right). Photos by O. Jennersten. Flowers of
Ipomoea habeliana (middle). Flower of *Opuntia
helleri* (lower left). Photos by P. R. Grant.
Cactus finches occasionally drink nectar and eat
the pollen from the three types of flowers.

Plate 6 Genovesa. The contrast between a very wet year, 1983 (upper), and a dry year, 1981 (lower). Photos by R. L. Curry.

Plate 7 Growth of grasses and sedges early in the wet season of 1987 (upper), and extensive growth of *Mentzelia aspera* later in the same year (lower). Photos by P. R. Grant.

Plate 8 Rampant growth of an *Ipomoea habeliana* vine in July 1983 (upper) compared with the remnants of dead stems twelve months later (lower). Photos by P. R. Grant.

Plate 9 Differences in the effects of rain on the vegetation in two El Niño years: 1983 (upper) and 1987 (lower). Photos by P. R. Grant.

Plate 10 The four species of Darwin's Finches on Genovesa. Upper: *Certhidea olivacea*, the warbler finch. Middle left: *Geospiza difficilis*, the sharp-beaked ground finch. Middle right: *Geospiza conirostris*, the large cactus finch. Lower: *Geospiza magnirostris*, the large ground finch. Photos by O. Jennersten.

Plate 11 *Cordia lutea* leaves stripped by sphingid caterpillars in early March 1987 (left) and regenerated two weeks later after the majority of caterpillars had been eaten or had pupated (right). Photos by P. R. Grant.

Plate 12 Caterpillars on *Cordia lutea* (upper) and *Croton scouleri* flowers (middle), a sphingid caterpillar (lower left) and a spider (lower right). Photos by P. R. Grant.

Table 3.2 Annual Survival of Birds of Known Age, Banded as Nestlings and Known to Have Fledged in the Years 1978–83

| | Total | | | Birds of Known Sex | | | | | |
| | | | | Males | | | Females | | |
Years	Numbers	Proportion Surviving	Age-Specific Survival	Numbers	Proportion Surviving	Age-Specific Survival	Numbers	Proportion Surviving	Age-Specific Survival
0	1244	0.080	0.08	(622)	(0.032)	(0.03)	(622)	(0.043)	(0.04)
1	100	0.040	0.50	20	0.027	0.85	27	0.032	0.74
2	50	0.020	0.50	17	0.019	0.71	20	0.021	0.65
3	25	0.011	0.56	12	0.013	0.67	13	0.010	0.46
4	14	0.005	0.43	8	0.006	0.50	6	0.003	0.33
5	6	0.002	0.50	4	0.003	0.50	2	0.002	0.50
6	3	0.001	0.50	2	0.002	1.00	1	0	0
7	1	0.001	1.00	1	0.002	1.00	0	—	—
8	1	0.001	1.00	1	0.002	1.00	—	—	—
9	1	0	0	0	0	0	—	—	—
10	0	—	—	0	—	—	—	—	—

Note: Year 0 is year of birth. Figures for males and females in year 0 are shown in parentheses because sexes were not known that year. An equal sex ratio among the 1244 fledglings is assumed.

probability of an individual of either sex surviving for another year decreased systematically over this period.

The estimates of survival are minima, because five birds born in the study areas in 1982 and 1983 (two males and three females) were still alive in 1988. For the purpose of constructing Table 3.2 they were assumed to have died in that year, whereas in fact two individuals of each sex were alive on our final visit to the island in February 1989. In addition, some birds may have survived outside the study areas after the last time they were seen in them. There is a possible bias here affecting the comparison of the sexes. Females occasionally change mates and territories, and they may leave the study area to breed elsewhere, whereas males almost always retain their breeding territory for life (Chapters 6 and 7). We doubt if this difference in residence is sufficient to account for the difference in estimated survival, however, because most females that change mates do so locally (Chapter 7). We have only once discovered a female breeding outside the two study areas after she had bred in them. Nevertheless, the annual influx of birds of both sexes from outside the study areas who stay to breed in them implies some movement in the opposite direction followed by breeding.

In summary, survival is very low initially; it is much higher after breeding age is reached; thereafter it remains constant or decreases slightly with age.

Birds of Unknown Age

The estimates of survival presented so far are average figures, obtained by combining data from all years because the sample (47 birds) is too small to allow a cohort-by-cohort analysis. Yet environmental conditions vary strongly among years (see Tables 2.1 and 2.4), and the particular year of birth may be an important factor in determining survival, as has been found in many studies of passerine bird populations (e.g., Perrins 1979; van Balen 1980; Bryant 1988). The sample of birds of known age gives an incomplete picture of survival patterns because it constitutes only 16.7 percent of the total number of banded birds breeding in the years 1978–87. It is worth attempting to use the rest of the breeding birds in an analysis of survival, even at the risk of making some mistakes in estimating their ages. We do so for birds in study area 1 only, because only they have been observed continuously and in every year. Survival patterns in area 2 after 1980 are the same, but the samples are smaller and some data are missing because part of the area was not visited in 1981.

To assess cohort effects, we must first estimate ages of individuals and then group birds of the same age together. This can be done in two ways. The first and simplest way is to combine those birds breeding in the study area for the first time in a particular year. For males, the second and better method is to estimate age by the degree of blackness in the plumage, using

Figure 3.2 Plumage categories. All birds start in brown plumage. Males acquire increasingly black plumage with successive molts, whereas females remain in brown plumage with successive molts. From Grant and Grant (1983).

an average rate of acquiring blackness as the basis for the estimation, and then to combine birds of the same estimated age.

Female ages cannot be estimated by plumage characteristics because females do not change plumage color with successive molts (Fig. 3.2).

ESTIMATING THE AGES OF MALES

Young birds molt from juvenal to adult plumage typically about 2–4 months after fledging, as has been described for related species of ground finches by Orr (1945) and Snow (1966). A limited degree of molt may also occur prior to each breeding season, but the standard pattern is one of a single, annual, post-breeding molt. At each successive molt, a male acquires the same amount of black in the plumage or more, but never less (Fig. 3.2).

Figure 3.3 depicts the plumage states of males of known age. It shows an average trend, with some variation at each stage. Thus young birds after their first molt may be in black 0 (i.e., completely brown), black 1, or black 2 states. Birds in their second year span the range from categories 0 to 4. Such variation may be caused partly by aseasonal molting of a few

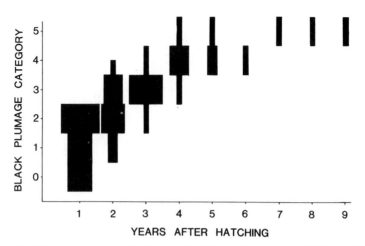

Figure 3.3 Plumage categories of males (see Fig. 3.2) in relation to age. Thickness of the bars is proportional to sample size. Only one male older than five years is present in the samples.

individuals. Errors of classification may contribute to it as well. There are limits to the variation, though, and the corollaries are important for estimating minimum ages of birds; for example, no bird is known to have acquired black 3 or higher categories in its first year, and only one bird is known to have acquired black 4 in its second year. The earliest acquisition of black 5 plumage is year 4.

These observations suggest the following working hypothesis or rule of thumb for estimating unknown ages: males in black categories 0–2 are likely to be in their first year (i.e., year of birth) or older; males in black 3 are in their second year or older; males in black 4 are in their third year or older; and males in black 5 are in their fourth year or older.

Many more males of unknown age have been observed to change from one state to another. The data from these are given in Table 3.3. The plumage transitions are generally consistent with the above age rule, but a few of the 102 birds might provide exceptions. Five birds passed straight from black 0 to black 3. Four were known to be at least two years old, but one could have been in its first year. Four passed directly from black 2 to black 4, and in two cases this could have happened in their second year. Finally, three more passed directly from black 3 to black 5, and two of them could have done so in their third year.

The ages of birds are more likely to be underestimated than overestimated. A substantial number of birds retain the same degree of blackness when molting, rather than advancing up the scale; notice, in particular, that the majority of males make no advance beyond black 4. Some individuals appear never to advance from black 4 to black 5.

Errors could also be incurred if rates of advance vary among years. We examined this possibility by comparing rates in the two dry years after the El Niño event of 1983 with rates in the previous, wetter years, to see if they were slower in the dry years. Separate χ^2 and Fisher's Exact tests, depending on sample sizes, were conducted with birds in each of five plumage categories, from black 0 to black 4, at the beginning of a year. None of the differences between wet and dry years in plumage advancement were significant ($P > 0.1$).

We have used this age rule to estimate minimum ages of males of unknown age, and then we have calculated survival curves for breeding birds.

The Initial Group of Adults

We start by considering just those birds that were present on the study area in 1978. They were the survivors of what was probably a drought. Very little rain had fallen on the island in the previous year (D. Day 1978, pers. comm.). *Bursera* trees had sprouted leaves in early January but had lost them by the end of the month, and the island remained grey in appearance for the rest of the year. To judge from the absence of males in black 0–2 plumage when we began the study in January 1978, there had been little or no successful breeding in 1977. It was known that a drought had occurred on Daphne Major in 1977; many finches had died, and none of the few fledglings produced that year survived to the following year (Boag and Grant 1981, 1984). Therefore the initial group of adults on the study area were almost certainly born in 1976, a moderately wet year in the archipelago (Grant and Boag 1980), or earlier.

The group comprises 44 individuals (Table 3.4) born at the latest in the years 1974–76 and hence belonging to a minimum of three cohorts. The

Table 3.3 Frequencies of Transitions between Plumage States in Successive Molts by 102 Males, Most of Unknown Age

To:	0	1	2	3	4	5	N
From:							
0	0.18	—	0.53	0.29	—	—	17
1	—	—	0.57	0.43	—	—	7
2	—	—	0.20	0.40	0.40	—	10
3	—	—	—	0.33	0.49	0.18	37
4	—	—	—	—	0.66	0.34	47
5	—	—	—	—	—	1.00	78

Note: N refers to the number of transitions. Plumage increases in degree of blackness on the scale of 0 to 5 (Fig. 3.2) or stays the same but never decreases.

Table 3.4 Survival of Birds Present on the Main Study Area (Area 1) as Adults in 1978

Years	1976 Cohort of Males			All Males			All Females		
	Numbers	Proportion Surviving	Age-Specific Survival	Numbers	Proportion Surviving	Age-Specific Survival	Numbers	Proportion Surviving	Age-Specific Survival
0									
1									
2	15	0.933	0.93	31	0.968	0.97	13	0.923	0.92
3	14	0.867	0.93	30	0.871	0.90	12	0.692	0.75
4	13	0.800	0.92	27	0.774	0.89	9	0.615	0.89
5	12	0.600	0.75	24	0.645	0.83	8	0.615	1.00
6	9	0.467	0.78	20	0.452	0.70	8	0.461	0.75
7	7	0.333	0.71	14	0.323	0.71	6	0.231	0.50
8	5	0.200	0.60	10	0.226	0.70	3	0.154	0.67
9	3	0.067	0.33	7	0.064	0.29	2	0.077	0.50
10	1	0.067	1.00	2	0.032	0.50	1	0.077	1.00
11	1	0.067	1.00	1	0.032	1.00	1	0.077	1.00
12	1	0	0	1	0	0	1	0	0

Note: The groups of all males and all females are presented as if they were born in 1976. Undoubtedly some of their members were born earlier. Plumage features (see text) were used to identify 15 males that were probably born in 1976; these are listed separately as the 1976 cohort. None were alive in February 1989.

1976 cohort of males can be identified by plumage traits with more confidence than can the rest, so these are shown separately in the table as well as in combination with older males. Patterns of adult survival in all groupings are quite different from the survival of birds born from 1978 onwards; the drought survivors persisted for much longer, on average, than did the birds of known age born later, as can be seen by comparing Tables 3.2 and 3.4.

A direct comparison can be made with the average life spans (Pyke and Thompson 1986) of members of the two groups beyond two years of age. Survival up to year 2 is not known for any of the initial adults, and complete survival thereafter is known only for those estimated to have been born in 1976 itself, except for one individual which we will assume died in its final (12th) year. Mean life spans after the age of 2 years were 4.3 years for the 15 males born in 1976 (Table 3.4) and 1.6 years for the 17 males born later (Table 3.2). The difference between them is significant (Mann-Whitney $U = 47$, $P = 0.002$). In other words, from their second year onwards the 1976 cohort of males lived for more than twice as long, on average, as did the males of known age born later. This pronounced difference in survival between the drought survivors and birds born after the drought is by itself enough to justify examination of birds of unknown but estimated age.

The 1975 cohort did just as well as the 1976 cohort from year 3 onwards. Survival after year 4 was higher in the 1975 and 1976 cohorts of males than in the 1974 cohort, which may mean that some of the birds classified as members of the 1974 cohort were really older. These differences between males were paralleled by differences between females. For example, 13 females grouped as if they were all born in 1976 (Table 3.4) lived on average for a further 3.9 years, whereas 20 females of known age born later (Table 3.2) lived after their second year for only 1.1 years on average (Mann-Whitney $U = 47$, $P < 0.002$).

Maximum longevity is also greater for the initial adults; it is at least 12 years. One male and one female that were born in 1976 at the latest were still alive in 1988, but not on our final visit to the island in February 1989. In contrast, the oldest birds in 1989 among the group born during the study were members of the 1982 cohort, 7 years old, little more than half the maximum age.

Comparison of the two groups is affected by uncertainty concerning the exact ages of the initial adults. Almost certainly some of the ages are underestimated, particularly among the females. Some of the birds may have been the same ones we banded in 1973 (Abbott et al. 1977; Smith et al. 1978), as they could have lost their identification by removing the pliable celluloid bands we used then. One bird in particular appears to have been a survivor from 1973. A male in black plumage banded by Ian Abbott in March 1973 had almost identical measurements to one we captured and banded at the same place in January 1978. If indeed it was the same bird, it must have been born in 1970 or earlier; as it survived to the latter half of 1982, it would have lived for at least 12 years, which is distinctly more than the 7 years we estimated on the basis of its plumage in 1978. If ages have been underestimated, the difference between the groups is greater than has been revealed.

Females appear to survive as well as males. This can be seen in Table 3.4 by comparing the age-specific survival of all females and all males. Mean annual survival from year 2 to 9 is 0.78 ± 0.17 (SD) for all females, 0.81 ± 0.10 for all males, and 0.80 ± 0.13 for the 1976 cohort of males alone.

An interesting feature of this group is that age-specific survival of males is not constant but decreases with age (Table 3.4), as was observed among the birds of known age. The mean rate at which it declines is 5.6 percent a year. This was determined by regressing the age-specific survival of males in the 1976 cohort on age from years 2 to 8 inclusive and finding the slope to be -0.056 ± 0.009 ($r = 0.94$, $P < 0.01$). The decision to truncate the curve after year 8 is arbitrary. The smallest number of birds, in the last age interval, may have had a disproportionate effect on the estimate of the slope. Therefore we repeated the calculation with the curve truncated after year 7. The slope (-0.049 ± 0.011) changed very little and remained significant ($r = 0.91$, $P < 0.05$).

Individual Cohorts

It is clear that drought (1977) survivors fared better over the next decade than did birds of known age born later. Each of these groups is heterogeneous with respect to year of birth and the particular years over which its members survived. This heterogeneity could have a bearing upon the observed difference in survival. Therefore we need to consider individual cohorts, identified by year of birth and considered from the time of attempted breeding onwards. Their members are recognized by their known year of birth or, where this is not known, by plumage features (males) or by the year in which they started to breed (females), on the assumption that breeding is first attempted at two years of age. The assumption of this breeding age for females is sometimes violated. Five out of twelve females of known age born in the years 1978–81 bred in their first year, six bred for the first time in their second year, and one started breeding in her third year. This pattern, incidentally, was almost identical for males ($N = 12$); six bred for the first time in their first year, five in their second, and one in its third year.

Five cohorts, in addition to birds born in 1976, have been followed since initial breeding for a minimum of six years. Figure 3.4a illustrates

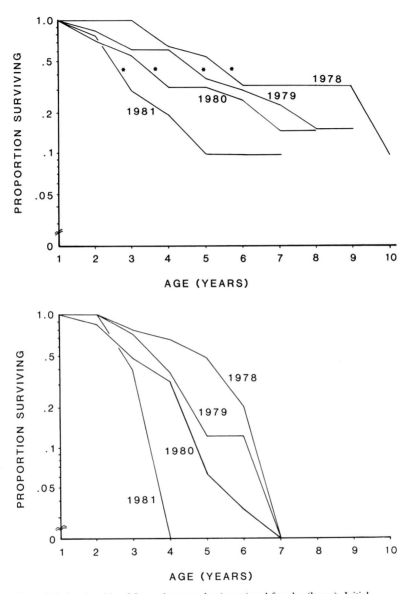

Figure 3.4 Survivorship of four cohorts; males (upper) and females (lower). Initial numbers of males were 11 (1978), 13 (1979), 20 (1980), and 11 (1981); initial numbers of females were 15 (1978), 10 (1979), 30 (1980), and 10 (1981). Stars identify survival in 1983.

adult survivorship of males of the first four cohorts produced after the 1977 drought; the fifth cohort (1982) is omitted because the heavy El Niño rains began exceptionally early in the middle of their critical first year, shortening the dry season and thus facilitating exceptionally high survival. The figure shows the striking feature that survival differs substantially and regularly among cohorts, with the first cohort (1978) surviving best and the last (1981) surviving worst. Differences become established in the first two or three years and are then maintained without further change (the 1981 cohort is a small exception). There is a tendency for the survival curves to get progressively steeper in a given year with increasing year of birth.

The initial adults fared better than all subsequent ones. This can be seen by considering only those data from year 2 onwards. Age-specific survival is not reliably estimated for most age groups because the samples are very small, so instead we use the median survival age, that is the age at which 50 percent of starting numbers have disappeared (D50), as a parameter characterizing the whole of the survival curve. There is an almost perfect rank ordering of this age with year of birth. The D50 point of those that survived to age 2 years was reached in the sixth year by the 1976 cohort, in the fifth year by the 1978 cohort, in the fourth year by the 1979 cohort, in the third year by the 1980 cohort, and in the second year by the 1981 and 1982 cohorts. For most of the cohorts the D50 point was reached in 1983. Despite anomalously high survival of members of the 1982 cohort in their first year, for reasons given above, survival was low in the second year and the D50 point for the cohort was reached in that year.

To compare the cohorts statistically, we have contrasted the proportions of each that survived for more than four years beyond year 2 with those that did not. This testing framework allows us to include the 1976 cohort, with a minimum age of two years, and the 1982 cohort, with a maximum age of an additional five years. There is a significant heterogeneity among the cohorts ($\chi^2_4 = 14.64, P < 0.02$). At the extremes, 80 percent of the 1976 cohort survived for at least four more years after reaching age 2, whereas only 11 percent of the 1981 cohort and 20 percent of the 1982 cohort survived for that long.

Cohorts of females show a similar pattern of survival in relation to year of birth (Fig. 3.4b), but they survive generally less well than males of the same cohort. The median survival age is younger for females than it is for males of the same cohort, except for the 1981 cohort. Similarly the last female of a cohort disappears at an earlier age than does the last male. Partial life table analysis of the largest cohort, the 1980 cohort, shows the same pattern (Table 3.5), with the mean expectation of further life being consistently higher for males than for females at every age from year 1 to year 8. The pattern of higher male survival was repeated by the 1983 cohort. Birds born in 1983 bred for the first time in 1986. Sex-specific survival up to 1986 was not known, but survival to the following year (i.e., during their fourth year) was 0.77 for males ($N = 19$) and 0.70 for females ($N = 20$).

There is more uncertainty in the age estimates of females than of males, and this is possibly a factor contributing to the difference between sexes in survival patterns. However, two potential biases work in the opposite direction, making females appear to survive better than males. Undetected early mortality is likely to be higher in females than in males, since sex identity is known by plumage and song in males before they breed, whereas it is not known in females before they breed. Therefore early survival of females appears to be higher than it really is, and this effect is greater than in males. Notwithstanding this potential bias, there were more females than males in the cohort (Table 3.5). Second, making the assumption that females breed for the first time in year 2 when in fact they do so on average at less than two years of age also has the effect of elevating the estimate of early survival.

Table 3.5 Partial Life Table of the 1980 Cohort

Age Class (years) x	Number Alive Each Year N_x	Proportion Surviving at Start of Age x l_x	Proportion Dying between x and $x + 1$ d_x	Mortality Rate q_x	Mean Expectation of Further Life at x e_x
Males					
0					
1	20	1.000	0.300	0.300	2.80
2	14	0.700	0.145	0.207	2.79
3	11	0.555	0.305	0.549	2.41
4	5	0.250	0.000	0	3.70
5	5	0.250	0.050	0.200	2.70
6	4	0.200	0.050	0.250	2.25
7	3	0.150	0.000	0	1.83
8	3	0.150	0.100	0.667	0.83
9	(1)	0.050	(0.050)	(1.000)	(0.50)
10	(0)	(0)	—	—	—
Females					
0					
1	30	1.000	0.133	0.133	2.33
2	26	0.867	0.400	0.461	1.61
3	14	0.467	0.167	0.358	1.57
4	9	0.300	0.200	0.667	1.17
5	3	0.100	0.033	0.330	1.50
6	2	0.067	0.034	0.507	1.00
7	1	0.033	0.033	1.000	0.50
8	0	0	—	—	—

Note: Starting numbers in the year of birth (year 0) were not known. One male was alive after the study had been completed, in February 1989. It was assumed to die in that year. Figures in parentheses show which entries in the table are affected by this assumption.

Possible errors in age estimation are unlikely to be entirely responsible for the difference between the sexes. None of the females in the four cohorts charted in Figure 3.4 survived beyond the sixth year, whereas males of all cohorts survived to the seventh year at least, and some to the tenth. Moreover, the difference is in the same direction as that shown by males and females of known age (Table 3.2); for this reason we think it is reliable. A similar observation has been made on Daphne Major where *fortis* females survived less well than males in two dry years out of three (Gibbs and Grant 1987a). In general, where there is a sex difference in adult survival in birds, females usually survive less well than males, possibly because they suffer greater energetic stresses during breeding (Ricklefs 1973).

In the previous section it was shown that birds of known age born after the drought survived less well than did the 1977 drought survivors. It can now be seen that this difference is attributable, partly or wholly, to the different years in which the groups were born and their different ages when they encountered the same environmental conditions. The 1974–76 cohorts were compared with the 1978–83 cohorts. Most birds of known age were born in the years 1980–82. Cohorts of those years, including as a minority those birds of known age, did not survive well.

Cohorts of birds born from 1978 onwards do not show the declining trend of survival with age exhibited by the initial adults. The slope of the regression for the largest single sample, the females of 1980, is -0.060 ± 0.062. It is not significantly different from zero ($P > 0.1$), although it appears to be similar to the slope for males of the 1976 cohort (-0.056 ± 0.009).

A Synthetic Survivorship Curve

To characterize the general trend in adult survival after the year of birth, we have integrated survival data from several cohorts of males in the following way. Survival from first to second year was calculated by averaging the values of the five cohorts studied directly—the 1978–82 ones. Survival from second to third year was calculated in a similar way, except that data for the 1976 cohort were included, after first scaling the numbers to the mean proportion of two-year-olds previously calculated. The procedure was repeated to include the 1975 and 1974 cohorts.

Although errors of age estimation undoubtedly render the calculated survival values imprecise, the resulting curve (Fig. 3.5) makes the fullest use of the data and is the best we can produce. It has the advantage of showing the pattern of survival over the full life span which the other figures do not. It is extremely similar to the survivorship of *scandens*, a congeneric and ecologically similar species on Daphne Major (Gibbs and

Grant 1987a), and it has the general features of standard survival curves of passerine bird species (Botkin and Miller 1974), but it is much shallower than the survival curves of temperate zone species of emberizine finches (sparrows) (cf. Tompa 1964; Ricklefs 1973; Baker et al. 1981), one of which is shown in the figure for comparison. Other comparisons will be made at the end of this chapter.

Environmental Determinants of Survival

Adult mortality rates of birds vary seasonally and are often highest in the breeding season when adults are believed to be most vulnerable to predators (Ricklefs 1973). Most adult *conirostris*, however, die between breeding seasons, not within them. The mean annual adult survival (sexes combined) calculated without regard to age is 0.78, varying from 0.58 (1985–86) to 0.94 (1980–81). This gives an expected average monthly survival of 0.98, and an expected survival over a typical breeding season lasting three months of 0.94. Survival of breeding birds during a breeding season was always higher than the expected value, 95–100 percent (mean 97 percent), except in 1982–83. In this exceptional breeding season a maximum of 11 percent of the banded breeders disappeared and may have died, but this is lower than the 15 percent estimated for an eight-month breeding season. Survival is therefore lowest, on a monthly as well as a

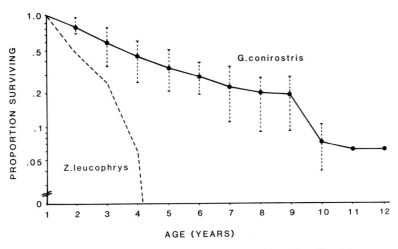

Figure 3.5 A synthetic survivorship curve for *conirostris*, with survivorship of the temperate zone white-crowned sparrow (*Zonotrichia leucophrys*) shown for comparison. Ninety-five percent confidence limits are shown for the *conirostris* estimates of mean survival by dotted lines. The number of cohorts (1974–82) vary, starting with four, rising to eight cohorts for years 4 to 6, and falling to one cohort for years 11 and 12.

seasonal basis, in the nonbreeding dry season, when food supply is lowest. This suggests that many birds die of starvation.

FOOD AND DENSITY

We do not have quantitative estimates of food availability to test the hypothesis that survival, or its complement mortality, is determined by food supply; behavioral evidence for the limiting effect of dry season food supply is discussed in Chapter 9. There is statistical evidence of food limitation of finch population sizes on Daphne Major (Boag and Grant 1981, 1984; P. R. Grant 1986b; Gibbs and Grant 1987a), where adult survival of *fortis* and *scandens* is correlated with the abundance of small seeds. An independent and negative effect of density upon survival was also found in both species (Gibbs and Grant 1987a).

Survival of *conirostris* also appears to be governed partly by density. The density of breeding birds increased steadily from 1978 to 1983, then declined until 1986 as a result of drought conditions. These changes are well represented by changes in the density of males only (Fig. 3.6), since the sex ratio did not vary much (Chapter 7). Variation among cohorts in survival, especially in the age interval of one to three years, was inversely related to density. The two trends are possibly related causally. The cohort trend can be explained by supposing that survival probability declines as density increases, and that young birds are more adversely affected by density than are old birds. We do not know if the density effects are direct, mediated perhaps by aggressive encounters, or indirect, through competition for food for example.

The D50 point was reached by four cohorts (1978–81) in the same year, 1983–84 (Fig. 3.4), the latter part of the El Niño disturbance and the

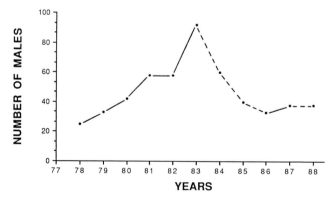

Figure 3.6 The number of breeding males on study area 1. Potential breeders are shown in two years when little (1984) or no (1985) breeding took place. Broken lines reflect the uncertainty in the numbers for those years.

aftermath. This raises the question whether mortality in general was unusually high then. For these four cohorts the answer is yes. In each case annual survival was lower during that year than in both following and preceding years. Food was abundant, at least in 1983, but overall density was also at its highest then, so once again a causal connection between survival and density can be inferred. Moreover, in the following two climatically stressful years when little (1984) or no (1985) rain fell and overall density declined, annual survival of the four cohorts increased. The 1982 cohort was exceptional, however, in experiencing lower survival in the first of the dry years (0.50) than either immediately before (0.92), when other cohorts fared relatively poorly, or immediately after (0.67), when the others did better. Finally, the survivors of the 1977 drought showed no change in survival in 1983.

In contrast to the poor survival following the El Niño event of 1983, adult survival was at its highest following the second El Niño event of 1987. This is apparent in Figure 3.4 which shows there was no mortality among males in three of the four cohorts depicted. In the fourth case, the 1978 cohort, numbers dropped from only three individuals to one, and one of those which disappeared was possibly eaten by an owl (see below) in the breeding season of 1987. Survival at this time was exceptionally high in all other cohorts of males. It was 1.00 for the 1976 ($N = 1$) and 1982 ($N = 4$) cohorts, and 0.93 ($N = 14$) for the 1983 cohort. Thus there are good years and bad years affecting all cohorts simultaneously and, to some extent, independent of age. The unusually high survival in 1987 can be attributed to a combination of plentiful food after a season of extensive production and moderately low density of adults (Fig. 3.6). The contrast between effects of different El Niño events on survival illustrates a general point: the consequences of unusual rainfall are determined, at least in part, by antecedent conditions.

PREDATION AND DISEASE

In addition to food supply and density, predation by short-eared owls is a factor influencing survival patterns, but a factor whose magnitude is unknown. Owls prey on nestlings. We have witnessed it rarely and have documented its frequency through the characteristic destruction of finch nests. Owls crush the roof when they alight on the top of the domed nest (Fig. 3.7) and then tear it apart to reach the trapped nestlings. On three occasions a breeding male finch disappeared at the time his nestlings were eaten. Possibly he was eaten at the same time while on a visit to deliver food to them. We have never seen an adult killed by an owl and have found skulls of only three *conirostris* in owl pellets (see also Curio 1965). To judge from the more frequent occurrence of *magnirostris* skulls ($N = 15$) in the pellets (see also Beebe 1924), owls feed more on this larger species,

60

Figure 3.7 Nests of *conirostris*. The middle one has been partly destroyed by an owl that has sat on the top to reach the nestlings inside. The cactus bush supporting the lower one has fallen to the ground during heavy rain and wind in 1987. Photos by P. R. Grant.

probably mainly. on fledglings. Therefore we can say little about owl predation as an adult mortality factor, other than we believe it to be relatively minor in importance. Predation of fledglings by lava herons and night herons (Chapter 2) appears to be rare, as we have witnessed it only three times in total. We have never seen either heron kill or eat an adult finch.

Many adult mockingbirds died on the island during the El Niño event of 1982–83, apparently due to avian pox (Curry and Grant 1989). The disease is known to affect finch populations on several islands (Grant 1986a), but we observed the symptoms—swellings on legs and head—on only three birds throughout the study. Fecal samples from 41 *conirostris* were examined for coccidea parasites because they are known to have debilitating effects in other birds (Ruff et al. 1974). None were found, although some were found in one out of six samples from *magnirostris* (T. McQuistion 1988, pers. comm.). No other indications of disease were encountered.

Age Structure of the Breeding Population

The unequal survival of different cohorts caused changes in the age structure of the breeding population. Figure 3.8 illustrates the changes among the males, whose ages are known more precisely than are the ages of females. An initially top-heavy age distribution (1981) was gradually transformed into a bottom-heavy distribution by 1983 as a result, in that year, of strong recruitment of young birds. The degree of top-heaviness is probably overestimated by using minimum ages for classifying the survivors of the 1977 drought, but it is safe to say that the age distribution did not become bottom-heavy before 1982. The extensive breeding in 1983 led to a more conspicuously bottom-heavy age distribution. Had 1983 been followed by climatically normal conditions this broadening of the base would have been even more pronounced, but high mortality occurred among young birds in the next two dry years, which reduced the potential contribution of the 1983 cohort to the breeding population. We calculate from the banding data that less than 2 percent survived to breed in 1986 (Table 3.1).

Our study ended with a repetition of these events. High survival of the 1987 cohort to 1988 (12 percent) broadened the base of the age distribution, but the drought that year threatened to make this effect short-lived.

Two other features of note in Figure 3.8 are: first, as many as nine age classes are present in the breeding population at any one time; second, some year classes are missing altogether as a result of the absence (1985) or failure (1984) of breeding in a particular year (Chapter 4). In the latter respect and in the presence of dominant age classes, the population

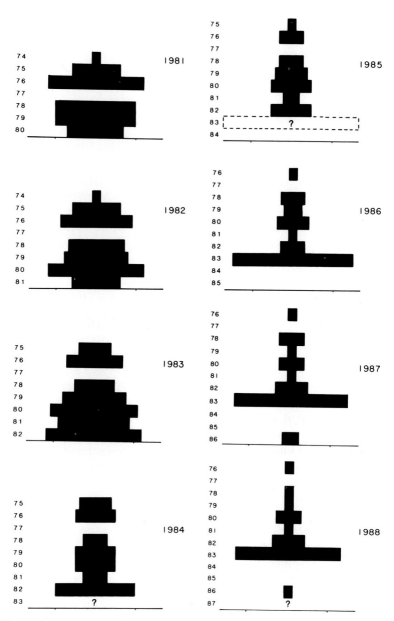

Figure 3.8 Changes in the age distributions of breeding males. The width of each bar indicates the size of each age group. For scale the distance between the two marks on the horizontal axis represents 10 individuals. Only a few pairs bred in 1984, one pair bred in 1988, and none did so in 1985. Age distributions in these years refer to potential breeders. The size of the 1983 cohort was not known until 1986. Question marks indicate uncertainty about the size of this cohort.

resembles long-lived tree species and fish (e.g., see Krebs 1985) more than typical passerines! Some interesting implications of the fluctuations in age structure will be considered in the chapter on mate choice (Chapter 7).

Comparative Demography

Land birds in wet forests of the tropics differ from their counterparts in the temperate zone in having low recruitment rates but in living longer (Cody 1966, 1971; Lack 1968; Ricklefs 1973, 1983). These differences are usually explained by the different mortality risks and different reproductive opportunities in the two regions. The Galápagos environment is interesting because it has features of both regions. Day length hardly varies throughout the year, and variation in temperature is moderate, as in other tropical locations at comparable altitudes. On the other hand, rainfall is both seasonal to a pronounced degree and annually unpredictable (Grant and Boag 1980), as in arid zones of the temperate region (e.g., Wiens 1974; Davies 1976; Maclean 1976). Given the mixed nature of the Galápagos climate, it is of interest to see how the survival characteristics of *conirostris* compare with those of other species.

LONGEVITY

Like tropical forest birds, *conirostris* live long. Our minimum estimate of the potential life span is 12 years, based on a sample of only 31 males and 13 females. On Daphne Major the fates of larger samples of two species of finches have been followed longer. Two male *scandens* were alive in 1988 at minimum ages of 14 years. Three other males were 13 years old, and two *fortis* males were also 13 years old. Since large birds are capable of living longer than small birds (Calder 1984), *conirostris* (~25 g) should be capable of living longer than either *fortis* (~17 g) or *scandens* (~21 g), hence to 14 or 15 years.

Unfortunately there are no data available for emberizine finches in continental Ecuador or neighboring countries to compare with Darwin's Finches. The oldest known finch from the Neotropics is a white-collared seedeater that lived for 11 years and 8 months in Belize (Klimkiewicz and Futcher 1987). Species of small land birds in other families in the Neotropics can live to 14 years (Snow and Lill 1974; Willis 1974, 1983), hence as long as Darwin's Finches.

In contrast, temperate zone emberizines of about the same size as Darwin's Finches rarely live to 10 years. Since estimates of potential life span are a function of sample size (Botkin and Miller 1974), it needs to be stressed that their longevities have been determined from impressively large samples. Klimkiewicz and Futcher (1987) have compiled recoveries of banded individuals of 31 species of North American emberizines. More than 1000 recoveries have been obtained for each of eight species, the largest single total being 16,330 song sparrows. The combined total for all eight species was 78,507. The oldest song sparrow was 11 years and 4 months; only one species exceeded this age: a white-crowned sparrow lived for 13 years and 4 months. Maximum ages for three of the eight species were less than 10 years.

These relationships are summarized in Figure 3.9. Longevity increases with body size in a group of fringilline and emberizine finches in the temperate zone. Tropical species live longer than is predicted from the relationship among temperate zone birds. Darwin's Finches do not differ from other tropical species in the degree to which they exceed expected longevity. *G. conirostris* has the proportionately lowest longevity among the group of tropical species, but this is almost certainly ascribable to the low sample size. Furthermore, the oldest individuals were still alive in 1988.

It is possible that the long lives of Darwin's Finches are brought about by low basal metabolic rates. There are no physiological data from the finches to support such a suggestion; it comes from two lines of reasoning. The first is from allometry. Many physiological processes scale with body size in the same way, and in the same way that life span does; size dependencies are all approximately the fourth root of body mass. Therefore all animals do the same things about the same number of times in their life spans (Calder 1984). Given this fixed maximum number, an increase in life span for an animal of a particular size is achieved by a lowering of the basal metabolic rate.

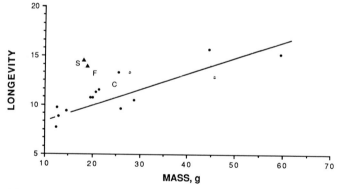

Figure 3.9 Maximum longevity as a function of size among 15 species of temperate zone finches and sparrows (solid circle). Longevities are taken from Klimkiewicz and Futcher (1987) and are based on a minimum of 1000 recaptures in each case. Weights are taken from Dunning (1984) and Pulliam (1985). Longevity = 0.56 weight$^{0.35}$ (r = 0.87, $P < 0.001$). Five tropical species not used in the analysis are indicated by symbols: C = *conirostris*, F = *fortis*, S = *scandens*, solid triangle = two species of rain forest species (not finches); white-bearded manakin and spotted antbird (data from Snow 1962; Snow and Lill 1974: Willis 1974, 1983).

The second reason is that lowered basal metabolic rates have been demonstrated in other finch species living in similar, hot, tropical environments (Weathers 1979; Weathers and van Riper 1982). In the Hawaiian archipelago the Laysan finch has been shown to have a daytime basal metabolic rate 20 percent lower than predicted by its body mass (31 g). The physiological interpretation is that thermal influences act as selective forces, favoring a reduced basal metabolic rate under conditions of strong heat loads from the sun and of difficulties in dissipating heat by evaporation owing to high ambient humidity (Weathers and van Riper 1982). If physical factors have been selective forces on Darwin's Finch metabolism in this way, their long life spans are a correlated effect. The long lives of tropical forest birds cannot be explained in the same way, because tropical species living in shaded habitats do not exhibit lowered basal metabolic rates (Weathers 1979).

ANNUAL SURVIVAL

Survival rates of adult tropical forest birds are almost twice as high, on average, as the survival rates of temperate zone birds of similar size. Fogden (1972) determined that mean annual survival of 34 species of forest birds in Borneo was approximately 86 percent, and Snow and Lill (1974) estimated that between 79 and 89 percent of white-bearded manakins in Trinidad survived each year. Willis (1974, 1983) studied three species of antbirds in Panama for more than a decade and found the mean annual survival to vary from 70 percent for the ocellated antbird to 81 percent for the spotted antbird.

Emberizines and other similar species in the temperate zone have substantially lower survival rates (Gibbs and Grant 1987a). They are estimated to be approximately 50 percent for the North American song sparrow (Nice 1937; Johnston 1956; Tompa 1964) and white-crowned sparrow (Baker et al. 1981), 49–59 percent for the indigo bunting (Payne 1989), 58 percent for the European house sparrow, and 35 percent for the Eurasian tree sparrow (Pinowski 1968).

Darwin's Finches resemble tropical forest birds in their high survival rates. Up to the age of eight years, adult survival is 69 percent for *fortis* and 82 percent for *scandens* on Daphne Major (Gibbs and Grant 1987a). The comparable estimate for *conirostris* is also 82 percent.

VARIATION IN ANNUAL SURVIVAL

Annual variation in rainfall causes food production to vary (Grant and Boag 1980; Boag and Grant 1984; Grant 1986a,b); as a consequence, adult mortality rates vary annually and age structure fluctuates (Gibbs and Grant 1987a). Since annual rainfall variation is exceptionally high on the Galápagos, adult survival varies to an exceptional degree. To compare Darwin's Finches with finches from the tropical mainland we need, but do not have, data from populations in the Santa Elena peninsula of Ecuador where climatic conditions are similar to those on the Galápagos (Marchant 1958, 1959, 1960; P. R. Grant 1985). Finches there are nomadic to varying degrees (Marchant 1960), and this fact, together with the degraded nature of the habitat (R. E. Ricklefs 1981, pers. comm.), makes it unlikely that comparable survival data will be obtained in the future. Nor are there comparable data for tropical forest species. But since annual survival varies more strongly in Darwin's Finches than in temperate zone forest passerines (Gibbs and Grant 1987a), it is reasonable to suppose that it varies in the finches even more strongly than in tropical forest passerines.

The most direct comparison of annual variation in survival between species is made with age-specific survival data, since this eliminates variation in survival associated with age and possibly different life spans. Gibbs and Grant (1987a) used data in Baker et al. (1981) to calculate the coefficient of variation for the survival of white-crowned sparrows from 1 to 2 years and from 2 to 3 years, and found it to be 11 percent in each case. Coefficients for *fortis* (40 and 51 percent) and *scandens* (39 and 43 percent) are much higher. The corresponding coefficients for male *conirostris* are lower: 13 and 23 percent, respectively. The number of cohorts used in these calculations were similar; 4 to 6 for *fortis* and *scandens,* and 5 and 6, respectively, for *conirostris.* Part of the reason for the lower coefficient for *conirostris* is that the 1983 cohort was not included, whereas it was included in the calculations for both *fortis* and *scandens.*

Thus adult Darwin's Finches, living in an environment that has both tropical and temperate zone climatic features, survive each year at a high average rate, which is a characteristic of tropical species, yet vary strongly in their annual survival, as do temperate zone species. *G. fortis* has the lowest mean survival but the highest variation among the three species studied. This pattern can be interpreted in terms of their food supply (Boag and Grant 1984; Gibbs and Grant 1987a). They are more dependent on small seeds as a food source in the dry season than are *scandens.* Annual variation in rainfall causes greater fluctuations in the supply of small seeds than in the supply of *Opuntia* cactus products, including seeds, upon which *scandens* specializes. Ecologically *conirostris* is similar to *scandens,* being strongly dependent on cactus (Fig. 1.1), and has similar demographic features.

AGE-SPECIFIC SURVIVAL AS A FUNCTION OF AGE

After reaching adulthood, birds in general have an approximately constant annual mortality rate (Deevey 1947; Botkin and Miller 1974). Put in different words, the risk of death is approximately constant on average over the adult life span (Nice 1937). Despite the approximate constancy, there are two arguments for expecting an increase in risk with age, and hence an increase in mortality rate.

The first is an evolutionary argument. If aging is a by-product of natural selection, favorable traits should be expressed early in life when the reproductive potential or "value" of an individual is high, and deleterious traits should be expressed later in life when reproductive value is lower (Medawar 1952; Williams 1957; Hamilton 1966). As a consequence of the latter, the probability of death in a particular time interval should increase with age and give rise to an increase in age-specific mortality (Emlen 1970). The second argument is more empirical. It derives from the calculation that observed longevities of birds are usually much lower than those expected from observed mortality rates when they are averaged over age groups and assumed to be independent of age (Botkin and Miller 1974). Calder (1984) has made the same point by showing that maximum life span and average survival (or mortality) do not scale with body size in the same way. Estimated longevity can be brought into line with observed longevity for a given sample size by making the mortality rate a slightly increasing function of age (Botkin and Miller 1974).

As with most birds, including *fortis* and *scandens* on Daphne Major, our estimates of age-specific survival (or mortality) are *approximately* constant. But unlike the Daphne data, ours do provide some indication of a rise in mortality rate with age. Age-specific survival decreased with age in each sex over the first four years of life among birds of known age, and decreased at a rate of 5.6 percent per year among the 1976 cohort of males from their third to ninth year. Loery et al. (1987) obtained a similar result, an estimated 3.5 percent annual decline in annual survival of black-capped chickadees. Likewise, Berndt and Sternberg (1963) documented a systematic decrease in age-specific survival with age of adult female pied flycatchers.

These observations can be construed as evidence for senescence, an intrinsic aging process. Hints of senescence have been seen in the data produced by several long-term studies of birds in the wild (e.g., Willis 1974; Woolfenden and Fitzpatrick 1984; Dhondt 1985; Koenig and Mumme 1987; Ratcliffe et al. 1988) and in captivity (e.g., Collias et al. 1986). The *conirostris* trends might be due to senescence or, alternatively, they could be the result of a gradual deterioration of extrinsic conditions associated with a rise in density. That is how we presented them earlier in this chapter. To expose an independent effect of senescence, which is likely to be small, we would need a better characterization of environmental quality in terms of food supply as well as density, and we would need larger samples of finches of known age.

Summary

About one in every ten *conirostris* fledglings survives its first year, and closer to one in twenty survives to breed. But survival thereafter is high, higher than in comparable temperate zone species, and higher in some groups among males than among females. The potential life span is also unusually long for a small passerine bird; the oldest birds reach 12 years or more. Thus the difference in success between the most and the least successful individuals, as measured by survival, is very great.

Mean annual survival from year 2 to year 8 was 81 percent for males and 78 percent for females among the adults present in 1978 when the study began. Annual survival rates tended not to remain constant but to decrease slightly with age over the major portion of the life span, and to drop sharply at the end. The mean rate at which survival declined with age was 5.6 percent a year for the 1976 cohort of males.

Monthly survival is higher in the wet season than in the dry season. Since food supply declines in the dry season, the high mortality at this time is probably the result of starvation. Adult survival varies among years in relation to food supply and density, and as a consequence, cohorts vary greatly in their survival patterns. Those whose members experience low density in the first two or three years of life fare better than those whose members experience high density. For example, the 1976 cohort had unusually high survival after the drought of 1977 under conditions of low density. Males of this cohort survived on average for 4.3 years beyond their second year, whereas males of known age born in the first five years after the drought survived on average for only 1.6 years beyond their second year. Increasing density may have contributed to the decrease in age-specific survival of the 1976 cohort. Survival could have been affected by density directly through aggressive interactions, for example, or indirectly through density-related competition for food.

The adult age structure of the population varies greatly among years, partly as a consequence of variation among cohorts in annual survival. The age structure becomes bottom-heavy as a result of strong recruitment following very wet years. It becomes top-heavy after dry years as a result of recruitment failure and high mortality among the young cohorts.

Survival patterns are similar to those exhibited by *fortis* and *scandens* on Daphne Major. For all three species they are explicable in terms of a variable food supply. Mean annual survival of adults is high, as it is in other tropical species of birds of similar body size living in forests. It is higher than in comparable temperate zone birds because finches do not experience an energetically stressful cold season or the hazards of migration. Annual variation in adult survival is also high, and in this respect the finches resemble temperate zone birds more than tropical (forest) birds. Thus *conirostris* and other Darwin's Finches, living in an environment that has both tropical and temperate zone climatic features, share demographic features with birds from both regions.

4

Demography: Reproduction

Introduction

Most birds die without reproducing. Several of those which attempt to breed fail completely, and although the majority succeed in raising their offspring to independence, few contribute to the next generation. The purpose of this chapter is to describe the principal features of reproduction and to explore some of the major reasons why success at reproducing varies so much.

Success varies primarily because suitable environmental conditions for breeding vary strongly and erratically. Some birds have the misfortune to be born just before those conditions deteriorate. Others have the good fortune to be born as those conditions are improving. Success varies for another reason; some birds are better than others at raising offspring. These two properties, of the environment and of the birds, serve to organize this chapter. The first part deals with patterns of variation in reproduction which characterize the population as a whole and which are associated with major variation in the environment: breeding phenology, the rate of reproduction, annual variation in the amount of reproduction, and recruitment. The second part is devoted to the performance of known individuals. It examines variation among individuals and the effects of age and prior experience on reproductive success. Then in Chapter 5 we will integrate the information on both reproduction and survival, and compare *conirostris* with other passerine species elsewhere.

It will be helpful to bear in mind the main climatic variation across the decade (Table 2.1 and Fig. 2.6). In particular, the years of 1983 and 1987 were extremely wet, 1984 was dry, and droughts occurred in 1985 and 1988.

Timing

Reproductive activity in the temperate zone is governed by a changing photoperiod, but in arid tropical and subtropical regions, where day-length variation is minimal, rain is of overriding importance (Lack 1950; Marchant 1958, 1960; Immelmann 1971; Kunkel 1974; Maclean 1976; Davies 1976, 1977). On the basis of a few observations on Islas Santa Cruz, San Cristóbal, Daphne Major, and elsewhere, Lack (1950), Snow (1966), and Grant and Boag (1980) suggested that rain which fell after a long dry season triggered the breeding of finches on the Galápagos. Rain could act as a direct stimulus and as an indirect one through effects on vegetation and arthropods to which the finches might respond. After the first rains the food supply increases with the development of nectar-producing flowers in less than a week and the emergence of some adult arthropods. Typically caterpillars appear first on *Bursera* foliage after about 10 days. Finches are in a state of readiness to respond rapidly to the first rain and take advantage of favorable conditions that may be short-lived, because both the time of arrival and the quantity of rain are highly unpredictable (Grant and Boag 1980).

The general correspondence between rainfall and the onset of breeding on Genovesa conforms to this simple scheme. Breeding does indeed begin soon after the first rain falls in the wet season (Fig. 4.1). The wet season of 1983 began exceptionally early, actually in November 1982, and breeding (egg laying) began within a couple of weeks. In maximal contrast, no rain fell in 1985 and breeding was not even attempted. In all other years, the first rain fell between early January and late February, and the first egg was laid 7–17 days later, with two exceptions discussed below.

When the wet season opens with a whimper, the breeding season does so too. In 1981, for example, the first egg was laid on day 17 after a total of 14 mm of rain had fallen. Only three females laid eggs, so the rain appears to have been an insufficient stimulus for the rest. Later in the season much more rain fell and all females bred. A repetition of the early events occurred in 1988 when 3 mm of rain fell in the last 10 days of January and 7 mm fell on February 11. One female laid eggs 4–6 days later. In contrast to 1981, only three more millimeters of rain fell and no other female laid eggs. In other years at least 20 mm of rain had fallen prior to the first egg being laid, and many females laid eggs. It therefore seems that 15–20 mm of rain is needed to induce a full reproductive response from the finches. The same conclusion has been reached from a study of patterns of breeding in relation to rainfall on Daphne Major (Boag and Grant 1984; Price 1985) and in the coastal region of mainland Ecuador (Marchant 1958, 1959; see also Maclean 1976).

Despite this general correspondence, the pattern of rainfall does not explain all variation in the onset of breeding. Other factors must be involved, some of which are intrinsic to the finches and unrelated to immediate rainfall, like physiological and endocrinological states, while others may be extrinsic and imperfectly correlated with rainfall, such as ambient temperature (e.g., Marchant 1959, 1960; Davies 1977) or atmospheric pressure.

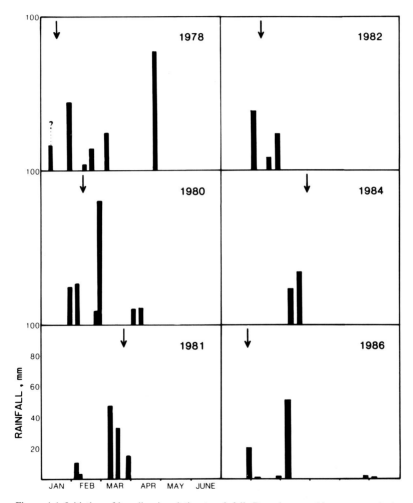

Figure 4.1 Initiation of breeding in relation to rainfall. Bars show weekly amounts of rain and arrows indicate the week when eggs were first laid. Years of exceptionally little and exceptionally extensive rains have been omitted.

For example, in 1986 breeding began *before* the first rainfall for the first time during the study. The first egg was laid on January 25, two days before the first rain (13 mm) of the year. This was not an isolated instance of a female out of synchrony with the rest, but the beginning of a population-wide breeding effort that reached a peak one week after the first rain and was completed on day 17 after less than 20 mm of rain had fallen. A few *Bursera* were in leaf, but caterpillars were not observed at this time on the four major plants with leaves: *Bursera, Cordia, Waltheria,* and *Cryptocarpus.* In fact the first caterpillars were not seen until February 8. Possibly a heavy garúa (misty rain) occurred in December 1985, and the effects on the vegetation were a sufficient if protracted stimulus for the birds to breed.

The same happened in 1987. The first rain of more than one millimeter fell on January 15 (6 mm), and the first heavy rain (35 mm) fell two days later. Four females began laying well before this, on January 5 and 6, whereas the rest began only after the heavy rain (Fig. 4.2).

Breeding in 1984 was also at variance with a simple explanation based on rainfall. In that year breeding (egg laying) began on day 17, which is equal to the longest previous delay, in 1981. In contrast to 1981, however, more than 20 mm of rain fell (on day 4), and by the time the first egg was laid 57 mm had fallen. Yet despite this apparent sufficiency of rain, only 12 females laid eggs, and chicks fledged from only one nest. Apart from the drought years of 1985 and 1988, this was the worst year for breeding.

The key to these unusual observations may lie partly in the pattern of breeding in the preceding wet season. The sluggish reproductive response to an apparently sufficient rainfall in 1984 followed a year of prolific breeding. The exceptionally early breeding in 1986 followed a year of no breeding. Responsiveness to external factors may be conditioned by a set

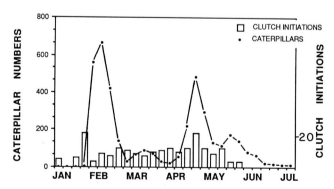

Figure 4.2 Clutch initiation in relation to caterpillar abundance in 1987. Data grouped into weekly units.

of internal factors, presumably at least partly hormonal, that are influenced by prior breeding experience. Whatever these unknown factors are, it is difficult to avoid the conclusion that birds were in a much greater state of readiness to breed at the end of the dry season early in 1986 than they were in 1984, and at the same time of year. The early breeding of a few individuals in 1986 and 1987 and of a single pair in 1988, following years of normal breeding, shows that other factors must be at work. The other factors probably include food supply, specifically the pollen and nectar of *Opuntia* cactus (Figs. 4.3 and 4.4). On Daphne Major, a few pairs of cactus finches (*scandens*) usually breed before the rains and at a time when the principal food supply is cactus pollen and nectar (Millington and Grant 1983, 1984; Boag and Grant 1984). The single pair of *conirostris* which bred in 1988 had the richest supply of cactus flowers in their territory.

THE END OF THE BREEDING SEASON

Breeding ceases in a much less concerted manner than it begins, and its end is much less obviously influenced by rainfall. In 1983, for example, the last chick fledged 9 days after the last rain, whereas in 1986 the comparable interval was 80 days, with two broods being produced after the last rain. Generally the interval is 35–49 days.

The major factor determining the end of breeding is probably food supply, in particular the supply of caterpillars and spiders, which are the principal foods given to nestlings (Table 4.1). Figure 4.5 shows a decline in the abundance of caterpillars toward the end of the breeding season in the three years in which we counted them. Breeding was almost at an end in 1983 when rain was still falling heavily and some plants were still growing vegetatively (P. R. Grant and B. R. Grant 1985, 1987). Just why the caterpillars declined before the end of the wet season in that year is not known. They were heavily exploited by increasing numbers of finches, mockingbirds, warblers, and doves; they were subject to some attack from insect parasitoids; and they may have had to cope with increasing levels of plant chemical defenses. On Daphne Major the caterpillar supply remained high for longer in 1983, and the finches (*fortis* and *scandens*) bred for one month longer (Gibbs and Grant 1987b). In all other years on Genovesa, the breeding of *conirostris,* and of *magnirostris, difficilis,* and *Certhidea* as well, ceased when weekly indices of caterpillar abundance on 12 major plant species fell to low values (but not to zero). The same has been observed for populations of *fuliginosa* and *difficilis* on Isla Pinta (Schluter 1984a).

A decline in caterpillar abundance is sufficient to explain the end of breeding in 1983, but heavy rainfall may have been an additional factor, reducing the amount of time for foraging. Parents are under the double stress of being unable to forage in the heaviest rain and having to meet the

Figure 4.3 *Opuntia helleri* flowers. The flower above has just opened and is intact; the one below has lost pollen and nectar to a cactus finch which has removed the stigmas. Photos by O. Jennersten (above) and P. R. Grant (below).

Table 4.1 Nestling Diets: Frequencies of Occurrence (%) of Food Items in the Crops of Nestlings in February to May 1978

Arthropods		Plants	
Spider	50.0	*Croton* seed	60.3
Caterpillar	74.4	*Bursera* aril	24.4
Moth	1.3	*Bursera* seed	11.5
Beetle	7.7	*Opuntia* pollen	9.0
Beetle larva	7.7	*Opuntia* fruit fiber	1.3
Fly	2.6	*Opuntia* seed	1.3
Fly larva	10.3	*Chamaesyce* seed	2.6
Fly pupa	32.0	Green seed	1.3
Chitin	6.4	Grass leaf	1.3

Miscellaneous	
Egg shell	10.3

Number of family crop samples	78
Number of nests	32
Number of food types	19

Source: Grant and Grant (1980).

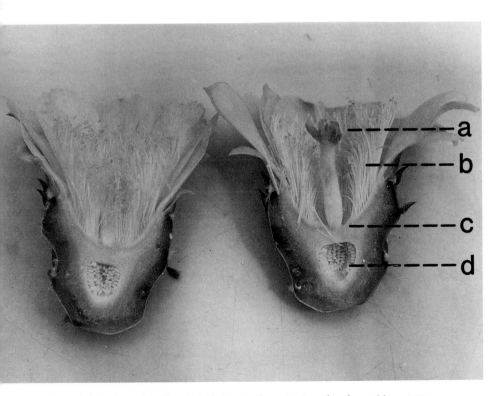

Figure 4.4 Sections of an *Opuntia helleri* cactus flower to show the stigmas (a), stamens (b), nectar-secreting pad around the base of the central style (c) and ovules (d). Photo by P. R. Grant.

increased energy needs of their nestlings when they get wet and cold. Most rain fell in the last two months of the 1983 breeding season (Fig. 2.7).

The Rate of Reproduction

Between the start and finish of the breeding season, the rate at which reproduction occurs depends on the length of three intervals: the intervals

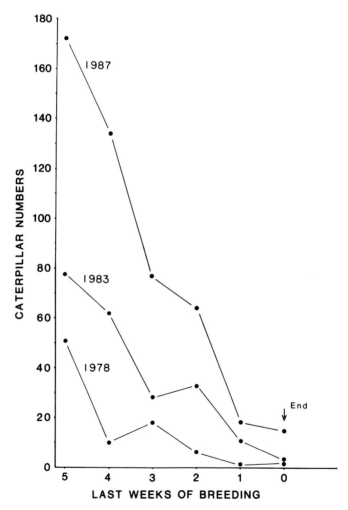

Figure 4.5 The decline in caterpillar abundance at the end of three breeding seasons. Caterpillars were counted every week on 100 leaves on each of 10 plants of the 12 most common plant species.

between egg laying and hatching, between hatching and fledging, and between hatching and the initiation of a new clutch.

Eggs are laid one per day on successive days until the clutch of usually three to five eggs is complete. The female (only) begins to incubate the eggs before the clutch is complete, and the incubation ends when nestlings hatch on the same or on two successive days. There is very little variation among clutches in the duration of the incubation phase. The interval between the day the last egg is laid and the day the last nestling hatches is usually 12 days, very rarely 11, and occasionally 13 (Grant and Grant 1980). There is likewise little variation in the nestling phase. From the day the last nestling hatches to the time the last one fledges, 14 days usually elapse. Extreme values of 12 and 17 days may be the result of disturbance by mockingbirds or owls on the one hand, and poor feeding conditions and growth on the other. Thus the period from the first egg entering the nest to the last nestling leaving it is 30 days.

The major component of variation in the rate at which reproduction takes place is the interval between successive broods. Females do not begin a second clutch before the first set of eggs has hatched, but neither do they always wait until the first nestlings have all fledged. Sometimes their successive broods overlap. Overlap occurs when the interval between first egg dates of successive clutches is less than 30 days. The shortest interbrood interval we have measured is 19 days, that is, a second clutch was started 4 days after the first hatching from the previous one, while the nestlings of the first were being fed by both parents. The longest interval was 49 days; the usual range is 25–37 days.

The incidence of brood overlapping varies substantially among years. It was most frequent in 1978 (44 percent) and least frequent in 1983 (11 percent). It also varies within seasons, tending to be greater early in the season (Table 4.2). The comparison of mean interbrood intervals shows us the same (Table 4.3). In 1980, 1981, 1982, and again in 1987, intervals between first and second broods were shorter than intervals between second and third broods (paired t tests, $P < 0.05$ in each case). The difference between early and late broods occurred in 1983, but over a longer breeding season ($F_{4,66} = 2.95$, $P < 0.05$). In 1986 the reverse occurred, with first intervals being longer than second ones (paired ($t_9 = 3.27$, $P < 0.01$).

Thus the rate of clutch production is faster at some times than at others. We suspect that the temporal pattern of food availability has the largest influence on this variation, although we have few measurements to support this idea. Intervals lengthen toward the end of the season (Table 4.3) when, in general, food supply declines. In 1986 the average interval shortened instead, because the birds first bred when little rain fell and then produced two more broods after much more rain fell, which may have induced a

Table 4.2 Incidence of Brood Overlapping

| Months | Number of Broods | | % Overlapping |
	Overlapping	Separate	
1978 January	3	5	37.5
February	5	3	62.5
March	3	6	33.3
1980 February	13	10	56.5**
March	0	11	0.0
1981 March	11	10	52.4*
April	1	13	7.1
1982 February	12	16	42.9
March	6	13	31.6
1983 December	3	6	33.3
January	4	16	20.0
February	0	9	0.0
March	0	14	0.0
April	1	18	5.6
1986 February	1	15	6.3
March	6	13	31.6
1987 January	11	6	64.7**
February	5	12	29.4
March	3	18	14.3
April	7	7	50.0

Note: Months refer to those in which the first (successful) of two successive clutches were started. Overlap occurs when intervals between first eggs of successive clutches are less than 30 days. Proportions differ between months in 1980, 1981, and 1987 only (χ^2 tests, $P < 0.05$*, and $P < 0.01$**).

Table 4.3 Average Intervals in Days between Successive Broods When the First Is Successful

Year	Dec	Jan	Feb	Mar	Apr
1978	—	31.4 ± 1.92 (8)	29.2 ± 2.32 (8)	32.3 ± 1.86 (9)	—
1980	—	—	29.0 ± 0.97 (23)	35.7 ± 1.06 (11)	—
1981	—	—	—	30.1 ± 1.15 (21)	34.6 ± 1.60 (14)
1982	—	—	31.4 ± 1.20 (28)	34.1 ± 1.49 (19)	—
1983	33.0 ± 2.25 (9)	32.0 ± 1.07 (20)	38.0 ± 1.74 (9)	36.2 ± 0.80 (14)	35.4 ± 1.25 (19)
1986	—	—	36.1 ± 1.26 (16)	30.4 ± 0.80 (19)	—
1987	—	28.2 ± 1.21 (17)	32.8 ± 1.20 (17)	33.7 ± 1.05 (21)	30.6 ± 1.12 (14)

Note: Intervals are measured by the time between the dates on which the first eggs were laid. Means and standard errors are shown, with sample sizes in parentheses.

greater response from both the vegetation and arthropod populations. The average interval shortened at the end of 1987. For this year we do have estimates of caterpillar abundance to compare with breeding phenology. Caterpillar numbers reached a peak in the second week of February and again in the last week of April (Fig. 4.2). Caterpillar biomasses probably reached maxima in the following one or two weeks. New clutches were initiated at these times after short intervals (Table 4.3); intervals were shortest between first (January) and second (February) clutches, and between fourth (April) and fifth (May) ones.

In addition to the supply of energy (food), the demand on the supply by nestlings has an influence on the timing of renesting. The larger the number of nestlings to be fed, the longer is the delay in renesting. This is indicated by a comparison of the interclutch interval and the number of nestlings from the first clutch. Successive interclutch intervals within a breeding season increased seven times and decreased six times (Table 4.3). Ten of the 13 changes were paralleled by changes in the number of nestlings that fledged. When the mean number of nestlings increased, the mean intervals lengthened; when the number of nestlings decreased, so did the time taken to start another clutch (sign test, $P < 0.05$, one-tailed). The three exceptions are interpretable in terms of the pattern of rainfall and hence the probable pattern of food supply.

Thus an interaction between food supply and nestling demand determines the time a female renests. Finches breed as quickly as possible, and intervals between successive clutches are shortest when food is most plentiful in relation to demand.

THE FEEDING OF FLEDGLINGS BY THE MOTHER

If the female has not begun another clutch by the time the nestlings fledge, the likelihood of renesting can be influenced by the physiological condition and behavior of her current fledglings, and hence the need to help the male feed them. The female can devote energy to feeding fledglings or to producing and incubating another clutch of eggs, but not to both activities, except to a very small extent. Thus these activities are in conflict. Resolution of the conflict varies annually. This variation is another indication that food supply governs the rate of reproduction, because the maternal contribution to the feeding of fledglings varies in association with rainfall. The evidence is as follows.

First we need to introduce the phenomenon of brood division. There is a tendency for the parents to specialize in the feeding of their fledglings, with the father feeding one to three of them and the mother feeding the remainder. Throughout the study a total of 235 fledglings were observed being fed by a parent only once in the ratio of 78 by the father to 22 by the mother. Fifty-two fledglings were observed being fed on two separate occasions. On a random (nonselective) basis, $0.78^2 = 0.61$ is the proportion that should have been fed both times by the father, $0.22^2 = 0.05$ should have been fed both times by the mother, and the remainder (0.34) should have been fed once by each of the parents. Observations differed from these expectations ($\chi^2_2 = 9.65$, $P < 0.01$). Observations of fledglings being fed by mothers only were overrepresented, and observations of fledglings being fed by both parents were underrepresented. Thus parental specialization and brood division, though not absolute, tend to occur.

The degree to which brood division occurs varies annually. Most of our observations of fledglings being fed were made in three years: 1980, 1986, and 1987. In 1980 and 1986, two years of moderate rainfall in which breeding occurred three times, 12 out of 49 fledglings were seen to be fed on two to four different days by the mother only, and an additional 11 were fed by both parents. In 1987, a year of extensive rainfall in which breeding occurred up to six times, only 1 out of 15 fledglings fed more than once was fed by its mother, and none was seen to be fed by both parents. The distribution of parental feedings did not differ between 1980 and 1986, but it did differ between the two years combined and 1987 ($\chi^2_2 = 8.14$, $P < 0.02$). The difference cannot be explained by a greater opportunity or need for brood division in the early years arising from a larger number of fledglings being produced, because on average more were produced in 1987 than in the early years (Table 4.4).

Price and Gibbs (1987) observed a similar annual shift in maternal contributions to the feeding of *fortis* and *scandens* fledglings on Isla Daphne Major in relation to rainfall and measured food supply. They interpreted brood division as a parental means of maximizing the efficiency of feeding the fledglings when food is limited. Further reproduction is forfeited or postponed under these conditions but not when food is plentiful, for then the energy needs of the fledglings can be met by a single parent. The mother, freed from the necessity of contributing to the feeding of fledglings, is able to lay and incubate another clutch of eggs. The father does neither, so he enhances his reproductive success by feeding the fledglings. From this reasoning, a positive association is expected between interbrood intervals and the incidence of maternal feedings. It was present in our study but was not strong. In 1987, when food conditions were favorable and mothers fed their fledglings rarely, intervals were not as long on average as in 1980 and 1986 (Table 4.3).

The Amount of Reproduction

How much breeding occurs is determined mainly by the amount of rain (Fig. 4.6) and the length of the period over which it falls. As mentioned previously, in drought years (e.g., 1985) breeding does not occur, whereas in wet years (e.g., 1982–83) breeding occurs uninterruptedly for as long as seven to eight months. Birds breed as long as conditions remain favorable, apparently, and that period can vary enormously from year to year.

Tables 4.4, 4.5, and 4.6 summarize the breeding statistics for each year. The best measures to compare among years are the average figures for eggs, nestlings, and fledglings produced by those pairs studied fully throughout a season (Table 4.4). The sample sizes are smaller than the actual totals because of uncertainties about the data or known incompleteness for many pairs, especially in 1983.

We start with the number of eggs in a clutch, which can be thought of as simultaneously representing two features of reproduction: initial effort and potential success. The first is measured in the currency of energy, the second in the currency of numbers of independent offspring. Completed clutches vary from 2 to 5 eggs (Figs. 4.7 and 4.8), with an average of between 3 and 4 (Table 4.4). Each egg weighs about 2.7 g (Grant 1982), which is a little more than 10 percent of an average female's weight, a typical value for finches (Newton 1972; Grant 1983a). Therefore an average clutch is a commitment of energy equivalent to 30–40 percent of her body weight. To judge from data for other passerines (King 1973), it is a commitment of daily energy 15 percent above the normal requirements

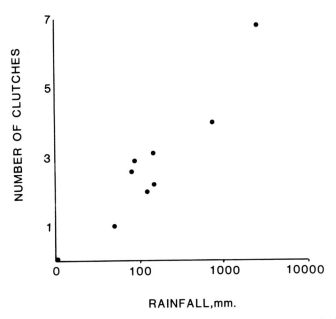

Figure 4.6 The mean number of clutches per pair as a function of the amount of rain in nine breeding seasons.

Table 4.4 Production of Eggs, Nestlings, and Fledglings by Those Pairs Studied Fully

Year	Number of Pairs	Mean Number per Clutch			Number of Clutches	Number of Fledglings/Pair
		Eggs	Nestlings	Fledglings		
1978	15	4.1 ± 0.1	3.1 ± 0.2	2.7 ± 0.2	3.1 ± 0.2	7.9 ± 0.7
1979	0			not known		
1980	35	3.4 ± 0.1	2.6 ± 0.2	2.1 ± 0.2	2.2 ± 0.1	4.6 ± 0.4
1981	46	3.5 ± 0.1	3.2 ± 0.2	2.2 ± 0.2	2.0 ± 0.1	4.5 ± 0.4
1982	44	3.3 ± 0.1	2.3 ± 0.1	2.0 ± 0.2	2.6 ± 0.1	4.9 ± 0.4
1983	15	3.4 ± 0.1	1.7 ± 0.2	1.3 ± 0.2	6.8 ± 0.3	8.1 ± 1.2
1984	12	2.9 ± 0.1	1.2 ± 0.4	0.2 ± 0.2	1.0	0.2 ± 0.2
1985	0	0	0	0	0	0
1986	17	3.6 ± 0.1	3.2 ± 0.2	2.3 ± 0.2	2.9 ± 0.1	6.9 ± 0.8
1987	13	4.1 ± 0.1	3.0 ± 0.2	2.4 ± 0.2	4.1 ± 0.3	9.9 ± 1.0
1988	1	3.0	3.0	3.0	1.0	1.0

Note: Birds who changed mates within a season were not included. Means and one standard error are shown.

Table 4.5 Nesting Success (Minimum of One Fledgling)

Year	Number of Nests Studied	Number Predated	Number Unsuccessful for Other Reasons	Percentage of Nests Successful
1978	41	3 (7.3)	4 (9.8)	82.1
1979			not known	
1980	77	12 (15.6)	7 (9.1)	75.3
1981	92	2 (2.2)	13 (14.1)	83.7
1982	109	20 (18.3)	34 (31.2)	68.8
1983	237	82 (34.6)	38 (16.0)	49.4
1984	12	5 (41.7)	6 (50.0)	8.3
1985			no breeding	
1986	62	8 (12.9)	3 (4.8)	82.3
1987	172	40 (23.3)	16 (9.3)	67.4
1988	1	0 (0.0)	0 (0.0)	100.0

for a period of three to four days. This is the initial reproductive effort. The number of eggs sets the potential success to be gained from a single breeding episode, a success that can be achieved only by devoting much more energy to parental care than the initial energy required in egg formation (e.g., see Ricklefs 1974; Hails and Bryant 1979).

Population figures for clutch size in Table 4.4 and also Figure 4.7 show annual variation in means. Figure 4.8 illustrates a characteristic tendency for mean clutch sizes to decrease as a breeding season progresses. Mean clutch size was approximately the same in most years, but deviated (significantly) from the general trend in three years (Fig. 4.7); it was

Table 4.6 Some Measures of Reproductive Success in Relation to Reproductive Potential Expressed in Percentages

Year	Hatching Success		Fledging Success	
	Total Clutch Failures		Fledglings/Nestlings	Fledglings/Eggs
	Included	Excluded		
1978	80.0	80.0	87.1	65.9
1979			not known	
1980	76.5	90.8	80.8	61.8
1981	91.4	94.4	68.8	62.9
1982	69.7	86.2	87.0	60.6
1983	50.0	75.3	76.5	38.2
1984	41.4	73.3	16.7	6.9
1985			no breeding	
1986	88.9	95.5	71.9	63.9
1987	74.4	84.9	82.8	61.6
1988	100.0	100.0	100.0	100.0

Source: Data from Table 4.4.

elevated when breeding density was low in 1978 and 1987, and depressed in the dry year of 1984 (Table 4.4). Clutch sizes of many passerine species vary among years with the amount of food available at the time of egg formation (Hussell and Quinney 1987, and references therein), so the magnitude of variation in clutch size reflects the amount of annual variation in food supply. The coefficient of variation of mean clutch size is somewhat lower in *conirostris* (11.40, $N = 8$ years) than in *fortis* (16.11, $N = 8$; from data in Gibbs 1988) and *scandens* (18.91, $N = 6$; Millington and Grant 1984) on Daphne Major, but all three species vary much more than two continental finch species that have been studied for similar periods of time, the scarlet rosefinch (3.87, $N = 8$) and bullfinch (2.73, $N = 10$) in Europe (Stjernberg 1979; Bjilsma 1982; Millington and Grant 1984). Mean clutch size of *conirostris* did not vary in association with annual

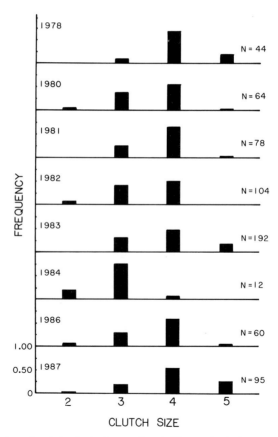

Figure 4.7 Annual variation in the frequencies of different clutch sizes. Sample sizes (*N*) are shown on the right. The modal clutch size was four in all years except 1984 when few birds bred.

rainfall ($r_s = 0.43$, $z = 1.13$, $P > 0.1$), however, unlike the situation on Daphne Major with both *scandens* (Millington and Grant 1984) and *fortis* (Price 1985; Gibbs 1988), possibly as a result of modifying effects of density.

The mean number of fledglings produced by a pair in a season is a measure of how much of the reproductive potential embodied in eggs is translated into reproductive output or success. The means fall into three groups: very low or zero in the dry years of 1984, 1985, and 1988; high values in 1978, 1983, 1986, and 1987; and intermediate values in the remainder. The main determinant of this heterogeneity is the number of clutches of eggs produced by a pair, which is a function of the amount of rain (Fig. 4.6) and the length of the breeding season. There is a very strong rank order correlation ($r_s = 0.98$, $z = 2.58$, $P < 0.005$, one-tailed)

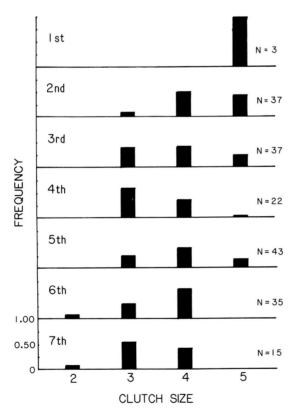

Figure 4.8 Seasonal variation in the frequencies of different clutch sizes in 1983. Sample sizes (*N*) are shown on the right. A slight seasonal decline is evident. Note in particular the frequencies of 2-egg and 5-egg clutches.

between number of clutches and total number of fledglings per pair among the eight seasons of breeding (see Table 4.4). Each of these is separately correlated with annual totals of rainfall ($P < 0.05$, one-tailed).

One other source of variation, the number of fledglings produced from each nest, completes the picture of an annually varying reproduction. In most years this number is an approximately constant proportion of the number of eggs laid: 63 percent (Table 4.6). In the extreme years of 1983 and 1984, fledging success was markedly reduced (Table 4.6). Predation of eggs and nestlings by mockingbirds and owls was more frequent than in any other year (Table 4.5). The frequency of abandonment of nests by the parents was also high in both years. In 1983 nests and the bushes they were placed in were damaged by storms, and five of them fell to the ground. Although the supply of caterpillars was plentiful, the time available for the parents to forage was repeatedly reduced by strong and continuous rain for spells of more than one hour. Rain fell on 11 consecutive days three times during the season. Some nestlings starved, and some nests were deserted following long periods of continuous rain. In 1984 the food supply itself declined while nestlings were being fed, and many of them starved. As a footnote to this comparison, not all El Niño years are the same nor are their consequences, as we found with adult survival (Chapter 3). In 1987 nests and bushes fell down (8 in total; see Fig. 3.7), and nest predation was a little high, but the combined losses did not result in a deviation from the general trend of about 63 percent of potential fledgling production (Table 4.6). Periods of rain did not restrict foraging to the same extent in that year as in 1983.

RECRUITMENT

Very few fledglings survive to breed (Chapter 3). Only 23 males and 32 females, born in the two study areas in the years 1978–86, bred or attempted to breed in the study areas in the years 1980–87. With such small numbers it is not easy to identify the causes of successful recruitment.

Not surprisingly, conditions favoring a high density of breeders promoted strong recruitment. The number of new recruits to the breeding population increased in direct relationship to total population density ($r = 0.90$, $P < 0.05$). The proportion of the total breeding population made up of new breeders of known origin did not remain constant, however. In the years 1980–86 the proportion varied significantly among years ($\chi_4^2 = 11.83$, $P < 0.02$). It was highest in 1983 (18.2 percent), breeding having begun after an unusually short dry season, and lowest in 1982 (5.3 percent). The proportion should have been low in 1980 (when banded recruits could only have been two years old because nestlings were not banded in 1979) and in 1981 (when only 1-year-old and 3-year-old banded recruits could have bred), but when the data for these two years are eliminated, there still remains a significant variation among years ($\chi_2^2 = 7.28, P < 0.05$).

The probability of surviving to breed (locally) appears to be generally random with respect to three factors of potential importance: the size of the clutch from which the individual originated, the brood of origin (first, second, etc., in a season), and the experience of the parents. Two exceptions occurred in 1983, however. First, birds born of inexperienced parents survived better, to breed eventually in 1986, than did the rest. The effect of experience of parents on their breeding success will be discussed in the next section. Second, birds born (hatched) early in the season also survived better than the rest. The seven banded recruits in 1986 were born in December 1982 (1) and January 1983 (6); none born in February to July 1983 survived to breed. Such a disparity is not expected by chance ($\chi_6^2 = 16.76, P \approx 0.01$). It presumably arose because early born birds were in better condition than later born birds when density increased and environmental conditions started to deteriorate in the latter part of 1983. The pattern of mortality of the 1983 cohort over the next three years indicates the same. Differential mortality with respect to month of birth occurred from 1983 to 1984 ($\chi_6^2 = 14.61, N = 420, P < 0.05$), but not from 1984 to 1985 ($\chi_6^2 = 9.62, N = 69, P > 0.1$), or from 1985 to 1986 ($\chi_4^2 = 4.40, N = 16, P > 0.1$).

Recruitment on Daphne Major has twice taken place remarkably early, in the year of birth. Young *fortis* and *scandens* individuals born in the first three months of the exceptionally long breeding seasons of 1983 and 1987 were breeding before the end of the season, in some cases with success (Gibbs et al. 1984; 1987 data unpubl.). We were alert to the possibility of a similar rapid maturation of *conirostris* juveniles, but we observed no signs of their intention to breed in either year.

Effects of Age and Experience

Variation in the population characteristics we have just described does not reveal how individual birds respond, possibly differently, to changing environmental conditions, or how individuals change as they grow older and gain experience. We will now consider the effects of age and experience on the timing and rate of reproduction, and on clutch size and breeding success. Age and experience usually cannot be distinguished as they generally covary. In the final part of this section we will attempt to separate their effects upon reproduction.

EFFECTS ON THE TIMING OF BREEDING

Under any one set of environmental conditions, females differ in their egg-laying dates; they differ in the rate at which they respond to rain or other factors and start to breed. A conspicuous contributor to this variation is age and experience. For example, three of the four females breeding before the rains in 1987 were at least 4, 5, and 11 years old, respectively; the age of the fourth was not known. Females with previous breeding experience in the study area (area 1) bred generally earlier than those breeding in the area for the first time. The differences were 8 days on average in 1980, 5 days in 1981, and 3 days in 1982. The first two were statistically significant (Mann-Whitney U tests, $P < 0.005$). In 1986 there was no difference, but the sample of experienced females was extremely small (4). The trend was most pronounced in the two El Niño years, when all experienced females bred at first brood time whereas the majority of new females started breeding two to four months later.

EFFECTS ON THE RATE OF REPRODUCTION

As birds age and gain experience, especially with each other, they speed up reproduction by shortening the interval between successive broods. For the years 1980–83 we made 11 comparisons between new pairs and old (established) pairs and found that the interval was shorter for the established pairs in 10 of the comparisons. We would not expect this to happen by chance (sign test, two-tailed, $P = 0.01$).

For the combined data there is a three-day difference between the means for established pairs (31, $N = 65$) and new pairs (34, $N = 76$). New pairs are heterogeneous; some are made up of birds breeding for the first time, others have members who have bred with a different partner before. The three groups differ significantly in interval lengths ($F_{2,125} = 4.61$, $P < 0.02$). The difference among them is greatest between old birds in old pairs and new birds in new pairs; old and new birds in new pairs are intermediate.

EFFECTS ON CLUTCH SIZE

Females differ in clutch size according to past breeding experience. In all years from 1980 onward, except 1986 when only four old females bred, and again in 1987, mean clutch sizes were higher among those who had previously bred than among those who had not (Appendix II, Table A1). In one year (1981) the difference was statistically significant ($t_{32} = 2.47$, $P < 0.02$).

The small differences in clutch sizes between experienced and inexperienced females breeding in the same year can arise in two ways; either (a) females increase their clutch size with age, or (b) inexperienced females that lay larger than average clutches have a higher than average chance of surviving to breed again as experienced females.

The second possibility of differential survival does not seem to be correct, because there was no difference in mean clutch size between survivors and nonsurvivors among females breeding initially in 1980 (t_{10} = 2.18), in 1981 (t_{13} = 0.45), or in 1982 (t_{19} = 0.57). The difference in 1980 approaches statistical significance ($0.1 > P > 0.05$), but it is in the opposite direction to that expected; the females which died had the larger clutches.

There is some support for the aging hypothesis, in that ten females had larger clutches in 1981, when experienced, than in 1980, when breeding for the first time (Wilcoxon $T = 4.5$, $P < 0.02$). Of course 1981 could have been a better year than 1980 for all females, but this does not seem to have been the case. Comparing the mean clutch sizes in 1980 and 1981 of eight females who had breeding experience prior to 1980, we found no difference (Wilcoxon $T = 7$, $P > 0.1$). Moreover, the mean clutch size of twelve inexperienced females in 1980 was the same as the mean of fifteen (new) inexperienced females in 1981 ($t_{25} = 0.44$). Therefore the increase in mean clutch size in 1981 of the group of ten females that had bred for the first time in 1980 can be attributed to age and experience and not to a year effect.

Although age and experience are important, they are not sufficient to explain, when combined with the different age compositions in different years, the annual variation in mean clutch size shown in Table 4.4. They do not explain the fact that in 1983 clutch sizes were generally higher, and in 1984 they were generally lower. For example, in 1983 three females were breeding for their third time (year), seven were breeding for their fourth time, and six were breeding for at least their sixth time. Collectively their mean clutch size for the whole of 1983 was higher than in 1982 (paired $t_{14} = 4.54$, $P < 0.001$). The nine who survived to 1984 and attempted to breed produced smaller clutches that year than in 1983 (paired $t = 3.72$, $P < 0.01$) or in 1982 ($t = 5.44$, $P < 0.001$). These differences demonstrate annual variation in clutch size among the same females, independent of age and experience. Price (1985) established the same for *fortis* on Isla Daphne Major.

EFFECTS ON REPRODUCTIVE SUCCESS

Using the same classification of pairs as experienced, inexperienced, and mixed, we have compared different aspects of reproductive performance to assess the role of age and experience in success. The statistics for each year are tabulated in Appendix II and summarized in Table 4.7.

Experienced pairs are superior to inexperienced pairs in all respects, with mixed pairs being intermediate. Among mixed pairs it does not seem to matter which member is inexperienced. There is a tendency for mixed pairs with a young female to do better than pairs with a young male, but not to a statistically detectable extent, so we have combined them. There is no detectable effect of additional experience (or age) after the second

Table 4.7 Reproductive Performance of Birds Who Have Bred Before (Experienced) or Not (Inexperienced)

	Pairs			% Difference between Experienced and Inexperienced Pairs
	Inexperienced	Mixed	Experienced	
Clutch size	3.68	3.71	3.81	3.5
No. clutches/season	1.87	2.75	3.07	64.2
% Nest failures	43.38	33.62	23.97	44.7
No. fledglings/brood	1.83	1.97	2.40	31.1
No. fledglings/season	3.27	4.77	7.38	125.7

Note: Pairs comprise two experienced birds, two inexperienced birds, or one experienced (of either sex) and one inexperienced (mixed pairs). The numbers are annual means and are based on values in Appendix II, Tables A1–A5.

year of breeding, so we have combined all experienced birds regardless of age.

Experience plays an important role in determining the number of clutches produced in a season; experienced pairs produced 64 percent more clutches on average than did inexperienced pairs (Table 4.7). The difference was consistent among years (Appendix II, Table A2) and statistically significant (t tests) in every year except 1986. Clutch size itself, although varying with experience as shown above, is a much less important contributor to success than is the number of clutches in a season (Table 4.7).

A factor of comparable magnitude to the seasonal number of clutches is nesting success. This refers to the success in raising at least one fledgling and encompasses the avoidance of starvation and predation. Experienced pairs were relatively more successful in this respect than inexperienced pairs in every year (Appendix II, Table A3), and overall their frequency of outright failure was approximately half that of inexperienced pairs. They abandoned clutches less often, fewer broods starved completely, and fewer were consumed by predators.

The number of fledglings produced from a given brood depends upon initial clutch size and subsequent losses through starvation, predation, or abandonment at either egg or nestling stage. Largely as a result of the inequalities described above, experienced pairs fledged more young from a single brood than did inexperienced pairs (Table 4.7). The differences, although not statistically significant in any one year, were consistent among the five years (Appendix II, Table A4). The relatively small difference in fledging success from individual nests becomes magnified, however, by the large difference in number of broods, so that the total production of fledglings per season was more than twice as great for experienced pairs as it was for inexperienced pairs (Table 4.7). The differences were statistically significant in four years out of five (Appendix II, Table A5).

Recruitment did not vary with parental age and experience, except in one surprising way. Five of the banded recruits in 1986 were born of inexperienced parents in 1983; the remaining two had experienced parents. In relation to the proportion of fledglings produced by experienced, inexperienced, and mixed pairs breeding in 1983, this disparity cannot be attributed to chance ($\chi^2_2 = 7.55, P < 0.05$). A possible explanation is that extended parental care was advantageous in 1983 because feeding was often interrupted by rain. Inexperienced birds in 1983 may have given long care to the offspring, as they produced fewer broods at greater intervals than experienced birds (Appendix II, Table A2).

AGE OR EXPERIENCE?

As birds age they gain experience in finding food, detecting predators, and breeding. The differences in clutch size and other aspects of breeding performance between experienced and inexperienced birds could arise from general experience associated with age or specific experience associated with breeding. To separate effects of age and prior breeding experience on reproduction we need to make comparisons between birds in such a way that one factor is controlled and the other one varied. For example, age could be controlled by comparing two-year-olds breeding for the first time with two-year-olds breeding for the second time (Nol and Smith 1987). Our samples of seven and one female, respectively, are inadequate for making such a comparison. Similarly three-year-olds breeding for the first time and second time could be compared. Many females bred for the first time in their third year in 1986, their first breeding having been delayed by the dry conditions of 1984 and 1985, but only one female of known age bred for the second time in her third year.

The alternative is to control experience but allow age to vary. This we can do by comparing the first clutches produced by 18 females that bred initially after one year with the first clutches of six others that started breeding after two years. Since they first bred in different years, and some started breeding at different times within a year, we first standardized clutch sizes to a common mean to eliminate the effects of seasonal and annual variation in average clutch size (Figs. 4.7 and 4.8). The older birds had larger clutches than the younger birds (Mann-Whitney $U = 84.5$, $z = 2.03, P < 0.05$, one-tailed).

A possible complication is that the older birds may have bred on better territories. As a check on this we compared the nearest neighbors of the two groups of females on the assumption that neighboring territories are similar in quality. The neighbors were at least two years old, and many were older. The two groups of neighbors did not differ in mean clutch sizes ($U = 64.5$, $z = 0.95$, $P > 0.1$).

We conclude that clutch size increases with age of female, at least from year 1 to year 2, in the absence of any effect of prior breeding experience. Prior breeding experience may have an additional and independent effect on clutch size, but we have been unable to assess this. We believe that prior feeding experience is likely to be an important factor underlying the effect of age on clutch size. It could account for the slowness in breeding as well as the small clutches of young females.

Studies of other passerine birds have shown a similar effect of age on clutch size (Stobo and McLaren 1975; Harvey, Greenwood, Perrins, and Martin 1979; Baker et al. 1981; Perrins and McCleery 1985; Nol and Smith 1987), and occasionally an independent effect of prior breeding experience (Harvey et al. 1985), but not always when it has been investigated (Nol and Smith 1987). The role of food-gathering skills in governing clutch size and the timing of egg laying can be inferred from correlations between mean clutch size (of birds of all ages) and food supply (Anderson 1977; Perrins 1979; Hussell and Quinney 1987), and from the experimental demonstration that abundant food increases clutch size and the number of breeding attempts as well as advancing laying date (Arcese and Smith 1988, and references therein). On Daphne Major clutch size increases with female age in *fortis* (Gibbs 1988) and increases with rainfall, which is positively correlated with food supply, in both *fortis* (Price 1985; Gibbs 1988) and *scandens* (Millington and Grant 1984).

Individual Variation

Individual females differ in the time at which they start laying eggs, the interval between successive clutches, and the size of their clutches. Age and experience account for some of this variation; older females reproduce more and faster. After allowance is made for these effects, however, there remains a large amount of variation in all three reproductive traits. Furthermore, individual females tend to be consistent from year to year, hence differences among them tend to be consistent. A prime example of individual consistency is the timing of breeding of No. 6980, the longest living female who bred from 1978 to 1987. She laid her first egg on day 1 of the egg-laying period in 1978, again in 1981 (the nest was destroyed in 1980), on day 3 in 1982, day 2 in 1983, day 4 in 1984, day 9 in 1986, then again on day 1 in 1987.

Consistent differences among females may reflect different genetic properties or nongenetic differences among them arising either from enduring effects of conditions during their growth or from immediate effects of the territories they occupy. If there is heritable variation among females, it should be manifested as a resemblance between mothers and daughters. For example, Perrins and Jones (1974) and van Noordwijk et al. (1980, 1981) have estimated the heritability of clutch size of great tits by regressing the clutch sizes of daughters on the clutch sizes of their mothers, obtaining significant values of 0.48 and 0.37, respectively. For our data the estimate is 0.74 ± 0.54, which is not significantly different from zero, but it is based on a sample of only 24 mother-daughter pairs, whereas the great tit studies had samples in the hundreds.

Nevertheless, consistency in the performance of *conirostris* females is statistically demonstrable. There is a tendency for females to breed consistently early, or consistently late, relative to others at the time breeding starts in successive years (Table 4.8). In like manner the interval between first and second clutches of individual females tends to be consistently long or short in different years, and their clutch sizes tend to be consistently large or small. All correlation coefficients in Table 4.8 are positive, and some are significantly different from zero.

For assessing the degree to which consistent differences among females reflect an underlying genetic variation among them, the appropriate technique is repeatability analysis. Repeatability is the proportion of the total variance in a trait, such as clutch size, that occurs among rather than within individuals (Lessells and Boag 1987). It provides an upper limit to the estimate of heritability (Falconer 1981) because the estimate may be inflated by nongenetic effects, such as consistent variation in territory quality, which could have a determining influence on both the rate and amount of reproduction.

Table 4.8 Correlations (r) between Years of the Date the First Egg Was Laid by Individual Females, the Interval between Their First and Second Clutches, and Their Mean Clutch Size

	Years				
	1978/80	1980/81	1981/82	1982/83	1986/87
Egg-laying date					
r	0.95*	0.20	0.62*	—	0.29
N	6	14	21	—	15
Interval between clutches					
r	0.13	0.73*	0.42	—	0.32
N	5	9	4	—	6
Clutch size					
r	0.52	0.48*	0.42*	0.19	0.22
N	9	17	24	25	22

Note: Statistical significance of the coefficients ($P < 0.05$) is indicated by an asterisk.

Table 4.9 Repeatabilities (*R*) of Egg-Laying Date, Interval between Successive Clutches, and Clutch Size

	All Females	All Males	Same Partner	Different Partners Females	Different Partners Males
Egg-laying date					
R	0.39**	0.08	0.55***	0.04	0.07
N	38	46	28	17	21
Interval between clutches					
R	0.32*	0.19	0.43*	−0.22	0.32
N	29	29	18	11	13
Clutch size					
R	0.33*	0.15	0.31*	0.35	0.01
N	41	41	27	14	14

Note: Statistical significance, as determined by *F* tests, is indicated by * ($P < 0.05$), ** ($P < 0.01$) or *** ($P < 0.001$).

To minimize age effects in the calculation of repeatabilities, we used data from the last two breeding seasons of a female's life. Data from different clutches within a season and in different years were standardized to a common mean before analysis. Repeatability estimates of the three reproductive traits are given in Table 4.9. There is a consistency (constancy) in the performance of individuals as well as a consistency in the results themselves. The three traits are repeatable among females but not among males.

For the set of males that stay with the same female, all three traits are repeatable, presumably as a result of a female influence and not as a result of a male influence. If the traits were repeatable for environmental (territory) reasons, and not for genetic reasons, repeatabilities for males that change partners should be as high as those for males which retain their partners, because in both cases the environment (territory) is unchanged. This is not the case, as can be seen in the table. All repeatabilities for males that change partners are indistinguishable from zero, in contrast to the repeatabilities for males that retained their partners. This is a further reason for believing that it is the properties of the female, and not of the male (or his territory), that determine the repeatability of the traits among pairs that stay together.

If environmental influences at the time of breeding played no part in modifying the tendency of females to differ consistently in these traits, females that changed mates should have the same repeatabilities as those which did not. Clearly this is not so (Table 4.9). In one case, that of clutch size, the repeatability appears to remain unchanged, but statistical significance is lost in the drastic reduction in sample size. We cannot comment further.

In the other two cases, egg-laying date and the interval between clutches are not consistent properties of females that change mates. The reasons for this are probably at least partly behavioral, hence environmental. Breeding involves coordinated activities. Partners familiar with each other are likely to be better coordinated in those activities, and hence quicker in breeding, than those unfamiliar with each other. Our data are consistent with this. Breeding of most females that retained their mates was relatively early in the second of the two years (12 out of 17), whereas most females that changed their mates bred relatively late (17 out of 28), although the difference in proportions is not statistically significant ($\chi^2_1 = 2.99$, $P > 0.1$). To remove this possible influence from the estimates of repeatability, we would use data from females in their second year of breeding with each of two different partners—if we had the data. Data limitations also prevent us from applying heritability analysis to the time of breeding and the interval between successive clutches.

Another environmental factor that influences the interval between clutches is the number of nestlings in the first brood (p. 79). Females may renest at similar intervals in successive years, and hence show a high repeatability for this aspect of the rate of reproduction, because they consistently produce large or small broods of nestlings. Brood size in turn might follow from clutch size, which is another repeatable trait. This possibility is supported by the following correlations. The intervals between first and second clutches were significantly correlated in one pair of years (Table 4.8). In this pair of years the numbers of nestlings reared to fledging from the first brood were also correlated ($r = 0.69$, $N = 9$, $P < 0.05$). The average number of nestlings produced by the nine females from first broods in the two years was correlated with the average clutch sizes at those times ($r = 0.73$, $P < 0.05$). Clutch sizes in the two years were correlated less strongly ($r = 0.60$, $P \approx 0.08$). Therefore the interval between clutches is a repeatable trait as a consequence of its connection to clutch size, which is possibly a heritable trait.

In the great tit studies, estimates of heritability and repeatability were very similar for each of two traits, clutch size and the date of clutch initiation (van Noordwijk and van Balen 1988). Van Noordwijk et al. (1980, 1981) found little reason to suspect an important influence of correlated environments of mothers and daughters on the heritability estimates, or of enduring effects of conditions during growth upon either trait. They argued from these results that female genotype affects clutch size and time of breeding, whereas male genotype is not important in either feature. Sample sizes in the two great tit studies were in the hundreds, ours are in the tens. Most of their results are statistically demonstrable, whereas ours, although similar in magnitude, are not and must therefore be considered as only suggestive. Another important difference between the studies is that clutch size varies much more in the great tit (5–15 eggs) than in *conirostris* (2–5 eggs), making it much easier to detect family resemblances in the former. On Daphne Major *fortis* shows similar variation (2–6 eggs), but despite having similar samples to those in the great tit studies, Gibbs (1988) was unable to detect a significant heritability or repeatability of clutch size.

Our conclusion is similar to the one reached in the great tit studies, but more tentative. Clutch size and clutch initiation date are possibly influenced by female genotype.

Lifetime Reproductive Success

The very small number of recruits from the banded fledglings implies that most breeding birds make no contribution to the next generation. As can be seen from Figure 4.9, this is true. Among breeding males, 78.8 percent produced no recruits; among females, 77.9 percent did not. Moreover the successful adults do not contribute equally to the next generation. While most contribute only one, a few contribute two or three. The outstandingly successful parent was a male, No. 6108, born in 1974 or earlier, who bred in the years 1978–82 and produced five recruits. Not only that, his offspring produced local recruits, which in turn produced local recruits, etc. (Fig. 4.10). Just how successful No. 6108 was can be appreciated by the comparison with the only other known genealogical tree that spans three generations. A male, No. 6251, was born in 1980; the parents were No. 6678 (♂) and No. 6198 (♀) of unknown origin. The male bred with No. 6675, also of unknown origin, and produced a single, local, female

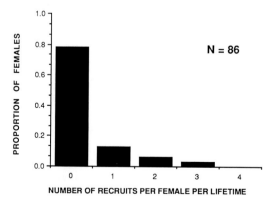

Figure 4.9 Recruitment from 86 females.

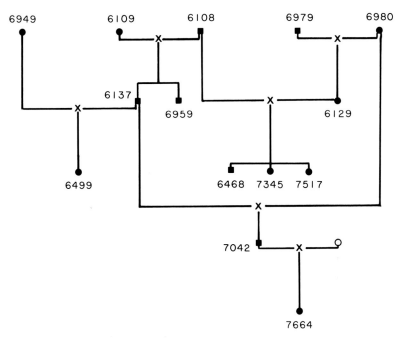

Figure 4.10 A genealogical tree. Males are indicated by squares, females by circles. The open circle refers to a female that was not banded.

recruit (No. 7873) who bred in 1983, 1986, and 1987. Her offspring were not seen after fledging, so the tree ends with her.

Some of the young banded as nestlings must have settled to breed outside the study area and were missed in our searches, although we doubt if we missed many (p. 43). Local recruitment is therefore an index of total recruitment, but it is an index of unknown accuracy. Possibly the disparity among breeders in their contribution to the next generation is overestimated by the use of only local recruitment. Nevertheless, the disparity is very pronounced in two other aspects of reproduction: lifetime production of eggs and fledglings (Fig. 4.11). Since females and males display virtually identical patterns, we have used only female data to illustrate them. For example, median values for the two sexes are the same; 11 eggs and 6 fledglings.

The most successful female, No. 6980, was also the longest lived. Between 1978 and 1987 she is known to have produced 58 fledglings from 110 eggs laid in 31 clutches in eight breeding years. Breeding was not followed in 1979, although we do know that three or four broods were raised that year, and that No. 6980 bred at least once (D. Day 1979, pers. comm.). She may have bred in 1976 and earlier. She may breed in the future, as she was still alive in 1988. Therefore the potential lifetime production of an individual female is at least 50 fledglings from 100 eggs. These are 8–9 times greater than the *median* values and 5–6 times greater than the *mean* values! Contrast these figures with the least successful producers. The modal group of females produced less than 6 eggs and fledglings, and 14.5 percent of the females failed to produce a single fledgling from the eggs they laid.

Summary

Breeding activity varies greatly from year to year and is largely governed by the pattern and quantity of rainfall. Breeding usually starts shortly after the first heavy rain that ends a 6–8 month dry season, and it ceases when caterpillars and spiders, which are the most important nestling foods,

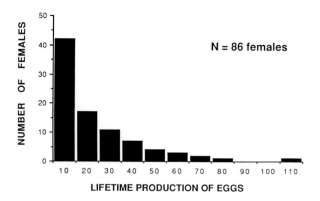

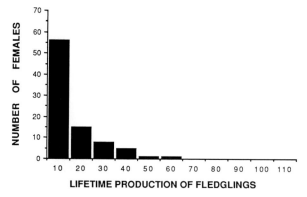

Figure 4.11 Lifetime production of eggs and fledglings by 86 females. The most productive female is included even though still alive at the end of the study in 1988.

become scarce. The first eggs are usually laid between one and two weeks after the first rain, although in two years egg laying began a few days before the first rain. Typically a minimum of 15–20 mm of rain is needed for successful breeding to take place. The length of the breeding season varies directly with the amount of rain. In an average year, 75–150 mm of rain falls and the breeding season lasts for 3–4 months. Extreme years deviate markedly from the norm. More than 2400 mm of rain fell from November 1982 to July 1983, and finches bred continuously for 7–8 months. Two years later (1985) no rain fell, and breeding did not occur.

Clutch sizes vary from two to five eggs ($\bar{x} \approx 3.5$). Incubation and nestling periods each span about two weeks, so in years of extensive rain finches breed once a month. Breeding is speeded up by brood overlapping; females start a new clutch when nestlings are still being raised from the previous clutch. When this happens, the interval between successive broods is reduced from 30–35 days to as few as 19 days. The interval between a female's successive clutches, her clutch size, and the time she starts laying relative to other females may all be heritable traits, since females tend to be consistent in successive years.

Seasonal production of young varies annually, and it varies strongly; the maximum was an average of 9.9 fledglings per pair in 1987, and the minimum in a year when breeding was attempted was 0.2 fledglings per pair in 1984. Production is also affected by age and experience. Old and experienced breeders breed more times in a season, suffer fewer total losses, produce slightly larger clutches, and raise more offspring from them than do young and inexperienced breeders. Their total seasonal production of offspring is more than twice that of inexperienced pairs. Pairs made up of one experienced and one inexperienced member have intermediate success.

The percentage of fledglings that are recruited to the breeding segment of the population in subsequent years is very low (usually 5–10 percent). The probability of an individual fledgling surviving to breed appears to be generally unaffected by the size of the clutch from which it came, when it was born in the breeding season, or the experience of its parents. The majority of parents contribute no recruits to the next generation.

Perhaps the most striking feature of reproduction is the great disparity between the most successful and the least successful individuals, even after allowance is made for the large numbers who do not survive to breed at all. Among breeders, the median lifetime production is 11 eggs and 6 fledglings. The most successful breeder (a female) was at least 8 times as productive as this. She bred in nine seasons and produced a minimum of 110 eggs and 58 fledglings; in contrast, 14.5 percent of females that laid eggs failed to produce a single fledgling.

5

Survival and Reproduction

Introduction

To understand the demography of *conirostris* we need to put together the findings on survival and reproduction and then, having determined the principal characteristics, compare them with those of other species living in different environments. This is an application of the comparative method to the task of identifying the most important environmental variables that influence a biological trait. In the present case the task is to determine which features of the environment have been responsible for the long life spans and the high reproductive potential of *conirostris*.

Survival and Reproduction

PRODUCTION OF OFFSPRING AS A FUNCTION OF LIFE SPAN

Conditions for survival vary from year to year, but most birds do not survive their first year regardless of which year it is, hence most do not survive to breed. If we take fledgling success to be 63 percent of eggs laid, on average (Table 4.6), and survival of fledglings for one year to be 8.0 percent (Table 3.1), then 1000 eggs give rise to only 50 one-year-old birds, about half of which die before breeding. If they are able to survive to breed, usually after one or two years, their subsequent survival is good if breeding density is low, but the higher the density when they first attempt to breed the lower are their chances of surviving to breed again.

The ability of an individual to stay alive appears to be the major determinant of how many offspring she or he will produce in a lifetime. In comparison, variation in reproductive effort at each episode is trivial. Some correlations bear out the importance of survival.

Lifetime production of eggs by a female is strongly correlated with the total number of clutches produced ($r = 0.99$, $N = 86$) and with the number of years of breeding ($r = 0.93$). Lifetime production of fledglings is strongly correlated with lifetime production of eggs ($r = 0.93$) as well as with total number of broods ($r = 0.92$) and with the number of years of breeding ($r = 0.88$). Lifetime production of recruits is less tightly correlated with total production of fledglings ($r = 0.45$), with total number of broods ($r = 0.43$), and with the number of years of breeding ($r = 0.42$). Nevertheless, all of these correlation coefficients are highly significant ($P < 0.001$). Correlations for males are almost identical.

Since total number of broods and number of years of breeding covary so strongly ($r = 0.94$), it is impossible to disentangle their effects upon reproduction. It is not even necessary to make the attempt, as the results can be summarized as follows: the longer a female (or male) lives beyond initial reproductive age the more times she breeds, in terms of both number of broods and number of breeding seasons; hence, the more eggs she lays, the more fledglings she produces, and the more recruits she contributes to the next generation. To simplify: whatever determines survival also determines lifetime reproductive success.

THE EFFECT OF REPRODUCTION ON SURVIVAL

Although long lives enable birds to breed many times, breeding many times could shorten the life span. According to standard life-history theory, there are genetic trade-offs or negative correlations between survival and reproduction that have been molded by natural selection (e.g., Williams 1966; Stearns 1976; Curio 1983; Reznick 1985). Those individuals that are capable of reproducing particularly well, for genetic reasons, survive less well on average than the rest, and those that survive well reproduce relatively poorly. In practice it is usually observed that birds who reproduce particularly well also survive particularly well (Högstedt 1981; Smith 1981, 1988; van Balen et al. 1987; Hötker 1988; but see Nur 1984, 1988; Ekman and Askenmo 1986). As an example from our own data, the average clutch size of experienced females breeding in 1982 was positively correlated with their life span ($r = 0.40$, $N = 22$, $P = 0.06$). Positive correlations arise through uncontrolled effects of environmental variation that influence survival and reproduction in the same way (Högstedt 1980; Horn and Rubenstein 1984; van Noordwijk and de Jong 1986). Genetic trade-offs can only be exposed by experimentally removing the obscuring effects of environmental variation.

In addition to genetic trade-offs, there may be phenotypic trade-offs when the environment for all birds changes in the same way. Survival may be enhanced when reproduction is not attempted because there is no rain. This effect, if present, is not likely to be detected because it will be nullified by the simultaneous opposite effect of a diminishing food supply. On the other hand, survival may be reduced after birds have bred repeatedly. If there is this type of survival cost of reproduction, it should have been manifested in 1983 and the aftermath of the El Niño event as a result of the stress of repeated breeding. The evidence bearing on this point in Chapter 3 is mixed. Four cohorts of males survived relatively poorly after the long breeding season of 1983, one (the oldest) was apparently not affected, and another (the youngest) survived extremely well. Reduced survival of *fortis* was recorded on Daphne Major at this time (Gibbs and Grant 1987a, b).

We need, instead, to compare the fates of individuals. Twenty-seven breeding females survived from 1983 to 1984, and they produced no fewer or more broods in 1983 than the 20 who did not survive. Only four females survived to breed again in 1986. Their breeding in 1983 spanned the full range of one to seven broods. We have larger samples of males with which to compare the breeders in 1983 who survived to 1986 with those who did not survive. The result of this comparison is the opposite of that expected from the cost hypothesis; survivors produced more broods on average, not fewer, than nonsurvivors ($t_{52} = 2.22, P < 0.05$). The difference matches findings in other species of birds, such as the song sparrow studied by Smith (1981, 1988) and the magpie studied by Högstedt (1980, 1981). It probably arose because experienced (old) birds generally survive better than inexperienced (young) birds, and so they produce more broods. Thus 25 percent of experienced males in 1983 survived to 1986, whereas only 15 percent of inexperienced males survived that year. Removal of the inexperienced males from the above comparison eliminates the difference in number of broods between survivors and nonsurvivors ($t_{32} = 1.20$, $P > 0.1$). There is no detectable survival cost to reproduction in these years.

INTEGRATION OF REPRODUCTION AND SURVIVAL

Patterns of survival and reproduction over the life span are brought together in Table 5.1. As is customary in such demographic analyses, the population is characterized by females giving rise to females (Krebs 1985). The best (largest) cohorts for analysis are those produced in 1976 and 1980, despite uncertainty about exact ages of some individuals (see Chapter 3). To construct survival and fertility tables we have adopted some conventions and assumptions (for further details about demographic analysis, its assumptions, and examples, see Ricklefs 1973; Woolfenden and Fitzpatrick 1984; Krebs 1985; and Koenig and Mumme 1987). Birth is treated as hatching, therefore fertility is represented in the table by the average number of female nestlings produced by a female each year. For this we assume that half of the total nestlings are females. Survival figures over the first two years for birds of known age (Table 3.2) are used in place of the unknown early survival of the 1976 cohorts of females. These figures were adjusted for an average survival of 79.3 percent from hatching to fledging, a value taken from Table 4.6 after excluding the minimal samples from 1984 and 1988. For the 1980 cohort of females, we took the observed 71 percent of nestlings ($N = 256$) fledging that year and multiplied it by the 15 percent that survived to the next year (Table 3.1) to obtain the proportion (0.1065) in age class 1. The calculations are not altered much if, instead of following the passage of female nestlings to adulthood and their production of female nestlings, we consider female eggs giving rise to female eggs as Baker et al. (1981) did, or female fledglings giving rise to female fledglings as Woolfenden and Fitzpatrick (1984) did.

THE PRODUCT OF SURVIVAL AND REPRODUCTION

The columns of particular interest in Table 5.1 are $l_x m_x$ and V_x. The $l_x m_x$ column shows the relative contribution of offspring made by females at different ages. Contributions were maximal in the first three years of breeding, then fell rapidly (Fig. 5.1), in both cohorts. This is a typical pattern for passerine birds (Ricklefs 1973; Baker et al. 1981; Woolfenden

Table 5.1 Survival (l_x) and Fertility (m_x) Table for Females of the 1976 and the 1980 Cohorts

	Age Class (x)	Probability of Surviving to x (l_x)	Nestlings/ Season (m_x)	Product of Survival and Reproduction ($l_x m_x$)	Reproductive Value (V_x)
1976	0	1.000	0.0	0.0	0.6843
	1	0.0634	0.0	0.0	10.7942
	2	0.0317	5.6	0.1775	21.5883
	3	0.0292	5.8	0.1694	17.3572
	4	0.0219	4.8	0.1051	15.4096
	5	0.0195	3.7	0.0721	11.9154
	6	0.0195	3.5	0.0682	8.2154
	7	0.0146	4.1	0.0599	6.2979
	8	0.0073	0.8	0.0058	6.7932
	9	0.0049	0.0	0.0	8.9286
	10	0.0025	3.5	0.0087	10.5000
	11	0.0025	7.0	0.0175	7.0000
	12	0.0025	0.0	0.0	0.0
	13	0.0	0.0	0.0	0.0
			$R_0 = \Sigma l_x m_x = 0.6843$		125.4841
1980	0	1.000	0.0	0.0	0.5880
	1	0.1065	1.9	0.2023	5.5214
	2	0.0919	2.2	0.2022	4.1967
	3	0.0495	3.1	0.1534	3.7071
	4	0.0318	0.0	0.0	0.9450
	5	0.0106	0.0	0.0	2.8349
	6	0.0071	3.0	0.0213	4.2324
	7	0.0035	2.5	0.0087	2.5000
	8	0.0	0.0	0.0	0.0
			$R_0 = \Sigma l_x m_x = 0.5879$		24.5255

and Fitzpatrick 1984; McCleery and Perrins 1988), although the rates of decline do vary.

Two features of the $l_x m_x$ values are not typical. The values decline to zero, but then rise before falling again to zero. This pattern comes about through the absence of breeding in some years ($m_x = 0$), which is not a feature of the temperate zone birds that have been studied in detail. The second unusual feature is a sum of all the age-specific values (i.e., R_0) distinctly less than 1.0. In a stable population that just replaces itself from generation to generation, the net reproductive rate R_0 is exactly 1.0. An example is the population of white-crowned sparrows studied by Baker et al. (1981) where R_0 values of 1.025 and 1.042 were calculated for two cohorts. It is clear that the two cohorts of *conirostris* portrayed in Table 5.1 did not replace themselves. The population was certainly not stable, as we have already seen in the four- to five-fold fluctuation in density of males (Fig. 3.6; also Table 6.4).

It is likely that the degree to which the cohorts failed to replace themselves was exaggerated by a bias in the estimates of survival. Some female offspring born on the study areas settled to breed outside them. We know of one such individual, and there were probably more. Their exclusion from the table results in an underestimation of l_x. Apart from these individuals, survival is well known after year 2 for the 1976 cohort and after year 1 for the 1980 cohort, and the schedule of births is accurately estimated for both of them except in 1983. In no instance did a breeding bird disappear for a year. But early survival is not known directly, and possible errors in its estimation introduced by using figures from other cohorts ramify throughout the whole l_x curve; by underestimating early survival we underestimate the numbers alive at all ages. The net repro-

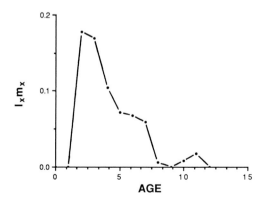

Figure 5.1 The product of survival and reproduction in relation to age ($l_x m_x$).

ductive rate of the 1976 cohort would be 1.0 if we allowed for survival over the first two years to be increased by a third.

The problem is not unique to *conirostris*. Net reproductive rates of female (0.624) and male (0.673) great tits (McCleery and Perrins 1988) are similar to our values. Like ours, the true reproductive rates are apparently lowered by dispersal and survival outside the study area.

REPRODUCTIVE VALUE

The last column in Table 5.1 is V_x, the age-specific expectation of producing more female offspring, or reproductive value at each age. This parameter is important in theories of life-history evolution. An optimal life history is produced when reproductive value at each age is maximized (Goodman 1982). The reproductive value at each age is a composite of current and future reproduction. In an unstable population such as ours the component of future reproduction (residual reproductive value) changes with population density. Some allowance can be made for this, but it is complex (Horn and Rubenstein 1984) and would not alter the peculiarities discussed below.

Owing to the component of future reproduction, the V_x curve does not fall to zero when there is no breeding, unlike the $l_x m_x$ curve, nevertheless the temporal pattern is the same. It is illustrated in Figure 5.2. The curve is bimodal for both cohorts, with a maximum reached in early life. The second mode is explained by two features of survival and reproduction: the absence of reproduction in the dry years of 1984 and 1985, and the large amount of reproduction in the following two years by the relatively few individuals that survived the dry period. Thus there was a strong benefit gained by individuals of both cohorts living a long time because their

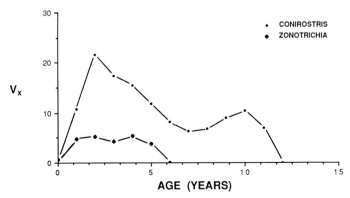

Figure 5.2 Reproductive value in relation to age (V_x); the temperate zone white-crowned sparrow (*Zonotrichia leucophrys*) is included for comparison.

reproductive value increased in old age. This is likely to be experienced by all cohorts in view of the tendency for very wet years to recur at intervals of four years or slightly longer (Grant and Boag 1980; P. R. Grant 1985), which is much less than the life span of a cohort.

Bimodality is not a typical pattern for birds. On the contrary, the V_x curve of temperate zone birds is unimodal, with a gradual and uniform decline from a peak to zero, whether the species is short-lived, like the white-crowned sparrow shown in Figure 5.2 or the great tit (McCleery and Perrins 1988), or is long-lived, like the acorn woodpecker (-15 years; Koenig and Mumme 1987) and the scrub jay (-20 years; Woolfenden and Fitzpatrick 1984). Given the different schedules of age-specific reproductive values among species, we have taken the maximum value during the lifetime of a cohort as a useful parameter for comparing them. Two cohorts of white-crowned sparrows studied by Baker et al. (1981) were calculated to have maximum values of 6.234 and 5.632 at age 1 year. These are close to the value of 5.521 at age 1 for the 1980 cohort of *conirostris* (Table 5.1). The maximum value for the Florida scrub jay, depicted in Figure 9.20 of Woolfenden and Fitzpatrick (1984), is a little more than 5 at age 4 years, and the comparable value for the acorn woodpecker is 5.94, also at age 4 years (Koenig and Mumme 1987). These fall far short of the 21.588 calculated for the 1976 cohort of *conirostris* (Table 5.1). We have been unable to find data on any other species that has such a high maximum reproductive value. Reproductive values may be as high for *fortis* and *scandens* on Isla Daphne Major. They have not yet been calculated.

To conclude, reproductive values of *conirostris* females are exceptionally high in some cohorts and exceptionally variable among cohorts. They are unusual in rising late in the life of a cohort as a result of the alternation of environmental conditions within the life span of the cohort; an alternation between very dry conditions, which are poor for survival, and very wet conditions, which are suitable for repeated, successful breeding.

Comparative Demography

REPRODUCTION

To place the reproductive features of *conirostris* in perspective, we have made comparisons with other species at several points in this chapter and in Chapter 4. Here we broaden the comparison. A logical starting point would be with the mainland descendants of the species which gave rise to Darwin's Finches. There are two difficulties with this. The first is that the close relatives of Darwin's Finches within the subfamily Emberizinae have not yet been identified with certainty (Grant 1986a). The second is that demographic data from species on the mainland are almost nonexistent. The best we can do is to use the reproductive data on five species of emberizines compiled by Marchant (1958, 1959, 1960) over four years

(1955–58) in the Santa Elena peninsula of southwest Ecuador. The comparison is appropriate because the area is similar to lowland sites on the Galápagos in both vegetation (Svenson 1946) and climate (P. R. Grant 1985).

Populations of birds on islands tend to have small clutches (Cody 1966, 1971; Lack 1968; Higuchi 1976; van Riper 1980; Crowell and Rothstein 1981). Summarized quantities in Table 5.2 show that *conirostris* and other species of Darwin's Finches have large, not small, clutches. They also have long nestling periods and high breeding success in comparison with mainland species. These differences are not expected from the scaling relationships of breeding characteristics and body size established for other species (e.g., Rahn et al. 1975; Calder 1984). *G. conirostris* is slightly larger than mainland finch species, on average. From simple allometric considerations, the mean clutch size should be lower, not higher, than those of mainland species (e.g., see Calder 1984), and the nesling period should be even longer than it is by 1–2 days (Grant 1982).

An obvious factor causing at least part of the differences between Darwin's Finches and mainland species is predation on the contents of nests (Grant and Grant 1980). At the mainland site there are predatory species of mammals, birds, and snakes that are entirely lacking on the Galápagos. Eggs and nestlings of Darwin's Finches are only taken by mockingbirds and owls, and on some islands not even these predators are

Table 5.2 Some Reproductive Characteristics of Five *Geospiza* Species and Five Species on the Santa Elena Peninsula of Continental Ecuador

Species	Duration		Production		
	Incubation Period	Nestling Period	Modal Clutch Size	Nesting Success	Breeding Success
Galápagos					
G. conirostris	11–13 (12)	12–17 (14)	4	85	69
G. magnirostris	10–14 (12)	12–17 (15)	4	68	60
G. difficilis	10–13 (12)	11–15 (13)	4	88	73
G. fortis	(12)	(14)	4	79	67
G. scandens	(12)	(14)	4	83	68
Mainland					
Parrot-billed seedeater	11	9–11 (10)	3	50	51
Chestnut-sided seedeater	11	9–10	3	37	36
Crimson finch	11	7–9 (8)	3	65	59
Collared warbling finch	11–12	8–9 (9)	4	47	38
Blue-black grassquit	10	9	—	—	—

Sources: Grant and Grant (1980) and Boag and Grant (1984) for the Galápagos, and Marchant (1960) for the mainland.

Note: Figures in parentheses are modal values. Nesting success is the percentage of nests that yielded at least one fledgling, and breeding success is the percentage of eggs that yielded fledglings.

a threat (Curio 1965, 1969). On Daphne Major, for example, mockingbirds are absent and owls do not attack nests there (Boag and Grant 1984), perhaps because the cactus where nests are placed are protected by stiff strong spines (P. R. Grant and K. T. Grant 1979). On Genovesa the spines are soft.

The number of breeding attempts made by mainland finches each year varies, as it does in Darwin's Finches. The maximum number may be small. Marchant's study included the El Niño year of 1957, when a large amount of rain fell on both the peninsula and on the Galápagos (P. R. Grant 1985). The breeding season spanned about four months in the peninsula, from February to June. With a nesting cycle lasting only 22–26 days (Marchant 1960), finches could have raised a maximum of four or five broods. Yet the mainland species bred apparently for only one to three times, whereas *conirostris* and other finches have been recorded breeding six times or more in other El Niño years (Table 4.4; P. R. Grant and B. R. Grant 1987; Gibbs and Grant 1987b). Whether predation plays a role in determining this difference, or whether plant and arthropod phenologies are responsible, is not known. Altogether it seems doubtful to us that mainland finches have reproductive values as high as those of *conirostris*.

For more comprehensive reproductive data we have to turn to temperate zone birds. Several studies have examined the relative importance of life-history stages to the determination of either lifetime or long-term reproductive success in birds (see Clutton-Brock 1988). This has been accomplished by partitioning variance among individuals in the number of recruits they produce or in the number of offspring reared to independence into three components (Brown 1988): the total production of eggs (fecundity), the number of chicks raised to fledging, and the life span of the parents. There are differences in the completeness of the data used in the nine studies described in the Clutton-Brock (1988) volume, but the following generalization emerges: the number of chicks raised to fledging makes the greatest contribution to overall success in short-lived passerine species, and life span makes the greatest contribution in long-lived nonpasserine species.

There are exceptions. Thus life span (< 8 years) appears to be the most important determinant of the success of meadow pipits in producing recruits (Hötker 1988). Life span does make a contribution in the other short-lived species, e.g., the great tit (McCleery and Perrins 1988; van Noordwijk and van Balen 1988), the song sparrow (Smith 1988), and the house martin (Bryant 1988). We have not used the same technique of partitioning variance, nevertheless the results of correlation and regression analyses lead us to conclude that the relatively long-lived *conirostris* is closer to the nonpasserine pattern. Lifetime production of eggs, fledglings, and recruits are all strongly correlated with each other, and with the number of breeding years of a female (p. 101). But in a multiple regression analysis, only the number of breeding years accounts for a significant ($P < 0.05$) amount of the variation in lifetime production of recruits.

In comparison with other passerines, *conirostris* is exceptional in two respects. Probability of reproductive success, measured as the percentage of breeding females that contribute at least one recruit to the next generation, is lower than in all other passerine species for which there are comparable data (Table 5.3). This is not simply the result of producing few fledglings. On the contrary, the maximum number of fledglings produced by a female is exceptionally high, apparently unmatched by any passerine species studied so far. Three of the 86 females fledged more offspring than the maximum (36) produced by the next most successful species (scrub jay). The average is also high ($\bar{x} \pm SD = 10.5 \pm 11.35$, $N = 86$ females). But in spite of this, recruitment success is exceptionally low, although this may be partly an effect of the small sample of *conirostris* females. Thus the difference between the few successful and the vast majority of unsuccessful individuals is very pronounced in *conirostris*.

EVOLUTION

The demographic features of *conirostris* that stand out in comparison with other species of passerine birds in continental regions are their long

Table 5.3 Indices of Lifetime Reproductive Success

Species	Percentage of Breeding Females That Produced One or More Recruits	Maximum Production by Females	
		Fledglings	Recruits
G. conirostris	19.8	58	3
Meadow pipit	50.0	19	4
Song sparrow	56.3	24	10
Galápagos mockingbird	29.2	14	4
Scrub jay	55.0	36	9
Collared flycatcher	47.5	35	4
Great tit[1]	36.0	—	7
Great tit[2]	42.5	—	7
Indigo bunting	—	27	—
House martin	—	28	—
Sparrowhawk	23.5	23	—

Sources: Hötker (1988) for the meadow pipit, J.N.M. Smith and W. Hochachka (pers. comm. 1988) for the song sparrow, R. L. Curry (pers. comm. 1988) for the Galápagos mockingbird, J. W. Fitzpatrick (pers. comm. 1988) for the scrub jay, Gustafsson (1989; pers. comm., 1988) for the collared flycatcher, McCleery and Perrins (1988)[1] and van Noordwijk and van Balen (1988)[2] for the great tit, Payne (1989) for the indigo bunting, Bryant (1988) for the house martin, and Newton (1988) for the sparrowhawk.

Note: All species except the last are passerines. The sample of breeding females is lower for *conirostris* than for all other species except the meadow pipit.

potential life spans in relation to their size, their large potential reproductive output, and large variation among cohorts in both of these features. They share these characteristics with other species of the ground finch group. We now conclude this chapter by discussing the factors that might be responsible.

Well-documented latitudinal gradients in clutch size, and to a lesser extent in nesting success and maximum longevity, have served to organize ideas about how natural selection has acted on these and other life-history traits (Moreau 1944; Lack 1954, 1968; Cody 1966, 1971; MacArthur and Wilson 1967; Ricklefs 1973). Life-history traits vary over latitudinal gradients both within and between species (Ricklefs 1973) because climatic conditions, food supply, and predation risks vary. There are two major theories to account for the life-history variation in these terms.

The first, r- and K-selection theory (MacArthur and Wilson 1967), contrasts the regimes of selection for high reproductive rate in strongly seasonal environments, where populations repeatedly increase and decrease, with selection for low reproductive rate in less seasonal environments, where population sizes are approximately constant. Density-dependent selection plays a pivotal role in the theory (Boyce 1984). The theory accounts for the latitudinal gradient by supposing that selection is persistently and strongly density-dependent in the tropics and occasionally and weakly so in the temperate zone.

The second, bet-hedging theory (Stearns 1976) contrasts the regime of selection in environments where survival of adults is highly variable with selection in environments where survival of juveniles is highly variable. In unpredictably varying environments, selection is postulated to favor adaptations that either spread expected reproduction over many seasons or concentrate it at early ages, depending on which age group is most vulnerable (Murphy 1968; Ricklefs 1983). Applied to latitudinal gradients, the bet-hedging theory could account for longer lives and lower reproduction in the tropics in terms of an assumed large variation in juvenile survival.

Neither theory is adequate to account for the demographic features of Darwin's Finches on the Galápagos, because the finches violate the pattern shown by continental birds. They live long *and* reproduce strongly without delay. Nevertheless elements of both theories are useful for understanding the finch traits.

Bet-hedging theory is applicable because the Galápagos environment varies strongly and unpredictably. Juvenile survival in *conirostris* is twice as variable annually as adult survival. Juvenile survival of *fortis* on Daphne Major varies much more, more than does adult survival. Therefore the conditions are right for expecting long life spans, which are observed, and low reproductive output in each year, which is not observed.

The high reproductive rate is explained by r- and K-selection theory, because such is expected in a strongly seasonal environment. The core idea, developed by Ashmole (1963) to explain patterns of reproduction in seabirds before the theory was formulated, is this: When the density of breeding adults is set by a low supply of food in the nonbreeding season, there is a substantial surplus of food produced in the breeding season, and in the absence of density constraints birds breed as fast as they can (see also Lack 1968; Ricklefs 1973, 1980). The theory is applicable to the ground finches (Grant and Grant 1980), even though density effects on the survival of adults are present in the nonbreeding season (Chapter 3; also Gibbs and Grant 1987a), because the basic conditions of a strongly fluctuating and temporarily abundant food supply are usually met (Grant 1986a, b). Given the high reproductive rate, however, the theory fails to account for long life spans.

The two theories need to be combined in some way, and Horn and Rubenstein (1984) indicate how this might be done. They point out, following Charnov and Schaffer (1973), that prereproductive (juvenile) mortality can be viewed as subtracting from natality to yield a lower, ultimately effective, natality of the parent. The effect of an increase in variation in juvenile mortality is somewhat analagous to a decrease in the average juvenile survival, as the end result is the same, selection for longer parental life. Viewed in this way, effects of variation in the survival of juveniles (bet-hedging theory) can be equated with effects of the mean survival rate (r- and K-selection theory). Horn and Rubenstein (1984) conclude that the dichotomy in selection regimes is between circumstances in which mortality falls more heavily on juveniles and those in which it falls more heavily on adults.

For small birds only one-half of the dichotomy applies, since it is almost inevitable that juveniles will survive less well than adults because they lack experience. Similarly, it is almost inevitable that juvenile survival will vary more than adult survival because juveniles are more sensitive to the mortality factors that vary annually (e.g., Klomp 1980). The few data available from studies of passerine birds support these statements (Table 5.4). In all cases in the table, adult survival is higher on average than juvenile survival, and in all cases, juvenile survival varies more from year to year than adult survival.

The significant dichotomy for small birds is between consistently high juvenile survival on the one hand and consistently low or high and erratic juvenile survival on the other. The former occurs (probably) in tropical moist forests as a consequence of the production of few young at any one time. Low reproductive rates have been selected by strong and persistent density-dependent factors associated with a fairly stable food supply. Reproductive rates are indeed lowest (manakins and antbirds: see Snow

Table 5.4 Variation in Annual Survival among and within Species

Species	N^a	Annual Survival	
		Mean	Coefficient of Variation
Galápagos			
G. conirostris			
Adult	10	77.8	14.4
Juvenile	7	11.4	29.0
G. fortis			
Adult[1]	7	69.0	64.0
Juvenile[2]	8	33.1	79.2
G. scandens			
Adult[1]	7	82.0	27.0
Juvenile	—	—	—
Galápagos mockingbird			
Adult	10	61.0	29.0
Juvenile	8	35.0	47.0
Tropical Forest			
White-bearded manakin[3,4]			
Adult	9	82.0	—
Juvenile	4	33.0	—
Spotted antbird			
Adult	11	81.2	—
Juvenile	11	36.0	—
Ocellated antbird			
Adult	11	70.0	—
Juvenile	11	53.0	—
Bicolored antbird			
Adult	11	71.0	—
Juvenile	11	42.0	—
Temperate Forest			
Great tit			
Adult	13	40.8	36.3
Juvenile	11	7.6	68.1
Pied flycatcher			
Adult	12	34.2	27.7
Juvenile	11	4.8	43.3

Source: Data from Gibbs and Grant (1987a)[1] and Gibbs (1988)[2] for *fortis* and *scandens;* Curry and Grant (1989) for the Galápagos mockingbird; Snow (1962)[3] and Snow and Lill (1974)[4] for the manakins; Willis (1974) for the antbirds, van Balen (1980) for the great tit, and Harvey et al. (1988) for the pied flycatcher.
[a]N = number of years.

1962; Willis 1974) where juvenile survival is highest (regrettably there are no data from tropical forest birds on variation). Low or erratic juvenile survival occurs on the Galápagos and generally in temperate zone habitats where high reproductive rates have been selected by factors acting either independent of density or occasionally in a density-dependent way. A congruent observation is that within the tropics clutch sizes tend to be higher in more seasonal habitats such as savannah than in less seasonal, moist, evergreen forest (Moreau 1944; Lack 1968).

Thus a combination of the two theories is better at explaining patterns of survival and reproduction in small birds than either is alone. The key to understanding the demography of Darwin's Finches, and of *conirostris* in particular, is the seasonality of the environment: primarily the degree to which feeding conditions differ between the seasons, and the unpredictable variation in that difference from year to year superimposed on a tendency for very wet conditions to recur at approximately four-year intervals. The reason why some finches live very long and produce many offspring is that the unpredictable conditions for breeding and for the survival of their offspring have selected for both.

Predation, in our view, plays a secondary but important role. Predation risk, though not measured, is probably much lower at most if not all life-history stages on the Galápagos than it is on the mainland. If we are correct in the supposition that predation risk is relatively low, then it has probably had the effect of permitting relatively large clutches of eggs, long nestling periods, and long life spans.

We do not believe predation has played a major role, for two reasons. First, clutch sizes and nestling periods do not differ between Darwin's Finches on islands with and without nest predators in the Galápagos (Grant and Grant 1980; Boag and Grant 1984). Second, predation does not explain why the single species of Darwin's Finches on Cocos Island has a clutch size of only 2 eggs (T. K. Werner and T. W. Sherry 1985, pers. comm.). Cocos Island has rain-forest habitat, and rain-forest birds studied by Snow (1962) and Willis (1974) have clutches of two eggs, but unlike rain forests in continental regions or even on Trinidad, where Snow studied manakins, Cocos Island has no native predators (Slud 1967; Sherry 1986). Therefore the small clutches on Cocos Island, possibly also those in continental rain forests, are likely to be the result of a relatively uniform food supply during the year. Interestingly, Cocos finches breed mostly at the drier times of the year (T. K. Werner and T. W. Sherry 1985, pers. comm.), suggesting that heavy rainfall has a deleterious effect on the activities of nesting birds, as on the Galápagos during an extreme El Niño year.

Summary

Survival and reproduction of *conirostris* are strongly associated. The longer a female (or male) lives beyond reproductive age the more times she breeds in terms of both number of broods and number of breeding seasons, hence the more eggs she lays, the more fledglings she produces, and the more recruits she contributes to the next generation. Therefore whatever determines survival indirectly determines reproductive success. There is no indication that by reproducing strongly in one year a female (or male) reduces her chances of surviving to the next year.

In comparison with other species of passerines, *conirostris* displays several exceptional features of reproduction. The maximum number of fledglings produced by a female in a lifetime is exceptionally high. The probability of reproductive success, measured as the proportion of females that contribute at least one recruit to the next generation, is lower than in all other passerines for which there are comparable data. Reproductive values of females are exceptionally high in some cohorts, and exceptionally variable among cohorts. Reproductive values of members of two cohorts increased toward the end of their life spans as a result of favorable conditions for breeding for the relatively few individuals that survived a preceding long dry period.

Thus the demographic features of *conirostris* that stand out in comparison with other species of passerine birds in continental regions are their long potential life spans in relation to their size, their large potential reproductive output, and large variation among cohorts in both of these features. They share these characteristics with other species of the ground finch group. The key to understanding them is the seasonality of the environment: the degree to which feeding conditions differ between the seasons and the unpredictable variation in that difference from year to year.

6

Song and Territories

Introduction

In Chapter 4 we discussed how intrinsic properties of individuals, age and experience, and the ability to survive through the dry nonbreeding season all contribute to the determination of reproductive success. Here we examine the breeding behavior in more detail. In particular we concentrate on two features of males that have a bearing on their success: the territories they hold and the songs they sing, as well as the interrelated functions of both in the context of breeding.

Birds vary considerably in their song characteristics and in their ability to learn and reproduce song. They range from the genetically controlled, single, repetitive songs of the European cuckoo and *Empidonax* flycatchers (Kroodsma 1984) to the complex repertoires of over two thousand songs of the brown thrasher (Kroodsma and Parker 1977; Broughey and Thompson 1981). Some species extemporize, mimic, and add to their repertoires throughout their lives (Laskey 1944; Thorpe and North 1965; Kroodsma 1974; Jenkins 1978; Payne et al. 1981), whereas others have shorter, sensitive periods for song learning confined to a few weeks after hatching, resulting in small repertoires which remain unaltered throughout life (Marler and Tamura 1964; Immelmann 1975; Nottebohm 1972; Marler and Mundinger 1971). Darwin's Finches belong to the latter group (Bowman 1979, 1983; Ratcliffe 1981).

Song is considered to play a role in: (1) the acquisition and maintenance of territories (Howard 1974; Yasukawa and Searcy 1985; Krebs et al. 1978; McGregor et al. 1981; Payne 1982); (2) the attraction and stimulation of females (Lott et al. 1967; Brockway 1969; Kroodsma 1976; Catchpole 1980; Catchpole et al. 1984; Eriksson and Wallin 1986; Searcy and Andersson 1986); and (3) species, individual, and kin recognition (Nicolai 1956, 1959; Weeden and Falls 1959; Immelmann 1969, 1975; Emlen 1971, 1972; Marler and Peters 1977; Peters et al. 1980; Searcy et al. 1981; Falls et al. 1988). In this chapter we will consider all these functions in the

context of territoriality, after first describing the structural characteristics of
song and the pattern of transmission.

Song Structure

Male *G. conirostris* sing one simple song, although there are two song
types in the population. Females vocalize near the nest but do not sing. The
two song types (Fig. 6.1) are discretely different, and the differences are
easily discerned by ear. They can be represented as *ch-ch-ch-ch* (song *A*)
and *chrrrrr* (song *B*). The most conspicuous differences between them are

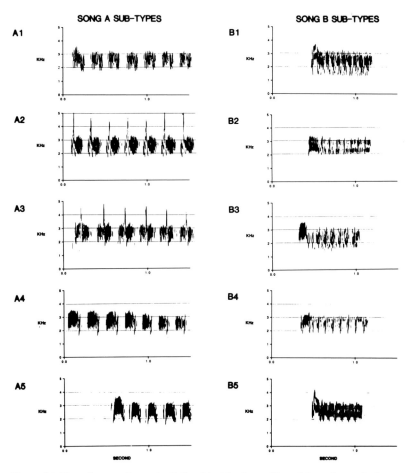

Figure 6.1 The subtypes of song *A*, distinguished by the position of the spike, and of
song *B*, distinguished by the introductory note.

Table 6.1 Measurements of Song Types (*A* and *B*) and Subtypes (*A*$_1$–*B*$_5$) from Sonagrams

Song Types and Subtypes	Number of Individual Birds	Number of Songs	Song Length
A	23	63	64.0 ± 1.41
A$_1$	7	14	69.6 ± 2.89
A$_2$	11	32	64.0 ± 2.06
A$_3$	5	17	59.1 ± 2.29
B	26	61	31.2 ± 0.51
B$_1$	14	37	31.5 ± 0.53
B$_2$	5	9	32.8 ± 2.06
B$_3$	4	8	31.1 ± 1.42
B$_4$	2	4	27.0 ± 1.00
B$_5$	1	3	26.7 ± 3.71

Source: B. R. Grant (1984).

Note: Units of 0.02 s; means ($\bar{x}$) and one standard error.

their lengths and degree of continuity. Song type *A* is two to three times as long as song type *B* and consists of a series of repeated notes or figures with silent intervals between each figure (Table 6.1; Fig. 6.1), whereas song type *B* has a short introductory section followed by a longer section of repeated syllables with no intervals among them. Both song types occur in the 1–5 kHz range and have a maximum energy band in the 2.8–3.0 kHz range, which is close to the optimum for long-distance sound transmission in most woodland habitats (Morton 1975; Bowman 1983).

Each type shows some variation (Table 6.1). Subtypes of song *A* can be recognized by the arrangement of four elements within a note, and subtypes of song *B* can be recognized by the pattern of the introductory section (see Fig. 6.1). These variations are not easily discernible to the human ear but are possibly perceived by the birds, since other species—for example, song sparrows, swamp sparrows, and meadowlarks—have been shown to be sensitive to changes in syllabic structure and temporal pattern (e.g., Searcy et al. 1981; Searcy and Marler 1981; Searcy, Marler, and Peters 1982; Searcy, Searcy, and Marler 1982; Stoddard et al. 1988; Falls et al. 1988), and to small changes in duration and pitch (Dooling 1982).

A male's song develops from an amorphous, poorly structured subsong, which is first produced as early as 100 days after hatching if some breeding is still occurring then. This development of song through a plastic phase, when birds seem to experiment with an overproduction of syllables, occurs in many species of birds (Marler and Peters 1981, 1982; Konishi 1985). A fully formed, crystallized song is produced in the bird's first breeding season and remains structurally unaltered for the rest of its life. We know songs remain stable from recordings of 56 individuals made repeatedly in the same and successive years for up to seven years. Bowman (1983) found the same constancy with captive birds of related species of Darwin's

Finches. What does vary is song length. Songs of the same individual, and songs of different individuals singing the same subtype, differ in the following respects: (a) the number of notes; (b) the length of the notes, and the length of the interval between notes in song A subtypes; and (c) overall song duration in the B subtypes (B. R. Grant 1984). Songs are produced singly, in pairs, or in triplets, depending apparently on the motivational state of the singer and the time within the breeding cycle.

Transmission of Song

In many species of birds, song is learned (Nottebohm 1972; Kroodsma 1978; Konishi 1985; Marler and Peters 1988a, b). Song can be copied from neighbors (Kroodsma 1974; Jenkins 1978; Payne 1981, 1982; McGregor and Krebs 1982) or learned from the father, as in bullfinches and zebra finches (Nicolai 1959; Immelmann 1969; Bohner 1983). The model of Cavalli-Sforza et al. (1982) describes the learned transmission of information as vertical from father to son, as horizontal from neighbors if the neighbors are from the same generation, and as oblique if the neighbors are from a previous generation. The type of transmission determines whether the information tends to remain stable or if it changes from generation to generation. It is generally most stable when transmitted vertically (Cavalli-Sforza et al. 1982).

The transmission of song by *conirostris* is vertical. Sons copy fathers. Altogether we have heard 30 males of known fathers sing, and 26 of them sang only their father's song type. Of the remainder, one never sang in either of its two breeding seasons; one sang the opposite song type to its father's; and two sang both song types. Such a consistent following of father's song type is not expected by chance (binomial test, $P < 0.001$).

Fourteen of the father-son pairs were recorded, repeatedly, within the same and different years, and in thirteen cases the son sang the same song type as his father; in the fourteenth the son sang the alternative song, which we will refer to as heterotypic. Figure 6.2 shows the first four father-son pairs that we analyzed. Sons' and fathers' songs are similar in detail, that is, in the arrangement of elements within the notes in song A and in the pattern of the introductory section in song B. In six instances they differed in duration, in the lengths of the notes, and in the intervals between them; for example, see pairs 2, 3, and 4 in Figure 6.2. One bilingual son was recorded, and he sang the same subtype of the A song sung by his father (Fig. 6.3). Thus the pattern of the father's song is adopted in quite close detail, although the copies are not identical. This gives rise to individual differences. The same phenomenon has been found by Böhner (1983) in his study of the accuracy of song copying by zebra finches.

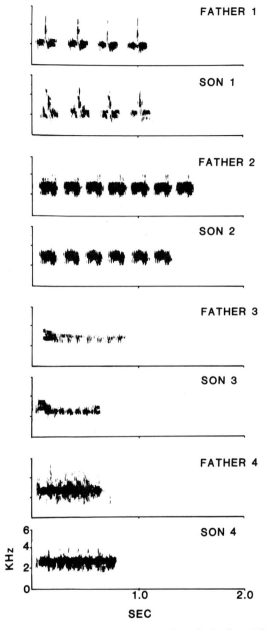

Figure 6.2 The songs of fathers and sons. From B. R. Grant (1984).

THE LEARNING CONTEXT

The identity of songs of father and son is probably due to learning and not due simply and exclusively to genetic factors. Laboratory experiments have shown that the songs sung by other species of Darwin's Finches are

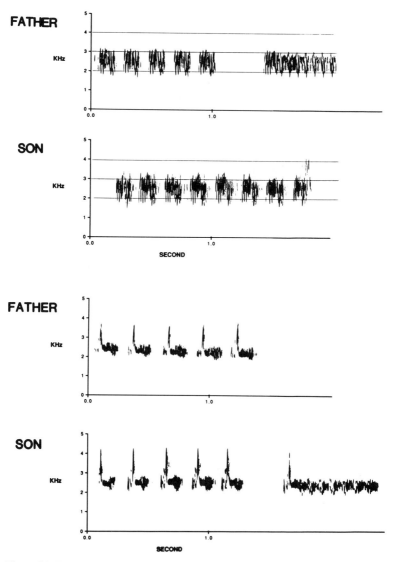

Figure 6.3 Two examples of bilingual birds. Upper: a father sang both song types, but his son sang only type *A*. Lower: a son sang his father's song (*A*) as well as song *B*. The sonagram of the lower son was supplied by R. I. Bowman.

not preprogrammed and unalterable. They are influenced by the songs the young birds hear in the period 10–40 days after hatching, as shown by exposing them to either conspecific or heterospecific tutors and finding that they reproduce those songs when fully grown (Bowman 1983). The short period of sensitivity for song learning corresponds to the time in the wild from the last few days in the nest prior to fledging to the point at which fledglings usually cease to be fed by their parents and leave the natal territory (Fig. 6.4). Most of our observations of the feeding of fledglings were made in 1980, 1986, and 1987. Combining these with a few made in 1978, 1981, and 1983, we obtain a total of 303 fledglings observed being fed, 262 of which (86.5 percent) were fed by the father. Most fledglings (77.6 percent) were observed being fed by a parent only once, and the majority of these (77.2 percent) were fed by the father. As with the feeding of nestlings, he usually sang immediately after feeding them and before he resumed foraging.

In view of the importance of song transmission from father to son, it might be expected that fathers would selectively feed them, providing they have some means of identifying them as sons. We do not know of any sex-specific cues in the fledglings' appearance and behavior, moreover the fathers do not restrict their feeding to sons. The sex of sixteen of the fledglings that were seen to be fed by one or both of their parents was determined in the following years. Five sons were fed once by their father, three were fed once by their mother, and two were fed by both parents. Daughters were fed in approximately the same proportions, 4:0:2. Thus neither parent feeds the fledglings selectively according to their sex.

Observations made for 30 minutes at 30 nests during the late nestling stage in 1987 suggest that the young birds quickly learn to associate their

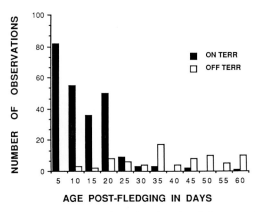

Figure 6.4 The location of fledglings on and off territory in relation to age in days after fledging.

father's song with feeding. At 29 of the nests the father followed a feeding visit to the nest by a bout of singing. On eight occasions the father was near the nest when the mother fed the nestlings, and at all of these times he sang after she had fed them. The pattern was repeated with fledglings. The association between father's song and food is logical as fledglings locate their father by his song and fly toward him, begging. When fledglings lose contact with the father and wander into an adjacent territory, they are never fed, even when they approach the male there and beg. Thus learning the father's song helps to ensure that the fledgling is fed. Fledglings probably learn morphological features of both father and mother at this time too. Social interactions in combination with song have been shown to be important for song learning in several species: bullfinches (Nicolai 1959), indigo buntings (Payne 1981), zebra finches (Böhner 1983), white-crowned sparrows (Baptista and Petrinovich 1984), and canaries (Waser and Marler 1988).

In many other species of passerine birds, songs are learned from neighbors, on territories adjacent to the one where sons were reared, or adjacent to the one where they breed (e.g., Jenkins 1978; McGregor et al. 1981; Payne 1981; Kroodsma 1983). In spite of the opportunity to learn from neighbors, however, cactus finches do so rarely (1 case out of 30). Nineteen of the 30 sons were born on territories where their neighbor(s) sang only the song of the opposite type to their fathers' (Table 6.2), yet only 3 of these 19 sang that song eventually; 2 of them were bilingual and sang their father's song type 10 times more frequently than the other song type. Fourteen of the sons established their own territories adjacent to at least one male who sang the opposite song to their fathers', and 6 of them had neighbors of only the opposite song type. If learning from neighbors took place at this time, we would have expected them to acquire the

Table 6.2 Territories of Sons Whose Paternal Song Type Is Known

Adjacent Neighbors	Number of Territories
	Natal
Only homotypic	1
Only heterotypic	19
Homotypic and heterotypic	8
None	2
	Breeding
Only homotypic	9
Only heterotypic	6
Homotypic and heterotypic	8
None	4

Note: Song types of adjacent neighbors are classified as homotypic when the same as the paternal song type, and heterotypic when different. Two bilingual sons and one who sang (briefly) but did not hold a territory have been omitted from the lower part of the table.

heterotypic song, but none did. Six of them had been heard singing before they established their territories.

The above indirect evidence from territory patterns suggests that the rare nonvertical transmission of song characteristics is oblique and occurs when young are still on their natal territory, as would be expected from the known sensitive period early in life. Sonagraphic analysis provides additional evidence that this is so. The son with the opposite song type to his father's sang a song (*B*) identical to that of his natal neighbor (Fig. 6.5), but of a different subtype from the song of his breeding territory neighbor. Likewise one of the bilingual birds was known to sing a heterotypic component identical to that of his natal neighbor, but different from his breeding neighbor's.

MISIMPRINTING

Most birds are genetically predisposed to learn the songs of their own species (Marler and Peters 1988b). This may be true of the closely related group of Darwin's ground finch species as well, but some do occasionally misimprint on the song of a congeneric species in the wild (Grant 1986a). If imprinting on father's song is the rule, how is the process perturbed to give rise through misimprinting to either heterotypic or bilingual singing? A developing bird, as nestling or fledgling, could confuse father's song and neighbor's song and learn both, or just the neighbor's, depending on when the confusion arose. A fledgling could lose contact with its father and be adopted by another male of the opposite song type upon which it then imprints. We have observed only two instances of a male feeding fledglings that were not his own. In one instance a male replaced another which disappeared, possibly killed by an owl, shortly after the arrival of the new male on the study area, and it inherited four fledglings when it paired with the mother. In the other instance a male courted and paired with a wandering female and adopted the single accompanying fledgling. We have never seen an adult feed nestlings other than its own as has been observed on Isla Daphne Major with *scandens* and *fortis* (Price et al. 1983).

The early history of the male who learned its neighbor's song type (Fig. 6.5) is suggestive of a process of adoption and imprinting. This bird (song *B*) was never seen to be fed by the father (song *A*), only by the mother. The neighbor (song *B*), an unmated male, was frequently seen courting the mother when she was feeding the fledgling both on her territory and on his. He was not seen to feed the fledgling but sang frequently; the association was repeated often enough for us to suspect that he did feed it. Interestingly, a male sib who was also not seen to be fed by his father subsequently sang his father's song (song *A*).

The early history of the bilingual birds is, unfortunately, not known, but it may be significant that the two who were raised on the study area were members of the final brood of the season. Song production rapidly declines after the last brood leave the nest, and the father in each of these cases may have been less persistent in singing than their neighbors were.

Despite the high frequency in the group of identified sons, bilingual birds are rare. We know of only 5 double-singers altogether among more than 250 males heard nearly daily and recorded repeatedly in one or more

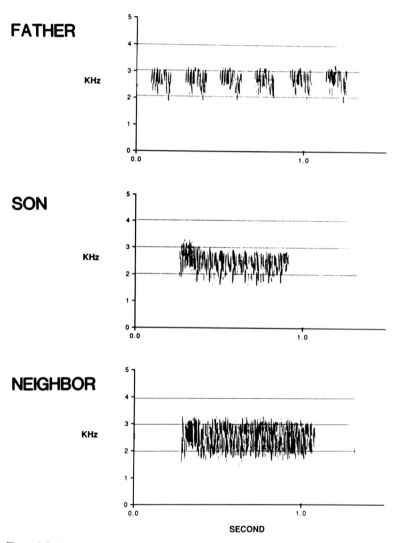

Figure 6.5 A rare example of nonvertical transmission of song. A son did not sing his father's song; his song was the same as that of the male on a territory adjacent to his natal territory.

breeding seasons. The number of times songs were heard from most of the males must have varied from the tens to many hundreds, so it is unlikely that we missed many, if any, of the rarer of two songs in an individual's repertoire. The father of one of the bilingual birds, for example, was probably heard several hundred times in each of the five years it bred, 1978–82, yet it was heard to sing only song *A*. Bilingual birds were at no apparent disadvantage, in spite of their rarity. Two of them had no difficulty in attracting mates and bred repeatedly in the study area from 1982 to 1987. Each produced one son that sang only the commoner of his two songs. The sons must have heard both songs, but for some reason sang only one of them. These two examples, together with the two bilingual sons of monolingual parents and the son who sang only a heterotypic song, show that sons do not always sing only what they hear from their fathers. Sons of the silent male might have further illuminated how songs are acquired, but unfortunately none were known to survive.

Territories and the Role of Song

Having described the structure of song and how it is acquired, we will now discuss the functions of song and the significance of singing a particular song. First we present background information on the establishment and occupancy of territories by males and their use by females, since songs are sung by males only on their territories.

ESTABLISHMENT OF TERRITORIES

Young birds leave their natal territory two to four weeks after they fledge, and wander. Usually they do not leave the territory voluntarily. They follow their parents off the natal territory, are fed for one or more days, and are then abandoned; or, if the parents are breeding again, they are chased off the territory by them at a time when the next brood is within a few days of fledging. In the dry season many that were born on the study area are still there, although a few have been seen temporarily as far as 3 km away. At the beginning of the following wet season one-year-old males establish territories, sometimes by initially staying within a small area of about 200 m^2 and remaining silent, but usually singing vigorously from one or a small number of cactus bushes. The typical but not invariant pattern is for the new territories to be established next to already established ones, rather than in isolated positions (Fig. 6.6), and to contain at least one cactus bush. Sometimes the attempt to establish a territory fails when the initiate is chased away by the neighboring territory owner. At other times the new bird abandons its territory and establishes another if it has failed to attract a mate, or if its breeding attempt with a mate has failed.

RETENTION OF TERRITORIES

Once having bred, a male holds its territory for life. The tenancy period can be as long as ten years. Territory boundaries may change from year to year, but usually on the order of only 10 to 20 m, rarely more. We have known two exceptions to life tenancy, with males abandoning the territories on which they had bred (successfully) and establishing new ones. The furthest distance moved was only 150 m. Two other males provide a different type of exception. In the last breeding season of their lives they abandoned their territories, wandered in the general vicinity singing very little, and did not breed. A fifth male provides yet another exception. In the middle of the 1986 breeding season it established a second territory less than 50 m from the first. With the same mate it produced the first and third brood on the first territory, and the second brood on the second territory.

Males spend most of their lives on their territories. They leave occasionally during the breeding season to feed in nearby areas, and more often in the dry season, but even at this time of general silence they return from a foraging trip elsewhere, sing weakly, and resume feeding.

The advantage to a male of retaining his territory throughout the year is that he is ready to breed at the onset of the first heavy rains of the year, without first having to re-establish ownership and negotiate the boundaries with his neighbors. Females visit and feed on many different territories during the dry season and are thus familiar with many males and their territories. As soon as it rains, pair formation is rapid. In all years males that retained their territories bred significantly earlier, raised significantly more clutches, and fledged significantly more young than those males that established new territories (Grant and Grant 1983). Early breeders are experienced, and age and experience are important factors in the reproductive success of a pair (Chapter 4, Appendix II), but they are not the only factors. The two experienced males who were exceptions to the rule of life tenancy and who changed territories bred a month later than those of similar age that retained their territories at the same time.

OCCUPANCY OF TERRITORIES BY FEMALES

Females are much less restricted than males to the territory on which they rear their offspring. They leave the territory at times when they are foraging for food for their nestlings. Occasionally the female, the male, or both lead fledglings off the territory for several days, particularly at the end of the breeding season, although usually the parents do not leave the territory for more than a few minutes at a time between successive broods.

In the early and middle parts of the dry season the males are most likely to be seen on their territories, whereas the females are most likely to be seen off theirs. If there is an unusually heavy shower at this time, the females quickly return to the territories. An example of this happened in 1982 after breeding had ceased in May. On July 10 a shower of 4 mm of rain fell. During the preceding three weeks, 16 of the 18 breeding males that had been seen were on their territories, whereas only 2 of 11 females seen were on territories. In the two days after the shower, 10 out of 12 females and 15 out of 16 males were on territory. The difference in proportions of females on territory before and after the shower is significant ($\chi_1^2 = 7.33$, $P < 0.01$).

Toward the end of the dry season, usually in the early part of the calendar year, some females appear to become increasingly restricted in their foraging to the territory on which they had previously bred, because they are never seen elsewhere. An example of site attachment is provided by female No. 6993. She bred with No. 6165 in 1978 in the central part of the study area. Even though she had changed mates in 1979 and bred 400 m away in the eastern part of the study area in 1979 and again in 1980, she was back on No. 6165's territory on 12 January 1981 and was seen repeatedly there in the next three weeks. It then rained and she returned to the eastern part of the study area to breed. But in 1982 she was not only back on the territory of No. 6165 but bred with him, after an interval of four years. Anecdotal evidence like this supports the view that familiarity with the territory is important in the choice of a mate by a female (Chapter 7).

TERRITORIAL DEFENSE

Territorial males sing repeatedly during the breeding season and defend their territories by chasing out intruding males. Not all territory owners are equal, however; there is some degree of age- and experience-related hierarchical behavior among them.

We made observations at all nests in study area 1 for periods of 30 minutes, three times during each breeding cycle in 1987. We found that males in black plumage (categories 4 and 5; Fig. 3.2) and males in partially black plumage (categories 0–3) intruded into neighboring territories to a proportionately similar extent ($\chi_1^2 = 0.01$, $P > 0.1$, $N = 94$,), but that responses of the territory owners differed according to their age as inferred from their plumage. Males in black plumage usually chased all other males from their territories, whereas males in partially black plumage chased intruders having a similar plumage to their own but hardly ever chased a male in black plumage ($\chi_1^2 = 10.98$, $P < 0.001$, $N = 94$). The partially black birds responded to black intruders by flying toward their mate or nest, remaining silent, staying low in the vegetation, or uttering ''weep-weep'' sounds. Thus young males breeding for the first time appear to be unable to drive away an intruding older male.

Both groups of males usually tolerate intruders in brown plumage, which are either females or wandering juveniles. Females are often courted, but both sets of brown birds are at some risk of being expelled from the territory during the dry season if they are directly competing for a food item, such as an *Opuntia* flower.

THE ROLE OF SONG IN TERRITORIAL DEFENSE

The generally silent intrusions and frequent chases between males suggest that at least one function of song is to broadcast the message ''keep out'' to other males. Song complements aggression and functions in territorial defense. The deterrent function of song has been demonstrated experimentally with other species of birds (Göransson et al. 1974; Krebs et al. 1978; Yasukawa 1981). The tape-recorded song of a great tit played back repeatedly in the territory of a male which had been temporarily removed was as effective as the singing owner itself in deterring intrusions. When the recorder was switched off the territory was soon occupied by a new male (Krebs et al. 1978).

THE ROLE OF *A* AND *B* SONGS IN TERRITORIAL DEFENSE

We carried out experiments with tape-recorded song to see if territory owners discriminate between the two types of song, to see if the particular song an intruder sings, be it type *A* or type *B,* makes a difference to the response of the owner. Songs of the two types were broadcast from a tape recorder in the center of 31 territories at a time when breeding had ceased but when territory owners were still singing frequently (June and July 1980). One song was repeated at 10-second intervals for 60 seconds, then the other song was played back in the same way after a silent interval of 180 seconds. Order of playback was randomized. The same song recordings (one *A* and one *B*) were used for all 31 experiments; the recorded birds themselves were not tested. Owners of the territory in which we broadcast the song responded to the playback of both song types by flying to within at least 3 m of the tape recorder (Table 6.3). Thus the males did not discriminate between the two song types in their own territories under these circumstances.

Another set of experiments, carried out in 1978 for a different purpose, gave the same result. The experiments were designed to test the ability of *conirostris* to distinguish between conspecific and heterospecific songs (Ratcliffe and Grant 1985). Regardless of whether it was type *A* or type *B,* the tape-recorded song elicited aggressive responses from the territory owner. Heterospecific songs of either *magnirostris* or *difficilis* were generally ignored.

Nevertheless males do have the ability to discriminate between *A* and *B* songs. During the *A-B* discrimination experiments, we had unexpected intrusions in 16 instances from a banded male occupying a territory adjacent to the one in which the playback experiment was being performed. The intruders did not influence the response of male territory owners on the experimental territory, because they did not sing and they always arrived after the initial response of the resident male. No intrusions occurred before playback. As soon as the territory owner saw the intruder, he chased him out of the territory. All 16 neighboring males entered the territory in response to playback of their own song type and flew to within 3 m of the tape recorder. None entered the territory in response to the opposite song type (Table 6.3). This difference in intrusion response by neighbors to playback of their own and of the opposite song type was significant (Fisher's Exact test, $P < 0.002$); the order of playback was not important ($P > 0.1$). The song type of the territory owner in the experimental territory was also not important to the neighbor, since a three-way G test of independence shows that neighbor responses depended only on playback of their own song type ($G = 21.22$, $P < 0.001$).

The chief implication of these results is that territory owners would be aware of a change in ownership of an adjacent territory when hearing the new song, and they would be more likely to investigate and possibly chase a new male away if the song type was the same as his own than if it was different. Thus males can discriminate between the two song types, and such discrimination could influence the spatial pattern of territories.

Song and Breeding

THE ROLE OF SONG IN MATE ACQUISITION AND RETENTION

Males sing vigorously and frequently in the breeding season, especially at the beginning of each breeding episode before they obtain a mate and after

Table 6.3 Responses of Males to Playback of Tape-Recorded Song

Song Type of Territory Owner		Number of Birds Tested	Number of Responding Birds			
			Playback Order		Playback Order	
			First Song *B*	Second Song *A*	First Song *A*	Second Song *B*
Territory owner	*B*	10	10	10	—	—
Territory owner	*B*	7	—	—	7	
Territory owner	*A*	8	8	8	—	—
Territory owner	*A*	6	—	—	5	6
Song *B* neighbor	*B*	6	1	0	0	5
Song *A* neighbor	*B*	3	0	2	1	0
Song *B* neighbor	*A*	4	3	0	0	1
Song *A* neighbor	*A*	3	0	3	0	0

Source: Grant and Grant (1983).

Note: Responses were scored as positive if a bird flew to within 3 m of the tape recorder.

pairing when the female is receptive. During the 30-minute nest observation periods, unmated males sang far more frequently ($\bar{x} = 163.6 \pm 13.72$, $N = 13$) than the mated males ($\bar{x} = 48.0 \pm 8.04$, $N = 23$; $t_{34} = 7.80$, $P < 0.001$), which indicates that song functions to attract potential mates as well as serving as a "keep out" signal to male would-be intruders. However among those 13 unmated males there was no correlation between the number of songs sung in 30-minute periods and the order in which they eventually attracted a female. This suggests that, although songs could signal to a female that a male was unmated, females do not use a comparison of song rates among males in choosing a mate.

Mated males sang more than twice as frequently at the time of clutch formation ($\bar{x} = 48.0 \pm 8.04$ SE, $N = 23$) as at the egg-hatching stage ($\bar{x} = 19.9 \pm 4.28$, $N = 21$). The difference is highly significant ($t = 3.00$, $P < 0.01$, two-tailed). An elevated rate of singing at the time of clutch formation could function to induce ovulation (Kroodsma 1976) and to repel intruders at the time when the female is most receptive (fertile). There was no significant difference in the number of songs sung by song type A and song type B males at any stage of the breeding cycle (t tests, $P > 0.1$), nor was there a significant difference between males of the two song types in the order of breeding (Mann-Whitney U test, $P > 0.1$).

FEMALE DISCRIMINATION BETWEEN SONG TYPES

When discrimination experiments were conducted with males, too few females were present to assess their responses to the playback of songs at the same time. In the middle of the breeding season of 1981, when another set of experiments was conducted, males chased responding females away from the tape recorder. When repeated in May, the experiments gave interpretable results. In 12 tests the female was alone on the territory at the time of playback. In 10 tests females responded to one playback only, and in the other 2 tests females responded to both. Nine of the 10 differential responses were elicited by playback of the song type *opposite* to that of the mate. This result is not expected by chance (two-tailed sign test, $P = 0.02$); it was not expected at all! Inclusion of the results of the other two tests, by assigning half to each group to give a ratio of 10:2, doubles the probability value ($P = 0.04$).

Since the responses were not elicited consistently either by the first or the second playback, nor by a particular song type, i.e., A or B (multi-way G test, $P > 0.1$ in each case), we conclude that females discriminated on the basis of song type in relation to their mates' songs.

Females were at the incubation stage of a nesting cycle during the experiments. David Anderson (1981, pers. comm.), who conducted the experiments for us, described the females as bursting forth from the nest during heterotypic song playback, but not during homotypic song play-back. The one exception was a female who had recently changed mates from a song *A* to a song *B* male. David Anderson observed females going to a conspicuous perch away from the nest, adopting a precopulatory posture, and rapidly producing "chick" sounds. He interpreted this "prostitution" behavior as serving to distract the intruding male from the nest. Copulations were not seen, and if they had occurred they would not have resulted in eggs unless the mate deserted the female, because the females were already incubating full clutches of eggs. The sexual response of the female could also indicate her readiness to mate at other times of the breeding cycle.

It is not clear why females did not discriminate between their mates' songs and tape-recorded songs of the same type (but different subtype), when males in other experiments (p. 129) did discriminate between tape-recorded songs and natural songs of the same type. Nevertheless the main point from the experiments is that females can discriminate between the two song types. The next question is whether they do discriminate when choosing a mate.

THE ROLE OF *A* AND *B* SONGS IN MATE CHOICE

Do females pair preferentially with males that sing a particular song in relation to the song type of their fathers? Assortative mating might arise from the effects of the same imprinting process in early life that has been implicated in the vertical (cultural) transmission of song from father to son, or it might arise because there are inherited mating preferences (e.g., Lande 1981; O'Donald 1983a). In contrast, females might pair with males that sing the opposite song type to their fathers' (disassortative mating), an outcome which could be rationalized as a means of avoiding close inbreeding, i.e., with father or brothers.

For 33 females we know the paternal song type and the song type of the mate at the first breeding attempt. Three of them were not helpful as their fathers were bilingual. The remaining 30 lead us to conclude that initial pairing by the female is random with respect to song type. Half of them had song *A* fathers; 5 paired with song *A* males and 10 paired with song *B* males. Half had song *B* fathers; these paired with males in the ratio of 9 song *B* to 6 song *A*. A Fisher's Exact test gives a probability value associated with the observed proportions of 0.28. Moreover the total frequencies of matings, 37 percent with *A* males and 63 percent with *B* males, are very close to the average frequencies of available mates over the whole study period (e.g., see Table 6.4).

By a small margin the majority of females (53 percent) paired with males that sang the opposite (heterotypic) song to that of their father. It is possible that all females prefer to mate heterotypically but that shortage of

Table 6.4 Frequencies of Males Singing Song A, Song B, and Song AB (Bilingual Birds) in the Main Study Area (Area 1)

Year	Mated			Unmated			Total	%A	Mated	
	A	B	AB	A	B	AB			% A	% B
1978	8	11	0	4	2	0	25	48.0	66.7	84.6
1979	10	16	0	4	3	1	33	41.2	71.4	84.2
1980	14	20	0	1	7	0	42	35.7	93.3	74.1
1981	16	31	0	2	9	0	58	31.0	88.9	77.5
1982	16	30	2	5	5	0	58	36.2	76.2	85.7
1983	36	51	2	2	1	0	92	41.3	94.7	98.1
1984	(4)	(5)	(2)	?	?	?	?	?	?	?
1985	?	?	?	—	—	—	?	?	—	—
1986	12	15	1	2	3	0	33	43.8	85.7	83.3
1987	12	19	1	0	6	0	38	31.6	100.0	76.0

the appropriate mates at the right time or some other factor prevents the preferences from being expressed. If this were the case, there are only two ways we would know it: if breeding success was lower among the homotypically pairing females than among the heterotypic ones, or if the females who first paired homotypically later re-paired heterotypically. There is no evidence of either. With regard to initial breeding success, females who paired homotypically produced as many fledglings on average $(2.0 \pm 0.54, N = 11)$ as did those who paired heterotypically $(1.9 \pm 0.36, N = 16)$. With regard to re-pairings, females that change mates (see Chapter 7) do not pair preferentially with males of a particular song type. Up to the beginning of 1982 there appeared to be a nonrandom tendency $(P = 0.02)$ for females $(N = 21)$ to pair heterotypically following the disappearance of a mate or after a nesting failure (Grant and Grant 1983). But by the end of the study we had data on 82 mate changes and there was no such tendency over the whole period $(\chi^2_1 = 0.01, P > 0.1)$.

Thus females do not show preferential mating with respect to song when pairing for the first time, even though they are capable of discriminating between the two song types.

Song Type, Mate Choice, and the Distribution of Territories

THE PATTERN OF TERRITORIES

Figure 6.6 shows the distribution of territories in each year. As population density rose, the total area occupied by territories increased. The peak was reached in 1983 when some territories were established in areas completely devoid of cactus. During the population decline that followed, the distribution relaxed back to one very similar to the earlier conditions in 1978 and 1980. Males that sang type A remained in the minority throughout, starting at a frequency of 48 percent and finishing at a frequency of 32 percent (Table 6.4).

In 1978 territories of mated song type *A* males and song type *B* males alternated in space: no two mated males of the same song type had adjacent territories (Fig. 6.6). In contrast, territories of unmated males were as likely to be next to males that sang the same song type as themselves (homotypic) as the opposite one (heterotypic). For the mated males alone, the ratio of 11 boundaries between heterotypic neighbors to 0 boundaries between homotypic neighbors cannot be attributed to chance (binomial test, $P < 0.002$). The distribution of territories of unmated males with respect to song type of neighbors is random ($P > 0.1$), but the difference between the unmated and mated male territory patterns is statistically significant (Fisher's Exact test, $P = 0.023$).

After 1978 the alternating pattern of territories of mated males with respect to song type changed. Table 6.5 gives the number of shared boundaries at the beginning of several breeding seasons. Not included are data for 1979 and 1983 when the initial distributions were not known, and 1984 and 1985 when little or no breeding occurred. Changes can be summarized as follows. The alternating pattern had disappeared by February 1979 (Grant and Grant 1983). It was present again at the beginning of the breeding season in 1980 and 1982, but not in the intervening and subsequent years.

The pattern changed as a result of the population being augmented by new males which established new territories adjacent to others, thereby creating more homotypic and heterotypic boundaries, and as a result of some of the new males with homotypic boundaries acquiring mates. It did not take place through wholesale shifts in territories. Prior to 1986 relatively few territory owners were replaced each year. An influx of new breeding males occurred each year during and after the time of the first brood. The most profound change took place in 1986 when the population was dominated by the 1983 cohort, and many new territories were established at the beginning of the breeding season.

MATE CHOICE IN RELATION TO SONG NEIGHBORHOOD

Throughout the study the frequency of shared heterotypic boundaries differed from expectations more among the mated males than among the total sample in all years except 1987, and to a significant extent in three years (Table 6.5). This could be the manifestation of female choice of mates that was influenced to some extent by the song of the territory owner and its neighbor(s). The data are not independent from year to year, however, because males that have bred remain on the same territory throughout the year and from year to year. Nevertheless, when we restrict attention to males holding a territory for the first time, we find further

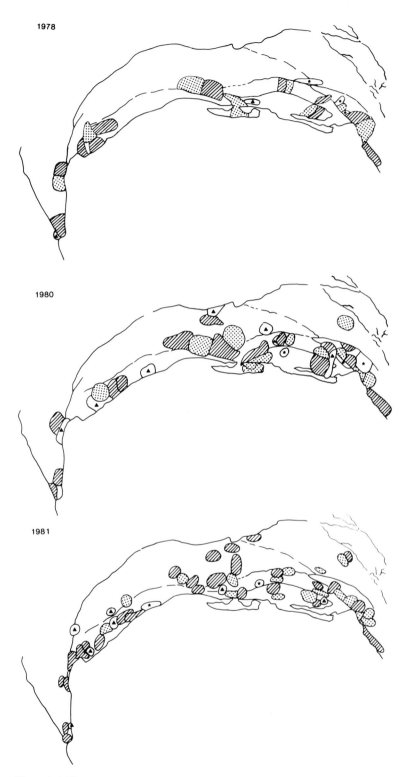

1978

1980

1981

Figure 6.6 Territories in study area 1, which may be compared with the distribution of cactus (bottom right; prepared by T. C. Will). Areas with dense cactus are preferred, but at the highest density (1983) areas almost devoid of cactus, especially the northeast

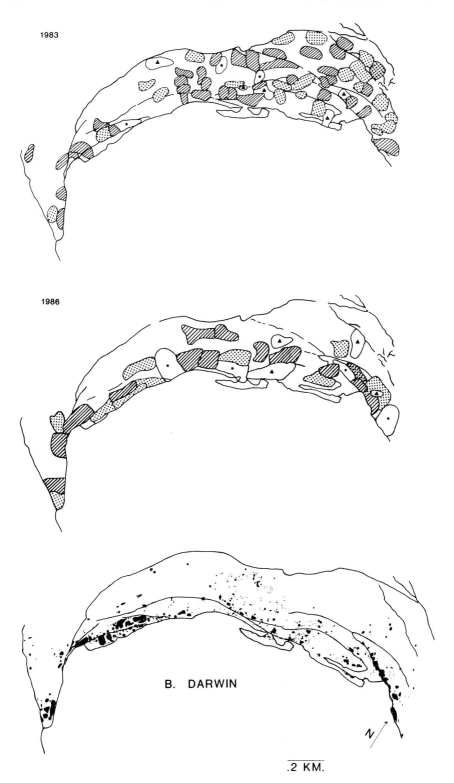

1983

1986

B. DARWIN

N

.2 KM.

section, were occupied by breeding birds. Symbols: stippled, mated song *A* males; lines, mated song *B* males; star, unmated song *A* males; triangle, unmated song *B* males. Faults in the lava are shown by continuous or broken lines.

Table 6.5 Number of Shared Boundaries between Males of Homotypic Song Type and Heterotypic Song Type at the Beginning of the Breeding Season

Year and Status	Number of Males		Number of Boundaries		χ^2
	Song A	Song B	Homotypic	Heterotypic	
1978: all males	11	13	4	13	4.89*
mated males	8	11	0	11	11.54***
1980: all males	11	19	10	18	3.58
mated males	9	14	3	11	5.37*
1981: all males	17	29	16	22	1.95
mated males	15	24	13	12	2.00
1982: all males	19	22	11	20	2.71
mated males	14	20	8	17	3.84*
1986: all males	14	18	13	11	0.05
mated males[1]	9	13	4	8	1.62
1987: all males	11	21	8	13	2.35
mated males[1]	7	16	4	7	2.04

Note: χ^2 values test whether the observed frequencies of boundaries can be expected by chance. *$P < 0.05$; ***$P < 0.001$.

[1]One bilingual bird with three territory boundaries (neighbors) has been omitted. χ^2 values are obtained by comparing boundary frequencies with expected ones derived from the frequency of occurrence of males with the two songs, given in Table 6.4 (Cliff and Ord 1973; B. R. Grant 1984).

reason for believing that pairing is partly influenced by song neighborhood. Most of our data come from 1980 and 1982. In these two years combined, proportionately more males with at least one heterotypic neighbor gained a mate (Table 6.6, $\chi^2_1 = 33.27, P < 0.001, N = 88$). A trend in the same direction was apparent in 1986 (Fisher's Exact test, $P = 0.09, N = 17$) but not in 1987 ($P = 0.33, N = 10$).

A further reason for suspecting that song neighborhood could play a role in pairing comes from the pattern of breeding. Using the date the first

Table 6.6 Mating Success of Males Establishing Territories between 1980 and 1982 in Relation to Song Neighborhood

Neighbors	Number of Males That	
	Gained a Mate	Did Not Gain a Mate
Only homotypic	9	16
Only heterotypic	28	1
Homotypic and heterotypic	17	0
None	10	7

Note: Proportionately more of those with at least one heterotypic neighbor gained a mate than did the rest (see text).

egg was laid as a measure of the start of breeding, we found with Mann-Whitney U tests that inexperienced, partially black males with at least one heterotypic neighbor began breeding significantly earlier than did partially black males with only homotypic neighbors or no neighbors at all in 1981 ($P = 0.008$) and in 1982 ($P = 0.03$). The ones with heterotypic neighbors also produced more fledglings in the two years ($P = 0.05$ and $P = 0.01$). In 1980 there was no detectable difference in pairing and breeding order. In 1986 we could not apply the test because all six males in partially black plumage had heterotypic neighbors, and in 1987 none had heterotypic neighbors; in other words, in neither year was a choice available.

THE CHOICE OF A TERRITORY BY A MALE

Since males maintain an approximately fixed territory location throughout their lives, while the song neighbors may change, the feature of primary importance to them when they establish their territories could be future resources (food, nest sites, etc.). Song neighborhood could play a secondary role since it appears that a young male will do best in obtaining a mate, at least in some years, if it can establish a territory next to at least one other male who sings the opposite song type. Results of playback experiments suggest that it will not be resisted in its attempts to establish such a territory as strongly as it would be if it had the same song type as the neighbor's. However, young males settling next to an experienced heterotypic male may run a high risk of being cuckolded. There are two reasons for making this suggestion. First, playback experiments show that a song which contrasts with the mate's song induces a precopulatory posture in females, whereas playback of the same song type as the mate's does not. Second, observations close to the nest show that partially black males do not chase intruding black males from their territory.

Thus there is a trade-off for a young male settling next to an older, heterotypic neighbor, as he may increase his chances of gaining a female, but he may also increase his chances of being cuckolded. This trade-off, along with considerations of habitat, would explain why we have no evidence of young males settling preferentially next to heterotypic singers in the years for which we have the most extensive and complete data (1981, 1982, 1986; χ^2 tests, $P > 0.1$ in each case). Further information on the subject of cuckoldry will be deferred to the next chapter.

ADVANTAGES OF HAVING A HETEROTYPIC NEIGHBOR

Nonrandom patterns of territory distribution and mating success, and the differential responses of both males and females to the two song types, can be understood in the light of two possible advantages of rearing offspring on a territory with a heterotypic neighbor. The learning of parental and species characteristics is enhanced, we suggest, by early (fledgling) exposure to heterotypic song.

Sound attenuates in a wooded environment (Morton 1975; Bowman 1979, 1983), so two songs of the same type are more difficult to distinguish 50–60 m away at the edge of the territory than are two songs of contrasting types. Therefore it is easier for fledglings to locate their father in a territory with a heterotypic neighbor and avoid making mistakes in identifying him. If they were to make a mistake and fly into the neighboring territory in response to a homotypic song, they would fail to receive food from the neighbor and possibly get lost and starve. This factor may have been important in 1980, the year when most fledglings were observed for the first four months, as there was a significantly higher proportional survival of young from natal territories with heterotypic neighbors than from territories with only homotypic neighbors ($\chi_1^2 = 10.07$, $N = 154$, $P < 0.01$).

Given that there are two song types in the population, males and females must be able to recognize both song types as *conirostris* songs and to discriminate between them and the songs of the two closely related sympatric congeners, *magnirostris* and *difficilis*. Both *magnirostris* and *difficilis* breed in the same habitat and hold territories overlapping those of *conirostris*. Fledglings raised on territories with a heterotypic neighbor can learn by associating auditory with visual cues to discriminate between the two *conirostris* songs, on the one hand, and the songs of their sympatric congeners, on the other, during the short sensitive period for song learning. Young born in natal territories without a heterotypic neighbor do not have the opportunity to associate *conirostris* morphology with the alternative conspecific song until after they have left their natal territory, when learning sensitivity is waning or has disappeared.

The Role of Song in Species, Individual, and Kin Recognition

In many species of song birds, males with large repertoires are favored by sexual selection. Large repertoires aid in obtaining and maintaining territories (Howard 1974; McGregor et al. 1981; Yasukawa and Searcy 1985) and in attracting and stimulating females (Kroodsma 1976; Catchpole 1980; Catchpole et al. 1984).

Species such as these have a long sensitive period for song learning (Marler and Tamura 1964; Nottebohm 1969). A sensitive period extending for life enables individuals to add to their repertoire continuously through copying others and extemporizing (Laskey 1944; Thorpe and North 1965; Nottebohm 1972). In some species, the canary for example, the song control centers of the brain diminish in the nonbreeding season but are reactivated the following breeding season (Konishi 1985). Males of these species learn new songs and change their repertoire from year to year (Konishi 1985). One consequence is an increase in the opportunity to obtain a territory through the deception of neighbors (Payne 1982).

Immelmann (1980) has demonstrated individual variation in the length of the sensitive period in zebra finches and suggests that it is largely under genetic control. Likewise, Kroodsma and Canady (1985) have shown that the differences between populations of marsh wrens in repertoire size and in the size of the song-controlling nuclei in the brain have a genetic basis. Thus long sensitive periods or large, active song-controlling nuclei in the brain can evolve under sexual selection where circumstances favor large repertoires.

Not all circumstances favor large repertoires, however, nor do all species have them or long sensitive periods for learning song. The great tit is unable to alter its repertoire after the first breeding season (McGregor et al. 1981; Lambrechts and Dhondt 1986, 1988), suggesting that the sensitive period for song learning is restricted to the first year of life in this species. An even shorter sensitive period, limiting song learning to the period prior to dispersal from the natal territory, restricts learning to the father's song in other species (Kroodsma 1974; Immelmann 1975; Konishi 1985). These are the most germane to our study because *conirostris* have a short sensitive period and a repertoire size of one.

These repertoire and learning features can be interpreted with reference to three potential disadvantages to a large repertoire or a constantly changing song: information in the song is less specific with regard to species, individual, and kin identity.

SPECIES RECOGNITION

All species have the problem of discriminating between songs of their own and those of other species. It is especially severe among those that are sympatric with many congeneric species which are similar to each other in plumage or behavior. Marler (1960) pointed out that recognition cues are selected to stand out against a background of other species' songs. Although different features of song are used by different species (Peters et al. 1980), the constant unvarying ones seem to be the most important (Emlen 1972). Peters et al. (1980) and Dooling (1982) have shown that swamp sparrows with a repertoire of 3–4 songs develop their species-discriminating abilities earlier than do song sparrows with their larger repertoire of 8–12 songs. Immelmann (1969) suggested that imprinting on father's song in zebra finches could prevent hybridization where many congeneric species breed in the same area.

Effective species recognition by song is important in *conirostris* because they breed in the vicinity of *magnirostris* and *difficilis*, often holding territories that overlap theirs and breeding at the same time (Grant and Grant 1980). Plumages of the three congeners are almost identical, as are many reproductive behavioral features such as precopulation displays and nest building (Lack 1947; and personal observations). The species differ morphologically only in body size and bill size and shape (Plate 10; Fig. 2.18). They differ also in song (Fig. 6.7). Ratcliffe and Grant (1983a, b 1985), using song playback and female mount experiments, demonstrated that song, body size, and bill shape are used by *conirostris* in species recognition. Playback of *magnirostris* and *difficilis* songs elicited virtually no response from *conirostris* territory owners in those experiments.

In species like *conirostris*, where song plays a major role in species recognition, individuals with an enhanced ability to transmit distinct

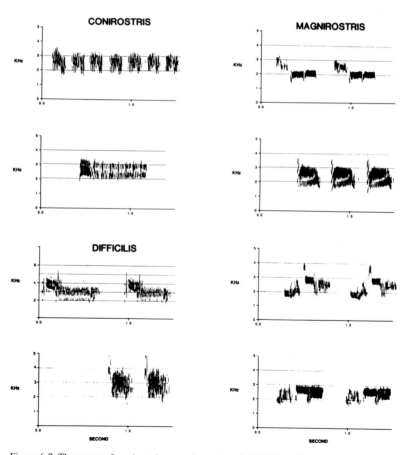

Figure 6.7 The songs of *conirostris*, *magnirostris*, and *difficilis* on Genovesa.

species-specific and stable songs and to recognize them have an advantage in obtaining mates. In the few cases where individuals sang two songs, those that sang both *conirostris* song types suffered no disadvantage in obtaining mates and rearing young. However, one *magnirostris* that sang both a *magnirostris* and a *conirostris* song did not maintain ownership of his territory or obtain a mate.

INDIVIDUAL RECOGNITION

Large repertoires cannot be used as efficiently as small ones for individual recognition. It has been found with song and swamp sparrows (Searcy et al. 1981) and with eastern and western meadowlarks (Falls and d'Agincourt 1981) that the species with the smallest repertoire exhibit the greatest difference in responsiveness toward neighbor song and stranger song. Our own playback experiments have demonstrated the ability of *conirostris* males to discriminate between neighbors and strangers by song.

In all species that use song in territorial defense, the features of the song used to identify individuals are expected to remain stable while the territory is occupied. Otherwise a change in song would indicate from a distance a change in territory ownership, induce heightened aggression in the neighbors, and result in unnecessary energy expenditure in renegotiating the territory boundaries, all of which could lead to a possible delay in the acquisition of a mate and the start of breeding. Since male *conirostris* hold their territories for life, they retain their song unaltered for life.

KIN RECOGNITION

A kin recognition system could be favored by selection if there is any advantage to be gained from nepotism or if there is a fitness disadvantage in mating with close relatives. We have no indication that *conirostris* engage in nepotism, in the form of cooperation between kin after dispersal, for example, or in lack of aggression between father and son when establishing territories. The one son which set up a territory next to his father's was no more tolerated than was his father's other neighbor. However there is a possibility that inbreeding between close relatives is avoided.

In small island populations where dispersal is restricted and individuals are long-lived, there will be some degree of relatedness between individuals due to common ancestry in the past, and there is a high probability that the gametes from two parents will contain alleles identical by descent (Baker and Marler 1980; Falconer 1981). Furthermore, if the dispersal distances from natal to breeding site are low, and if they are the same for males and females, the opportunity for sibs to mate with each other by chance is high (Greenwood and Harvey 1982), which further increases the probability that the gametes will have alleles identical by descent.

Inbreeding depression, which is the loss of fitness caused by the expression of deleterious recessives or by developmental instability due to increased homozygosity (Simmons and Crow 1977; Lande and Schemske 1985; Charlesworth and Charlesworth 1987), is a possible consequence.

Kin recognition systems that facilitate inbreeding avoidance have been found in several species of animals (see Holmes and Sherman 1982; Lacy and Sherman 1983). Bateson (1982) has demonstrated with a laboratory population of Japanese quail that individuals can discriminate between plumage variants, and that they prefer to court birds that differ from their parents although not by much. Bullfinch siblings learn their father's song (Nicolai 1959) and form social bonds after leaving their natal territory, but then later they become antagonistic toward each other as the breeding season approaches. They have never been known to mate with each other (Nicolai 1956). Kin recognition cues cannot be confined to information obtained from *conirostris* nest mates, because many sibs that were raised in different broods are available as potential mates. However, as with quails (Bateson 1982), signals associated with the parents could be matched against those of potential mates. A song that is stable and copied in detail from the father could signal "possibly kin." The signal itself need not be a genetically fixed property; this is in contrast to the phenotypic matching models of Beecher (1982) and Lacy and Sherman (1983), where the signals are assumed to be under genetic control.

The genetic effective size of the *conirostris* population is low, less than 200 (see Appendix IV). There is a moderately high probability that sibs will mate with each other by chance because the average dispersal distances from natal to breeding sites are relatively short, on the order of only 4 to 5 territories. Moreover there is no differential dispersal of the

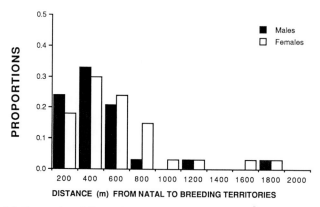

Figure 6.8 Frequencies of dispersal distances from natal to breeding territories of 33 females and 30 males.

sexes (Fig. 6.8). The median distance for males (340 m, $N = 29$) does not differ significantly from the median distance for females (420 m, $N = 33$; Mann-Whitney U test, $P > 0.1$). Despite the likelihood of matings between close relatives by chance, none of the 55 birds born in the study areas who subsequently bred there mated with a known relative. This is not strong evidence of inbreeding avoidance because many kin relationships among breeders were not known.

If inbreeding is avoided, the detailed structure of the father's song could be used to identify close kin. Although mating is random with respect to song type, it may not be random with respect to song subtype. Individuals could recognize kin by the fine structure of the song and avoid breeding with them by an active process of discrimination. Other cues seem unlikely. For example, as far as we know plumage features do not convey kinship information.

None of the 21 females whose father's and mate's song subtypes were known mated with a male of the same song subtype as her father. Of these 21 females, only 8 mated with males of the same song type as their fathers', and by so doing could have mated with close kin. These numbers were accumulated over six breeding seasons with different sets of mates available among the eight song subtypes recognized in the population. The samples are too small to assess the probability of observing by chance the absence of a female mating with a male of the same subtype as her father's. We conclude that the evidence is only suggestive of an avoidance of breeding with kin through the identification of kin by the fine detail of the song.

In summary, circumstances that favor a distinct, precisely copied, and stable song are the long life of individual males, life tenancy of their territories, and the proximity of congeneric species that are similar in appearance. Advantages of individual recognition and possibly kin recognition may also contribute to those circumstances. Collectively these circumstances have selected for a shorter sensitive period limiting song learning to the time prior to dispersal from the natal territory.

Origin and Maintenance of the Two Song Types

All populations of finches studied in the genus *Geospiza* have one, two, or a few song types (Bowman 1983; Ratcliffe and Grant 1985), therefore *conirostris* on Genovesa is not unusual in having two songs. The first colonists of the island may have comprised both *A*- and *B*-singers. Alternatively, a song type could have been introduced by a closely related species immigrating and breeding with local residents. Song *A* on Genovesa has a strong resemblance to a song of *scandens* on Daphne Major (Fig. 6.9) and Santa Cruz, whereas very different songs are sung by birds in the other two populations of *conirostris* on Española and nearby Gardner, far away in the south of the archipelago (these songs are illustrated in Fig. 9.4, Chapter 9). As a third possibility, one song may have arisen through a substantial error in the copying of the other.

Why are there only two song types in the population? A factor limiting the increase in diversity of song types in any population of Darwin's Finches is the tendency for a new song to become extinct, partly because it (or the singer) is discriminated against, and partly because it is rare. The new song, arising perhaps through a copying error, could differ sufficiently from the original to constitute a third type. However, the more it differs from the original the less likely it is to be recognized as a *conirostris* song by individuals other than the neighbors of the singer. For example, Smith (1976) has shown that abnormal songs in the red-winged blackbird are recognized by neighbors but not by other individuals in the population. Such recognition probably involves associative learning, as suggested by

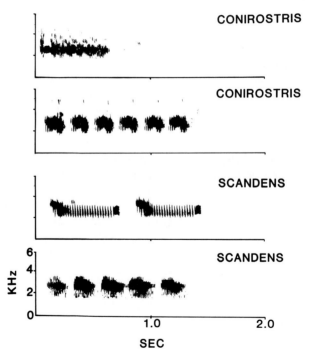

Figure 6.9 Songs of two species of cactus finches compared. The upper two sonagrams show the *B* and *A* song of *conirostris* on Genovesa; the third one shows a *scandens* song from Marchena; the fourth one shows a *scandens* from Daphne Major. The Daphne *scandens* song resembles the *A* song of *conirostris*. The Marchena sonagram was supplied by R. I. Bowman. From Grant (1986a).

the work of Searcy et al. (1981) and Yasukawa et al. (1982). Lack of recognition, therefore, reduces the chances of a male with a new song type obtaining a mate, so new song types will be incorporated into the population infrequently. The extinction rate of a new song type at low frequency will be high owing to random processes as well, especially at times of population crashes associated with droughts. Therefore, with a low recruitment rate and a high extinction rate of new song types at low frequency, there is never likely to be more than a few song types in the population at any one time.

In contrast, new song subtypes should be more frequently incorporated into the population because, arising as small copying errors within the framework of a song type, they should be easily recognized as *conirostris* songs.

The two song types in the population tend to be maintained in a frequency-dependent manner. Males with the rarer song (type *A*) more often had a heterotypic neighbor and in every year fledged more offspring (Appendix III, Table A7) to a statistically significant extent in 1981 and 1982 (*t* tests, $P < 0.05$). This frequency-dependent breeding advantage tends to produce a 1:1 ratio of the two song types, but random events, particularly in the relative proportions of experienced and inexperienced breeders in the two groups, cause stochastic fluctuations (see Appendix III for detailed demographic characteristics of the two groups).

Summary

Males sing a single, short, and simple song, whereas females, although they vocalize near their mate or nest, do not sing. Sons learn their song, even subtle structural features of the song, from their father and retain it unaltered for life. Song learning takes place in the late nestling and early fledgling stages. There are two song types in the population, and a very small number of males (1–3 percent) sing both types. Song plays a role in the acquisition and maintenance of territories, the attraction and stimulation of mates, and the recognition of species, individuals, and possibly kin.

Males acquire a territory in their first or second year, and once they have bred they keep it for life. They respond equally to the two song types played back from a tape recorder in their territories, but they discriminate between them when songs are broadcast experimentally from an adjacent territory; they intrude only when their own song type is played back, regardless of the territory owner's song. These results indicate that a male is aware of a change in ownership of an adjacent territory by the song the new bird sings, and that he would be more likely to investigate the newcomer and perhaps act aggressively toward it if the song type was the same as his own than if it was different.

Females pair randomly with respect to song type, but possibly not with respect to song subtype, when breeding for the first time. There is no association between the song type of the father and the song type of the mate. Nevertheless females can discriminate between song types, as shown by results of playback experiments in the breeding season when they responded to the song type opposite to that of their mates by assuming a precopulatory posture.

In 1978 territories of mated males of the two song types alternated in space; that is, no two mated males of the same song type shared a territory boundary. Territories of unmated males did not show this pattern. The alternating pattern broke down in subsequent years as a result of recruitment of new males into the breeding population, but it reappeared at the time of the first broods in 1980 and 1982. The distribution of unmated males remained random throughout the study period. In some years more males with a heterotypic neighbor gained a mate than those without a heterotypic neighbor, and those breeding for the first time with a heterotypic neighbor bred significantly earlier and raised more offspring than those without.

These data suggest that a female may prefer to mate with a male in a territory with a heterotypic neighbor. If this is true there is a mating advantage to be gained by a male holding a territory with a heterotypic neighbor. There are three other possible advantages, the last two being shared by the female. First, older males may increase their chances of an extra-pair copulation with the female of a young heterotypic neighbor, since playback experiments show that females respond sexually to songs contrasting with their mate's. Second, song attenuates in a wooded environment, making contrasting songs easier to distinguish from each other than homotypic songs. This facilitates location of the male by the fledglings and aids fledgling survival. Third, young must learn to recognize both song types as *conirostris* songs and to distinguish them from *magnirostris* and *difficilis* songs. Young with heterotypic neighbors are favored because they can use visual and auditory cues to learn both song types during the short sensitive period for song learning.

Species recognition by song has been demonstrated by playback experiments. This is important because *conirostris* breed in the vicinity of and at the same time as two closely related congeners. Song is species-specific and is stable from generation to generation as a result of transmission from father to son by copying. Individual recognition by song could be important in the maintenance of territories. Since territories are maintained for life, there is an advantage in song being stable for life. Kin recognition by song is possibly important for inbreeding avoidance. The low effective population size of less than 200, the relatively short dispersal distances from natal to breeding sites, and the lack of differential dispersal of the sexes all indicate a high probability of kin pairing together by chance. In spite of that, no female is known to have mated with a male of the same song subtype as her father's. Detailed characteristics of the song may have been used to avoid mating with kin.

In most species of song birds, males with large repertoires are favored by sexual selection. Large repertoires aid in obtaining and maintaining territories and attracting and stimulating females. In *conirostris*—and, we suggest, in other species where song plays a major role in species, individual, and possibly kin recognition—individuals with a distinct, precisely copied, and stable song have a fitness advantage. This type of song is most likely to be present when song learning is restricted to the short sensitive period before dispersal from the natal territory.

7

Mate Choice

Introduction

Sexual selection theory attempts to account for the evolution of traits through mating preferences (Darwin 1871). Recent studies have helped to clarify the genetic and selective mechanisms by which the traits and the preferences for them evolve (e.g., see Maynard Smith 1985; Bradbury and Andersson 1987). Nevertheless some issues remain unresolved. The most important one concerns the nature of the cues used in mate choice. The issue is whether the cues used are condition-dependent and thereby reliably indicate underlying genetic constitution, as argued by Andersson (1982a, 1986), Hamilton (1982), and Hamilton and Zuk (1982), or whether the trait or traits are arbitrary, in the sense of being unconnected with fitness, and maintained solely through a mating advantage (Fisher 1930; O'Donald 1980; Lande 1981; Arnold 1983a; Kirkpatrick 1982, 1986).

Experimental studies of captive, monogamous birds have shown that the choice of mates can be influenced by many factors, including early experience (Immelmann 1975; Cate and Mug 1984), genetic relatedness (Bateson 1982, 1983), and an exaggeration of natural colors and plumage features (Burley 1981, 1986; Burley et al. 1982; Andersson 1982b; Møller 1988). These experiments clearly demonstrate a female's ability to discriminate among males on the basis of appearance and behavior. Burley (1986) has documented the fitness consequences of particular matings of captive zebra finches. In nature additional factors, such as song rate or quality (McGregor et al. 1981; Catchpole et al. 1984; Gottlander 1987; Lambrechts and Dhondt 1986) and territory size or quality (Millington and Grant 1983; Davies and Lundberg 1984; Price 1984a; Harper 1985; Alatalo et al. 1986; Nagata 1986), can be important in mate choice and in fitness. They may be of overriding importance.

In our discussion of *conirostris* up to now we have considered only song as a factor in the choice of a mate. We showed in Chapter 6 that females exhibit no preference for a particular song type, although they do appear to be influenced by the song type of neighboring males to some extent in some years. In this chapter we extend the discussion to consider more broadly and directly the question of how and why particular mates are chosen. In view of the findings from studies of other species of birds, we have paid special attention to the previous breeding experience of cactus finch males, their appearance, and their territories as factors of possible importance governing mate choice.

We begin by summarizing the main features of the mating system and the reproductive behavior of cactus finches to provide the context for subsequent discussion of mate choice.

The Mating System: Mainly Monogamy

Cactus finches are monogamous; less than 5 percent of males on average pair bigamously with two females in different parts of the same territory (Table 7.1). Most bigamous matings occurred in 1983, but even in this year of extensive breeding all bigamous males bred monogamously more often than bigamously. In the next year of extensive breeding, 1987, no bigamous pairs were formed. One male was paired bigamously with different females in successive years. All 11 bigamous males were old and in black plumage; their minimal ages were estimated to be in the range of 3–8 years (5.2 ± 1.67, $\bar{x}$ ± SD). The relative shortage of females and of old males sets an upper limit to the potential number of bigamous matings,

Table 7.1 Breeding Status of Males: Number of Males That Paired Monogamously Throughout a Breeding Season, Numbers That Paired Bigamously for at Least One Brood, and Numbers That Remained Unmated for an Entire Breeding Season

| Year | Mated | | Unmated | Total |
	Monogamous	Bigamous[1]		
1978	19 (76.0)	0 (0.0)	6 (24.0)	25
1979	?	?	?	33
1980	32 (76.2)	2 (4.8)	8 (19.0)	42
1981	46 (79.3)	1 (1.7)	11 (19.0)	58
1982	46 (79.3)	2 (3.5)	10 (17.2)	58
1983	79 (86.8)	9 (9.9)	3 (3.3)	92
1984	(12)[2]	(0)[2]	?	?
1985	—	—	—	—
1986	28 (84.8)	0 (0.0)	5 (15.2)	33
1987	32 (84.2)	0 (0.0)	6 (15.8)	38
$\bar{x}$	(80.8)	(3.3)	(15.9)	

Note: Percentages are given in parentheses after the numbers.
[1]All males were in black 4 or 5 plumage.
[2]In parentheses because only a few pairs attempted to breed.

but this was far from being reached in any year as there were always many more old males paired monogamously than bigamously. Unlike the situation in some other species (cf. Smith et al. 1982), bigamy is not caused by shortage of males.

Males can, but rarely do, achieve high reproductive success by pairing bigamously. Only one male fledged more offspring from the two nests combined (7) than the maximum (5) produced from a single nest of a monogamous pair, and only six of the eleven males produced more fledglings per episode on average when paired bigamously than when paired monogamously. Thus in general there appears to be little reproductive gain or loss in breeding bigamously. There is no detectable survival cost (B. R. Grant and P. R. Grant 1987), and in this respect, in the rarity of bigamy, and in the small reproductive advantage gained from it, cactus finches resemble many other passerine birds (e.g., Lack 1968; Nolan 1978; Mock 1985; Dhondt 1987).

Courtship, Pairing, and Parental Care

Immediately after the first heavy rainfall of the year, males sing frequently and build one or more display nests. Visiting females are led to the nests, fed, and sometimes followed on long acrobatic flights. In view of their relatively short and broad wings, these flights must be energetically costly (Norberg and Norberg 1989). Females have been seen to visit many territories and be courted by several males before pairing. Once paired, a female is accompanied closely on the territory most of the time by the male (mate guarding), especially up to and during the early incubation stage, when she is receptive and the likelihood of cuckoldry is highest. She chooses one of the display nests, lines it with feathers while accompanied by the male, and lays a clutch of eggs. Only females incubate the eggs. Both parents feed the nestlings, approximately equally, with food gathered mainly from the territory, but males do the larger share of feeding the fledglings for a period of two to four weeks (Chapter 6). Participation of both parents in the feeding of young nestlings appears to be essential because death or disappearance of one parent in the first half of the two-week nestling phase has always resulted in abandonment of the young.

The Choosing Sex

The pattern of courtship behavior just described can be interpreted as showing a choice of males as mates by females. Females are usually the limiting sex. In seven breeding seasons the proportion of territorial males without mates varied from 3 percent to 24 percent (Table 7.1). Only in 1983 were there about as many breeding females as males. Both members of a prospective pair are expected to contribute to the joint decision to form a pair because both raise the offspring, but females have the greater opportunity to exercise a discriminatory choice. Our subsequent discussion will be framed in terms of female choice.

Cues Used in Mate Choice

If females do choose among several potential males, which cues do they use? In view of the effect of age and experience on reproductive success, discussed in Chapter 4, they could be expected to use some estimate of age and experience, either behavioral or morphological (plumage blackness). They could use two other traits that have been implicated in mate choice among medium ground finches (*fortis*) on Isla Daphne Major: beak size (or body size) and territory size (Price 1984a). Thus properties of the male or properties of his territory may be important to females when choosing a mate.

TERRITORY SIZE AND QUALITY

Territory area was measured directly or from territory maps in all years from 1978 to 1987. In 1982 only, an attempt was made to determine territory quality. We concentrated on the amount of *Opuntia helleri* cactus and the number of *Opuntia* flowers and fruits produced in each territory as a measure of its quality, because 90 percent of nests are placed in cactus bushes and more than 50 percent of foraging time throughout the year is spent on *Opuntia* flowers and fruits, and on arthropods in *Opuntia* pads (Fig. 1.1; see also B. R. Grant and P. R. Grant 1979, 1981, 1982, 1983). The territories of mated and unmated males did not differ significantly in total area, or area of cactus, or number of flowers and fruits (t tests, $P > 0.1$). Therefore territory area in all years and quality in 1982 had no detected effect on mate choice by females.

It is possible that females use other unmeasured cues of territory quality in choosing their mates, such as the abundance of *Croton scouleri* whose seeds are fed to some nestlings (Table 4.1) and fledglings. However, small variations in territory quality do not seem to be very important to *conirostris* because once males have bred in their territories they retain them for life. Over the years, features of the territory change; for example, the relative quality of *Opuntia* resources on the various territories changes from one year to the next. If the changes in quality were large and important, males of this long-lived species would be expected to track the resources and change territories between seasons. It may be that certain habitat features and familiarity with them, as well as the advantages of retaining a territory for early breeding, are more important than minor fluctuations in quality.

SIZE

We compared males who acquired a new mate with those who did not in each of the years of breeding and found no difference between them in beak or body size traits in any of the years (t tests, $P > 0.1$).

Even though beak size and body size are not important in determining whether or not a male acquires a mate, they might be important among the subset of males that do acquire mates. Females could choose males that resemble themselves in size, or they could choose mates that resemble their fathers, either because they inherit a phenotypic preference from their mothers (e.g., Lande 1981; O'Donald 1983a; Price 1984a, b) or because they acquire a preference through imprinting on their fathers (e.g., Immelmann 1975; Cooke 1978). We examined these possibilities with correlation analysis but failed to find evidence to support either of them. There was no correlation ($P > 0.1$) between mates in 1980 for any of the standard six measures (see Chapter 1) of body and beak size ($N = 20$ pairs). We repeated the analysis in 1987, when all breeding birds except two were new, and obtained the same result ($N = 44$ pairs). Finally, none of the dimensions of a female's first mate was correlated with dimensions of her father ($N = 21, P > 0.1$) or her own ($N = 17, P > 0.1$). In short, there is no evidence that females are influenced by these morphological cues when choosing mates. If females really do have preferences for mates based on the size traits we have measured, they must be weakly expressed as a result of being overridden by other factors. It is not surprising that they would be overridden because there is a limited supply of unmated males at the time of pairing, and time constraints may prevent preferences from being expressed (see also Janetos 1980). A possibility remains that some individuals may have mating preferences and express them, while others do not (e.g., O'Donald 1983b).

PLUMAGE

Among the morphological traits that are possibly important in mate choice, only plumage state is associated with mating success; males in black plumage (categories 4 and 5; Fig. 3.2) succeed better than younger males in partially black plumage (categories 0–3) in acquiring mates (Table 7.2).

Ignoring males who retained their mates from one year to the next, we combined the data from the years 1981 and 1982 and compared the males who acquired a (new) mate with those who did not. The two groups differed in plumage state: proportionally more of those who did acquire mates were in black plumage (56 percent) than those who did not (17 percent; $\chi_1^2 = 7.90$, $N = 57$, $P < 0.01$). A trend in the same direction occurred in 1987. For example, at the time of the fourth brood all three unmated males in black plumage acquired mates, whereas only two out of eight in partly black plumage did so (Fisher's Exact test, one-tailed, $P = 0.03$). Data from other years have not been analyzed because prior pairing history was not known in 1978, breeding was not followed in 1979, some of the early breeding of 1983 was not observed, very few birds bred in the dry year of 1984, and none bred in 1985. Data from 1986 are exceptional and will be treated separately below.

AGE AND EXPERIENCE

Males in black plumage are successful in attracting mates, because females base their choice either on the plumage itself or on some other cue, such as courtship behavior, which is correlated with plumage state. As a male ages and acquires breeding experience, it changes both plumage state and courtship behavior. Under normal circumstances it is not possible to distinguish between their effects. In the years 1978 to 1983, and again in 1987, almost every black male had had previous breeding experience.

In 1986 plumage and the behavioral correlates of breeding experience became uncoupled for the first and only time in the study. This unprece-

Table 7.2 Mating Success of Males: Partially Black (Black Plumage Categories 0–3) Contrasted with Black (Categories 4 and 5)

Year and Plumage	Unmated	Mated	% Mated
1978			
Partially black	5	2	28.6
Black	1	17	94.4
1980			
Partially black	6	10	62.5
Black	1	25	96.1
1981			
Partially black	8	20	71.4
Black	2	28	93.3
1982			
Partially black	8	22	73.3
Black	2	26	92.9
1983			
Partially Black	3	52	94.6
Black	0	36	100.0
1984			
Partially black	(6)	(1)	(14.3)
Black	(15)	(12)	(44.4)
1986			
Partially Black	3	7	70.0
Black	3	21	87.5
1987			
Partially black	6	8	57.1
Black	0	24	100.0

Note: Figures for 1984 are given in parentheses because several potential breeders did not breed.

dented situation arose because insufficient rain fell in 1984 and 1985 for successful breeding to take place (Chapter 4), and as a result some birds born in 1983 were in black plumage by 1986 when their first opportunity for breeding occurred, whereas others in the same cohort were still in partially black plumage. In taking advantage of this situation, we were helped by a shift in the age structure that had occurred in 1983 (Chapter 3). Birds born that year were a significant component of the population in 1986.

If black plumage, by itself, is important in determining mating success, inexperienced males in black plumage should have had the same mating success as experienced males in black plumage, and inexperienced males in black plumage should have had an advantage over inexperienced males in partially black plumage. They did not meet either expectation. All but 1 of the 12 experienced black males initially lacking mates acquired new ones at the time of the first brood; the one exception had acquired a mate in 1983 but was the only experienced bird to have never raised a brood then. In contrast only 5 out of 11 inexperienced black males acquired mates (Table 7.3). Experienced males had the higher mating success (Fisher's Exact test, $P = 0.02$), and inexperienced males had the same success regardless of their plumage ($P > 0.1$). So in this year, previous breeding experience, signaled presumably by behavior, was a more important determinant of mating success than was either plumage state or age.

Similar patterns have emerged from studies of two related species on Daphne Major, although previous breeding experience was not as well known in those studies as in ours. Millington and Grant (1984) found that *scandens* females paired preferentially with old males among a group that were all in fully black plumage; some of these old males were known to have bred previously, whereas the younger black males probably had not. In Price's (1984a) study of *fortis*, mating success of males was associated with black plumage, which covaried with previous breeding experience to some extent, but among inexperienced birds (the 1978 cohort) mating success was independent of plumage type.

ORDER OF BREEDING

If we assume that all males become reproductively active at the same time, the order in which males obtain (new) mates can be taken as a reflection of their relative attractiveness to females. Darwin (1871) used this type of argument to explain how sexual selection could occur in monogamous birds. The assumption of simultaneous reproductive activity is at least approximately true; with few exceptions all male finches start defending territories and singing vigorously within the first two days of the first heavy rainfall of the wet season. The time of pair formation cannot be determined precisely, whereas the day the first egg is laid can.

Table 7.3 Mating Success of Males in 1986

		Mating Success	
Experience	Plumage	Mated	Not Mated
Experienced	Black	11	1
	Partially black	1	0
Inexperienced	Black	5	6
	Partially black	5	4

Note: Numbers of males that mated or did not mate are shown. Experience was more important than plumage in determining which males mated.

An early analysis of the data from 1978 to 1982 (B. R. Grant 1985) showed that, by the criterion of breeding order, the most preferred males were those in black plumage and those with at least one neighbor which sang the opposite song type (heterotypic) to their own. Plumage and this particular feature of territory arrangement covary, so females could choose mates on the basis of one or both, or on the basis of age-related courtship behavior that is correlated with both. Experiments provide the preferable way of disentangling correlated effects on female choice (e.g., see Alatalo et al. 1986), but experimental manipulation was not available to us. Therefore, to separate the influence of these correlated characters on breeding order into direct and indirect (correlated) effects, we have used selection analysis that employs the technique of partial regression (Lande and Arnold 1983). Only males that obtained new mates in a given year were used in the analyses.

Results are given in Table 7.4 under two headings. Entries in the selection gradient (β) represent the direct effects of selection on each trait, while the selection differentials (s) represent the combined effects of selection on the trait and on all correlated traits. Fitness was scored by the rank order in which males obtained their females, as indicated by the date the first egg was laid. The relative fitness was obtained by dividing by the mean, so that the mean relative fitness equaled one (Lande and Arnold 1983). Results of the selection gradient analysis revealed that females chose males in black plumage in three out of the five years, and in one of the others (1986), when the separate effects of plumage and experience could be determined, they chose on the basis of experience. Territory position had a significant effect on the order of breeding in two years, and beak size had an effect in one year. Thus plumage and behavior associated with experience emerge as the most important identified factors in mate choice. The net effect, measured by the selection differentials, is early breeding of males in black plumage.

Order of breeding was not correlated with territory size in any year from

Table 7.4 Selection on Characters Associated with the Order in Which Males Obtained New Mates

| Year | Character | Coefficients | |
		$\beta \pm$ SE	s
1980	Plumage	0.32 ± 0.08**	0.33**
($N = 23$)	Territory position	0.26 ± 0.09**	0.30*
	Bill depth	-0.19 ± 0.09	0.05
	Bill length	0.12 ± 0.10	-0.15
	Bill width	-0.21 ± 0.19	-0.18
1981	Plumage	0.27 ± 0.10*	0.32**
($N = 26$)	Territory position	0.19 ± 0.10	0.29
	Bill depth	0.02 ± 0.10	0.01
	Bill length	-0.17 ± 0.19	-0.01
	Bill width	0.08 ± 0.19	0.01
1982	Plumage	0.20 ± 0.10	0.23**
($N = 24$)	Territory position	0.29 ± 0.10**	0.31**
	Bill depth	0.05 ± 0.10	0.18
	Bill length	0.16 ± 0.19	0.13
	Bill width	0.23 ± 0.19	0.13
1986	Plumage	0.13 ± 0.14	0.04
($N = 26$)	Experience	0.31 ± 0.14*	0.29**
	Territory position	0.19 ± 0.13	0.11
	Bill depth	0.18 ± 0.12	0.15
	Bill length	0.17 ± 0.13	0.05
	Bill width	-0.17 ± 0.18	0.02
1987	Plumage	0.86 ± 0.26**	0.55*
($N = 20$)	Territory position	-0.33 ± 0.24	-0.11
	Bill depth	-0.02 ± 0.13	-0.32
	Bill length	-0.24 ± 0.11*	-0.30
	Bill width	-0.08 ± 0.11	-0.19

Note: Standardized directional selection gradients ($\beta \pm$ standard error) measure the direct effects of selection on each character. The standardized selection differentials (s) measure, separately, the net effects of selection (i.e., direct and indirect effects). N = sample size of males. Statistical significance of the coefficients is indicated by *($P < 0.05$) or **($P < 0.01$); see B. R. Grant (1985) for details of the analysis, and p. 155 for details of the characters.

1978 to 1987. Territory quality estimated in 1982 was not correlated with order of breeding in that year. If territory or habitat quality influences the choice of mates, as has been found for other species (Millington and Grant 1983; Price 1984a; Alatalo et al. 1986), we have been unable to detect it.

Fitness Consequences of Mate Choice

The early pairing advantage gained by experienced males in black plumage is translated into a reproductive advantage; early breeders produce more offspring during a breeding season than do late breeders (Grant and Grant 1983). The results of a second selection analysis, in which fitness of a breeding male was scored as the number of fledglings produced per season,

Table 7.5 Selection on Characters Associated with the Number of Fledglings Produced by Mated Males in a Season

		Coefficients	
Year	Character	β ± SE	s
1980	Plumage	0.44 ± 0.14**	0.51**
(N = 23)	Territory position	0.24 ± 0.14	0.39*
	Bill depth	−0.01 ± 0.16	0.22
	Bill length	−0.08 ± 0.34	−0.20
	Bill width	−0.09 ± 0.32	−0.21
1981	Plumage	0.60 ± 0.14**	0.63**
(N = 26)	Territory position	0.18 ± 0.14	0.40*
	Bill depth	0.17 ± 0.14	0.14
	Bill length	−0.16 ± 0.26	0.03
	Bill width	−0.09 ± 0.26	0.01
1982	Plumage	0.30 ± 0.19	0.47**
(N = 24)	Territory position	0.24 ± 0.18	0.45**
	Bill depth	0.08 ± 0.17	0.30
	Bill length	0.10 ± 0.36	0.12
	Bill width	0.08 ± 0.35	0.08
1986	Plumage	−0.08 ± 0.09	0.22
(N = 26)	Experience	0.33 ± 0.09**	0.32*
	Territory position	0.11 ± 0.08	0.13
	Bill depth	0.06 ± 0.12	0.05
	Bill length	0.10 ± 0.09	0.10
	Bill width	−0.01 ± 0.12	0.05
1987	Plumage	0.29 ± 0.32	0.19
(N = 20)	Territory position	0.06 ± 0.30	0.12
	Bill depth	−0.05 ± 0.14	−0.25
	Bill length	−0.17 ± 0.14	−0.35
	Bill width	−0.12 ± 0.15	−0.21

Note: Standardized directional selection gradients (β ± standard error) measure the direct effects of selection. Standardized selection differentials (s) measure the net effects. N = sample size of males. Statistical significance of the coefficients is indicated by *(P < 0.05) or **(P < 0.01).

are shown in Table 7.5. Once again only those males that acquired a new mate in a given year were used in the analyses.

In two years, 1980 and 1981, plumage enters significantly into the selection gradient; it is the only variable that correlates directly with fitness when the effects of other variables are held constant. In 1986, the only year in which the effects of experience could be separated from the effects of plumage, experienced males gained a fitness advantage through producing more fledglings than inexperienced males, whereas this measure of fitness was not associated with plumage.

Thus experience-related behavior seems to be the main feature to which females respond and with which reproductive success is correlated. Plumage usually covaries with experience, hence black males usually gain the fitness advantage. In some years males with a heterotypic neighbor gain a fitness advantage over those which do not have such a neighbor, as shown by the positive selection differentials. Presumably this is the result of a positive correlation with plumage, experience, or some other unmeasured factor. Reproductive success was not associated with beak size or, in a separate analysis, with territory size or quality in any year.

The fitness differences between experienced (black) and inexperienced (partially black) males that had arisen up to the time of fledging remained unaltered through to recruitment. Of the 1244 fledglings produced in the main study area in 1978–83, 39 survived and bred in the years 1980–86. Ten of the recruits (25.6 percent) had partially black fathers. This is almost the same as the number (11.5) expected from the proportions (29.5 percent) in the total sample of fledglings.

The fitness advantage gained by a female mating with an experienced male could arise in two ways. The first is through better parental care, because the male has learned from previous breeding experience. The most direct evidence of better parental care is the higher nesting success of experienced pairs than of inexperienced pairs (Chapter 4). The second is through superior quality of the male's genes. This possibility is far more difficult to investigate. The argument for it is as follows. Experienced males are old. In each year more of the young partially black males died than did fully black males. Old males have survived more dry seasons of food shortage and more threats from predators, parasites, and pathogens, and therefore they could be a selected group with higher than average genetic quality (Hamilton and Zuk 1982; Searcy 1982; Halliday 1983; Price 1984a; Manning 1985).

Thus black plumage and experienced behavior could be phenotypic signals of both the age and genetic quality of the males.

Courtship Behavior and Experience

The above results shed new light on the facts discussed in the previous chapter. It now appears that females do not respond mainly to song neighborhood when choosing mates. Plumage state emerges as the single most important factor in mate choice and fitness, although it probably functions in association with important behavioral variables that change with experience, which we have been unable to measure.

The important classes of behavior are courtship, nest building, and interactions with intruders and predators. They are modified by learning, improve with time, and hence vary with age and experience. Males change in their response to intruders with age (Chapter 6). They also increase mate guarding as they gain experience. Four partially black males were observed for 30 minutes during each of their first two broods in 1987 and again during each of their fourth and fifth broods, in all cases when their females were receptive. During the first two broods they spent less time on average guarding their females than during their last two broods (paired $t_3 = 3.92$, $P < 0.05$). In contrast there was no significant difference in time spent by black males guarding between the first two broods and the last two broods ($P > 0.1$).

Extra-pair Copulations

If there is any genetic gain to a female in mating with an experienced black male, some late-pairing females will be denied it by being forced to pair with inexperienced males, unless they engage in extra-pair copulatory behavior with old experienced males. By copulating with experienced males they could gain not only superior genes for their offspring, which might, for example, reduce the risk of laying infertile eggs, but also an increase in the genetic variation among their offspring, which could increase the chances of at least one offspring surviving to breed in this unpredictable environment (cf. Williams 1975). The females would not gain better parental care, because only the male associated with the nest helps to rear the offspring and he is intolerant of other males on his territory. Only two males other than the partner have ever been known to help in rearing fledglings in this population (p. 123), and in neither case was there any reason to believe they were the true father. In contrast to those paired with inexperienced males, females already mated to experienced males have little or nothing to gain genetically from extra-pair copulations, and they stand to lose some of the parental care on which they depend through a possible weakening of the pair bond (Gladstone 1979; Alatalo et al. 1984).

OBSERVATIONS

There are four reasons for believing that mating with nonpartners occasionally gives rise to broods of mixed paternity. First, females showed a sexual orientation to stimuli from nonpartner males in playback experiments (Chapter 6). Second, extra-pair copulations have been observed three times. They occurred 5, 7, and 11 days respectively before a new clutch of eggs was started, and thus were probably within the female's period of receptivity (fertility) in the first two cases (cf. Westneat 1987a). They could have resulted in fertilizations, although the likelihood must have been low in view of the much more frequent intra-pair copulations ($\sim$98 percent) that occur daily during the period of receptivity. A feature of these three observed extra-pair copulations which stands against our argument is the black plumage, and not partially black plumage, of all three cuckolds (the cuckolders, all neighbors, were also in black plumage).

Third, males visit the nests of neighboring females, or the females themselves, much more at the time of clutch formation than at other times in the nesting cycle. During 128 nest watches of mated males at the time of the first two broods in 1987, we observed a total of 65 intrusions into the territory by neighboring males. Intrusions occurred more frequently at the time of clutch formation (80.0 percent) than at the time of egg hatching (20 percent; $\chi^2_1 = 30.09$, $P < 0.001$). The timing, and the silent and stealthy behavior of the invading males, suggest that they were seeking to mate with the neighboring females at a time when they were receptive. They were not intruding to gain food, because during the breeding season intruders ignore food sources in the intruded territory, in contrast to the frequent feeding by intruders during the dry season. Thus they are reproductive and not ecological (food) competitors.

Fourth, females paired with partially black males have a greater opportunity to copulate with other males than do those paired with black males. This difference arises from the fact that partially black males, unlike black males, do not chase black intruders from their territories (Chapter 6), and they do not increase the guarding of their females during the receptive stage, at least during the first two breeding episodes in a season. Partially black males chase other partially black males away, but they respond to black intruders by staying low in the vegetation near their mate or nest. During four of our nest watches the black intruder sang, and on two other occasions when the female approached the intruder her mate did not intervene. Observations at nests also revealed that black males spend more time guarding their females when they are receptive than when they are not ($t_{33} = 2.72$, $P < 0.02$, two-tailed). The number of minutes spent within sight of the female or within 12 m of her in a 30-minute period was greater at the time of clutch formation, the receptive period ($\bar{x} \pm SE = 26.78 \pm 1.10$ minutes, $N = 18$), than at the time of egg hatching, the unreceptive period ($\bar{x} = 20.80 \pm 1.92$, $N = 17$). Partially black males guarded less than fully black males during the receptive period ($\bar{x} = 18.00 \pm 3.19$, $N = 4$; $t_{20} = 3.20$, $P < 0.01$, two-tailed). Mate guarding is also more frequent in the receptive period than in the nonreceptive period in swallows (Beecher and Beecher 1979; Møller 1987a).

We conclude that extra-pair copulations occur rarely, perhaps most often at dawn or dusk when observations are difficult to make, and that they even more rarely have significant reproductive effects. Partially black males are more likely than black males to be cuckolded, owing to their own behavior as well as their females' behavior. A useful exercise in the future would be to apply paternity exclusion analyses to biochemical data in order to estimate the frequency of fertilizations resulting from extra-pair mating (e.g., see Gowaty and Karlin 1984; Burke and Bruford 1987; Wetton et al.

1987; Sherman and Morton 1988), since the frequency may be higher than observations would suggest (Westneat 1987a, b). Regardless of its frequency, the tendency for females to copulate with other males is the probable evolutionary cause of mate guarding.

INDIRECT EVIDENCE

Heritability analysis of morphological variation can be used as a technique for detecting the occurrence of extra-pair copulations (e.g., Boag 1983; Alatalo et al. 1984; Møller 1987b). The reasoning in the present context is as follows. A black experienced male associated with the nest is assumed to be the father of all young fledging from it, because black experienced males are preferred by females and have the highest fitness. In some instances a partially black male could be the father of none or some but not all of the young fledging from its nest. Consequently, given enough females engaged in extra-pair copulatory behavior, the slope of the regression of offspring morphological values on male values will be higher, and the scatter of points will be smaller, when the males are black (always the true fathers) than when they are partially black (the apparent fathers). Alternatively the slopes will be identical if extra-pair matings do not occur, or if they do occur but are not restricted to one group.

Bill depth and bill width are the most useful traits to employ because they have low measurement errors (Grant 1983b). This and heritability analysis are discussed more fully in Chapter 8. Heritabilities of beak depth and beak width estimated by using 37 black male parents are both statistically significant, whereas neither is statistically significant when estimated from 22 partially black male parents. Furthermore, the scatter of points (the error attached to the estimates) is much smaller in the black male regression than in the partially black one. Both of these results are consistent with the hypothesis that some females paired with partially black males copulate with other males. A more critical test is to compare the slopes of the two regressions with each other, but this is doomed to failure when the variation around one is large, as is the situation here for interpretable reasons. In fact, by analysis of covariance there is no difference between slopes for either bill depth or bill width (F tests, $P > 0.1$).

REDUCING THE RISKS OF CUCKOLDRY THROUGH SYNCHRONOUS BREEDING

A male can minimize the risk of being cuckolded by establishing a territory next to another inexperienced male, by remaining in close proximity to his mate during her receptive period, and by breeding at the same time as a neighboring pair. There is a connection between the last two behaviors. With synchronous breeding, each male is likely to stay close to his female at the same, critical time of fertility and spend little time intruding into the neighbor's territory. Even though the reproductive gains to be made by a male through extra-pair matings are high at this time, the potential losses through cuckoldry are also high.

Synchrony of breeding is largely governed by rainfall (Chapter 4; Fig. 4.2). Birds respond independently and similarly to the same stimuli. The song of neighbors can have an additional synchronizing effect, as suggested by the results of laboratory experiments with ring doves, white-crowned sparrows, budgerigars, and canaries in which it has been shown that an increase in male song helps to bring females into a state of readiness to breed (Lott and Brody 1966; Lott et al. 1967; Brockway 1969; Nottebohm and Nottebohm 1971; Kroodsma 1976; Morton et al. 1985). The combined songs of neighboring males should have a greater stimulating effect on females than the songs of isolated males, and as a result neighboring females should tend to breed at the same time. According to this argument, a male has some influence over when a female breeds, and so does his neighbor.

Two analyses provide evidence that neighbors breed more synchronously than do isolated pairs. In the first analysis we compared the time of egg laying in 1987 of isolated and not isolated, partially black males. We compared these two groups because inexperienced males appear to be at the greatest risk of being cuckolded and stand to gain the most from local synchrony. As a measure of synchrony we used the difference in days between the time the first egg was laid by a female paired with a partially black male and the time the first egg was laid by a female paired with a black male on a contiguous territory or on the nearest, nonadjacent territory. Only those pairs which hatched at least one egg were considered. The breeding of partially black males was more synchronized with the neighbors' breeding when the neighbors were adjacent than when they were not (Mann-Whitney U test, $z = 3.03$, $P = 0.002$, $N = 18$).

In the second analysis we considered only experienced (black) males that had retained their mates from the previous breeding season and had bred (successfully) at the time of the first brood in the years 1980 to 1987. Data for these years were combined. As before, breeding was more synchronized with the breeding of contiguous neighbors than with the breeding of noncontiguous neighbors (Mann-Whitney U test, $z = 2.86$, $P = 0.004$, $N = 28$).

Results of both analyses show that rainfall does not account completely for breeding synchrony, since both isolated and non-isolated males received the same rainfall stimuli yet differed in their degree of synchrony. Results of the second analysis show that it is not just the partially black males which synchronize their breeding with the breeding of their adjacent neighbors; all males do. The timing of breeding is affected by the time when the pairs are formed and by post-pairing behavior and physiology. Since synchronized breeding of adjacent neighbors occurred among both previously paired and previously unpaired males, the important determinant of synchrony is the behavior of the birds after the pairs are formed.

A further possibility to consider is that neighbors with contrasting song types have a more stimulating effect on females than those which sing the same song, and as a result their breeding is more synchronous. We reanalyzed both sets of data with respect to song type of adjacent males and found no significant differences in degree of synchrony between homotypic and heterotypic neighbors. Therefore the particular song type of a neighbor plays no role in breeding synchrony.

The results are consistent with the hypothesis that the risks of cuckoldry are reduced by synchronous breeding. Biochemical assessment of paternity could be used to test the hypothesis directly. Our results have interesting implications for other species of passerine birds. In many studies it has been found that birds change their song to match a dominant neighbor's (e.g., Payne 1981). It would be interesting to examine these species more closely to determine if, by copying the song of their neighbor, they synchronize their breeding with his and thereby reduce the chances of being cuckolded. Avoidance of cuckoldry is likely to be of particular importance in species that are short-lived and have time to raise only one or two broods in a breeding season or in a lifetime.

Mate Changes

Males and females invest much time and energy in rearing offspring, and both are necessary for the offspring to be raised to independence. Their commitment to raising offspring together is likely to be weakened if one or both spend time courting other potential mates (Gladstone 1979). Desertion may ensue. The characteristics, causes, and fitness consequences of mate changes, including desertion, provide us with additional information on the basis and significance of mate choice.

CHARACTERISTICS

All long-lived females have bred with more than one mate. The maximum recorded number of mates is seven over a period of six breeding seasons (Table 7.6). Some short-lived females also breed with more than one mate, and a change of mates is not rare. One female paired with four males in succession during the extended breeding season of 1983.

Mate changes take place within and between breeding seasons. We have 112 records of known females breeding in successive breeding seasons. In 37 percent of the cases the female changed mates; half of them being induced by the death or disappearance of the previous mate, the remainder occurring

Table 7.6 Number of Mates of Nine Females That Bred in at Least Five Years

Female No.	Minimum Number of Breeding Years	Total Number of Mates	Changes in Mates	
			Within Season	Between Seasons
6111	7	2(1)	1	1
6196	5	2	0	1
6747	5	2	1	0
6753	6	3	1	1
6940	6	2(1)	0	1
6949	6	8(2)	3	4
6980	9	4(1)	1	2
6990	5	2	0	1
6993	6	4(2)	0	4

Note: In some cases the death of a previous mate caused the female to pair with another; the number of times this occurred is indicated in parentheses. Two females, No. 6111 and No. 6993, went back to a previous mate.

at an annual rate of 12.5–27.3 percent, rarely higher (41.6 percent in 1987). We refer to the remainder as desertions of the male by the female. Alternatively they could be caused by males accepting new females before their previous mates had returned to the territories to breed. This is unlikely as interactions between females are rarely seen, and females that change mates do not breed later than those who replace them. Mate changes take place less frequently within a breeding season ($\bar{x}$ = 6.1 percent per year), and they are much less frequently induced by the death of the previous mate (B. R. Grant and P. R. Grant 1987).

CAUSES OF DESERTION

The most obvious possible cause of desertion is complete reproductive failure (Harvey, Greenwood, and Perrins 1979; Coulson and Thomas 1983). In 47 cases we know the reproductive fates of breeding attempts before the change of mates. Since the data for within- and between-season changes did not differ, we combined them. Only 10 mate changes (21.3 percent) were preceded by complete failure, that is, eggs laid but no fledglings produced. Therefore, complete reproductive failure can account for only a small minority of mate desertions.

Females do not desert males because they are young. There was no difference between deserted and not deserted males in the proportion of partially black males when desertion took place either within (χ^2_1 = 0.15) or between (χ^2_1 = 0.22) seasons. Females that deserted and re-paired within a season did not have a history of breeding consistently rapidly (i.e., short interbrood intervals) or slowly (Chapter 4), or a history of producing relatively small clutches (cf. Perrins and McCleery 1985).

In all 56 cases of desertion the female re-paired and attempted to breed.

Desertion can therefore be viewed as a female preference for one mate and its territory (the new one) over another (the old one).

CHARACTERISTICS OF THE NEW MATE

As in other species (e.g., Wunderle 1984), a factor of importance to females in re-pairing is the location of the new mate. Females often re-pair with a neighbor (Fig. 7.1). Ten out of 21 that deserted their mates from one season to the next moved only to the adjacent territory and bred with the male there. Among those females that had to re-pair the next season because their previous mate had died, approximately the same proportion (11 out of 18) re-paired with a neighbor or stayed on the previous territory. This indicates a strong component of site tenacity in the choice of a new mate and is striking because females wander widely in the study area between successive breeding seasons.

Mate changes occurring within a breeding season show a different pattern. Deserting females usually do not re-pair with a neighbor; only 13 out of 34 (38.2 percent) were recorded doing so. In most but not all cases neighboring males were mated at this time. In contrast to this low frequency, 5 out of the 6 females (83.3 percent) that re-paired following the death of the previous mate did so with a neighbor or stayed on the previous territory and acquired a new mate. The difference between the two groups of females is pronounced but statistically on the borderline of significance (Fisher's Exact Probability, $P = 0.054$). Thus the death of a mate during a breeding season may constrain a female to re-pair rapidly with any available male nearby or with one who replaces the original male.

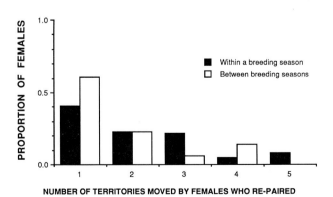

Figure 7.1 Frequencies of dispersal distances by females who changed mates. Females who re-pair with a neighbor are grouped with those who stay on the same territory, because in both cases the territory boundaries are sometimes redefined.

Desertion appears to reflect the opportunity for, and exercise of, greater choice.

A single anecdote indicates how and when desertions occur within a breeding season. In 1986 an unmated black 5 male from study area 1 was observed 500 m from his territory in a part of study area 2 not occupied by breeding cactus finches. He courted a female and fed her dependent fledgling. This is one of only two known cases of a mate other than the father feeding a fledgling (p. 123). Whether influenced by this or not, she paired with the male shortly after and bred with him on his territory in study area 1. The male fed the juvenile again during the ensuing nesting cycle. Unfortunately, her previous breeding history outside the study area was not known as she was not banded. Desertions may thus occur at the time females are feeding their dependent fledglings away from the breeding territory and encountering new males. Deserting females are often seen off their breeding territory before their re-pairings are known.

The basis of choice is partly plumage or a factor correlated with it, such as experience. Just considering the mate changes which occurred from one season to the next, we compared the plumages (partially black or black) of males who acquired a female through a mate change (the successes) with those without mates at the beginning of the second season when re-pairing took place (the failures). The combined data for all years yielded a highly significant difference ($\chi^2_1 = 19.37$, $P < 0.001$). Sixty-two percent of black males acquired a female whereas only 18 percent of partially black males did so. We made similar comparisons between successful males (obtaining a female through a mate change) and unsuccessful males at the same time within breeding seasons. In the aggregate the data give evidence again of nonrandom pairing ($\chi^2_1 = 10.23$, $P < 0.005$). In all seven breeding seasons, old males were proportionally more successful than young males, despite a preponderance of partially black birds among the unmated males. Thus altogether 21 out of 28 females (75 percent) that deserted a black male within a breeding season re-paired with another black male, and 8 out of 12 females (67 percent) that deserted a partially black male re-paired with a black male.

Females may choose males also on the basis of song type, or size of the males, or size and quality of the territory when re-pairing. A variety of tests showed no departure from random mating with respect to any of these features (B. R. Grant and P. R. Grant 1987), even though we had previously found with a smaller sample that females tended to re-pair with males of the opposite song type to their mates (Grant and Grant 1983).

In summary, females that change mates are influenced in their choice of a new mate by two factors: the age and location of a new male. They do not appear to be influenced by territory size, song type, or body size either when they pair for the first time or when they change mates.

CONSEQUENCES OF CHANGING MATES FOR
THE FEMALES

Why do females change mates? The obvious answer is that they do so to gain a reproductive benefit, but it is less obvious what the benefit might be and whether they obtain it or not. Females may gain a quantitative advantage by breeding sooner (Boag and Grant 1984), by breeding more successfully, or by producing a genetically more varied set of offspring than they would by breeding with only one male. The last advantage is extremely difficult to assess. We investigated the first two possible advantages by comparing females who changed mates with those who did not, on the assumption that those who did not would provide an estimate of the potential reproductive performance of the others had they not changed mates.

For mate changes taking place within a season, the interval between first egg-laying dates in successive clutches is the only precise measure we have for the rate of clutch production. Clutch or brood failures shorten the interval, and the dates of failure are not precisely known, so we concentrate on successful breeders.

Our analyses showed that females do not shorten the interval between clutches by changing mates (B. R. Grant and P. R. Grant 1987). Rather, there is a trend in the opposite direction. Females who changed mates in the 1980–82 seasons produced a clutch with a new mate after a significantly longer interval than females who did not change mates ($t_{118} = 4.31$, $P < 0.01$). In 1983 mate-changing females appeared to neither gain nor lose, but the data may have been biased by our failure to detect all broods; whereas in 1986 we can be certain that females neither lost nor gained in time by changing mates because all broods were found (B. R. Grant and P. R. Grant 1987), and the same is true for 1987.

Females that changed mates between seasons did not gain a time advantage either. In none of the years 1980–82, nor in 1987, was there a significant difference in order of breeding between those who had changed mates and those who had retained previous mates (Mann-Whitney U tests, $P > 0.1$ in each case). On average, females who changed mates bred slightly earlier in 1980 and 1982, and slightly later in 1981 and 1987, than those who did not, but the differences were less than one day in each case.

Thus females do not speed up their reproduction by changing mates. Nonetheless they may gain an advantage by fledging more offspring with a new mate if the new mate, or its territory, is superior. We can assess this possibility because we know the breeding success immediately before and after the change of 17 females who deserted their mates within a season, a total of 22 times in all breeding seasons combined. The number of fledglings produced after the change (2.4 ± 1.53, $\bar{x} \pm$ SD) does not differ significantly from the number produced before (2.1 ± 1.80)

(Wilcoxon test, tied ranks ignored, $T = 51$, $N = 16$, $P > 0.1$). This apparently straightforward comparison is confounded by a decline in the average fledgling production toward the end of the breeding season in some years. To eliminate this seasonal effect, we use the nearest neighbors who did not change mates as a control group to compare with those who did. In the breeding episode immediately after the changes occurred the two groups did not differ in their production of fledglings (Wilcoxon test, $T = 43.5$, $N = 14$, $P > 0.1$). In other species it has been found that mate-changing females reproduce less well (e.g., see Coulson and Thomas 1983; Rowley 1983; Perrins and McCleery 1985).

Therefore we have been unable to detect a benefit to females in changing mates, but we cannot rule out the possibility that they did obtain a benefit, for two reasons. First, those who bred successfully may have gained a benefit from the genetic diversity of their offspring. This benefit should show up in a higher frequency of their offspring among the recruits, but our sample sizes of offspring from these females and of recruits are too low to give us a chance of detecting such an effect. Second, some or all of the deserted males could have been unwilling or slow to breed again when they were deserted, which is why they were deserted, and so just by breeding with other males the females stood to gain an advantage over remaining with the original males. This highlights a weakness of the comparative method we have used. The proper comparison we need cannot be made: it is between a female breeding with the original male and the same female breeding with another male, at the *same* time! Experiments are the only means of circumventing that particular dilemma.

CONSEQUENCES OF CHANGING MATES FOR THE MALES

When desertion occurred between seasons, all deserted males ($N = 26$) acquired new mates at the beginning of the second season or soon after, except one with an injured leg, another who died shortly after the beginning of the breeding season, and two more in 1987 who nevertheless acquired mates at the time of the fourth brood. Those who were deserted and re-paired in the years 1980–83 produced with their new mates an almost identical number of fledglings in the first brood (2.55 ± 1.44, $\bar{x} \pm SD$, $N = 15$) as those who retained their mates from the previous breeding season and who bred at the same time (2.26 ± 1.62, $N = 28$). They bred no later (or earlier) than those who retained their mates, so they suffered no reproductive cost of being deserted (Mann-Whitney U test, $P > 0.1$).

In contrast, desertion occurring during a breeding season imposed a strong reproductive cost on the deserted males. Most (78.6 percent) did not acquire another mate during the season, although all survivors acquired a new mate in the next breeding season. Thus, for most, reproductive success after desertion was zero, unless they had fertilized their females' next set of eggs before desertion took place. Moreover, in most cases they were left to complete the feeding of the fledglings. The number of fledglings produced in the season of their desertion was an average of 1.0 in the years 1980–82. This is low because the deserted males generally bred only once, not because, when deserted prior to their offspring fledging, they had difficulty in feeding their nestlings. The corresponding figure for the production of fledglings by males that retained their mates was 2.5. Therefore deserted males, as a group, bore a reproductive cost of 1.5 fledgling or 60 percent of their potential production. In an El Niño year of extended breeding, the cost is lower because more females are available as a result of young birds coming into breeding condition in mid-season. For example, in 1987 we calculated the cost for deserted males to be only 28 percent of their potential.

To summarize, some breeding partners change from one season to the next, and neither the males nor the females lose time or reproductive potential by doing so. But when a change in partners takes place during a breeding season, the deserted males incur a reproductive cost, whereas females may gain a reproductive benefit.

Synthesis

We are now in a position to outline and interpret the rules of mate choice. Females are not influenced directly by beak size, body size, territory quality, or song neighborhood, but they are guided in their choice by male appearance and behavior. They pair preferentially with males who have bred before, which they choose on the basis of aspects of courtship behavior and possibly plumage as well. By mating with experienced males they enhance their fitness; they are able to produce a large number of fledglings, partly because their eggs or nestlings are less likely to be eaten by predators (Chapter 4), and hence make a large contribution of recruits to the next generation. Females make the same choice when changing mates. The majority of changes are desertions that reflect a preference for a new mate over a previous one. Females usually re-pair with fully black, experienced males who hold territories adjacent to the deserted males or nearby. They incur no reproductive cost by deserting, but they possibly gain a benefit when changing mates during a breeding season by producing more fledglings from the next brood than they would have done by remaining with their original mates. On the other hand, males who are deserted within a breeding season incur a cost because they have difficulty in obtaining another mate; in some years they lose more than 50 percent of their future potential production in the season of desertion. The primary immediate benefit that females gain by pairing with experienced males is likely to be enhanced parental care. The genetic quality of their offspring may be enhanced too, if experienced males have above average fitness for genetic reasons, and genetic diversity of the offspring is enhanced by changing mates. Observations at the nest and results of heritability analyses suggest that some females paired with partially black, inexperienced males may raise young fathered by other males, possibly experienced males, although extra-pair copulations appear to be rare as we have seen only three. Nevertheless they appear to be the key to understanding both mate guarding and intrusions into other territories by males.

Females choose experienced males on the basis of courtship behavior. From our limited analyses it appears that plumage plays a subsidiary role in mate choice by females. We suggest that in combination plumage and behavior give a reliable indication of the age and condition of a male (e.g., see Hamilton and Zuk 1982; Andersson 1986), and thus they are the best guide to his ability to provide parental care. A female could use conspicuous black plumage to identify a superior male at a distance, then use interactions with him through courtship flights, courtship feeding, and inspection of nests to modify her initial assessment. An experienced male in only partially black plumage may be at a disadvantage in competition for a mate with experienced black males. We have too few observations of such males to assess this possibility.

Courtship and nest-building skills are reliable indicators of age and experience because they develop through time, being modified by learning (Orr 1945; Searcy and Yasukawa 1981). Studies of other organisms show that mate choice can be based on energy-demanding behavioral displays (Searcy 1982; Gibson and Bradbury 1985; Borgia 1985; Watt et al. 1986; Diamond 1988). Thus the trait used by males to attract females in *conirostris* and some other species is not under genetical control. If there is genetic variation in such traits as susceptibility to parasitic infection, and if females can discriminate between signals that indicate age, vigor, and experience, then the signal itself need not be under genetical control. Watt et al. (1986) have shown that by selecting mates with enhanced behavioral displays, females choose males with superior genotypes. Behavioral signals could more accurately reflect the underlying genetic constitution of the male than genetically controlled signals, where many genes subject to recombination would be involved.

For black plumage to reliably signal age, there must be a selective disadvantage to the early acquisition of it. We can identify two disadvantages to acquiring black plumage early, or the corollary, advantages to retaining brown or only partially black plumage. First, there could be an advantage to retaining female-like plumage in the first dry season as it enables young males to feed in the territories of adult males with minimal harassment. In the dry season food is in short supply and many birds in their first year die (Chapter 3). Birds in brown plumage are more likely to be overlooked or tolerated while feeding in territories of adult males than are black-plumaged males. Birds in brown plumage are more cryptic than are those in black plumage against the prevailing background colors of greys, browns, and greens. This may be an advantage also in the presence of predators (Grant 1986a).

Second, if male plumage signals sex and social status (Nicolai 1956; Rohwer and Rohwer 1984, Rohwer 1985, Rohwer and Butcher 1988; Lyon and Montgomerie 1986), an inexperienced male in black plumage would signal dominance without the appropriate behavior to match the signal. It may suffer harassment from other males as a result (Price 1984a) in both the breeding and nonbreeding seasons. During the breeding season the harassment would place an extra burden on the time and energy of the male, who needs to give close attention to the female during her periods of receptivity and to provide food to the female during the incubation phase, to the chicks during the nestling phase, and to himself at all times. Thus young males in partially black plumage may have an advantage over those in brown plumage in attracting a mate, without suffering the disadvantage of excessive harassment from neighboring old males.

There is one factor which could be of importance to our understanding of mating decisions, and hence the mating system, but for which we have no data. That is variation in female quality. The analyses have assumed female quality to be constant, or to vary unimportantly. In the absence of information on the age of most females, we have no means of knowing if this assumption is true, although we know that experienced females reproduce more successfully than inexperienced females when paired with inexperienced males (Chapter 4). We do not know if males ever reject females that have chosen them. We have never witnessed such behavior.

Finally these results and interpretations, as well as the observations on the outcome of rare bigamous matings, help us to understand the monogamous mating system of the species. Given the potential reproductive gains to both sexes from polygamy, especially to the males (Trivers 1972), monogamy can be viewed as polygamy constrained (Wickler and Seibt 1983; Mock 1985). The potential for covert polygamy certainly exists, and it is known to be realized, in several species of birds that should be referred to properly as apparently monogamous, because males have been observed, or inferred, to copulate with females other than their partners (see McKinney et al. 1984 for examples). The most important constraint is the need for biparental care. Fitness is likely to be lessened if any activity causes a weakening of the pair bond and a reduced commitment to parental care. Another constraint is environmental. Variation in habitat quality among territories is not so pronounced as to make bigamous reproduction on the good territories distinctly superior to monogamous breeding. Females do rarely pair bigamously, and at a time when young males are without mates, but like their partners they do no better than when paired monogamously, and they do no better than other females paired monogamously with young males at the same time (B. R. Grant and P. R. Grant 1987).

Summary

The rules governing the choice of mates can be inferred from mating patterns and behavior. The predominance of males and the behavior of females suggest that females have the greater opportunity to exercise a discriminatory choice, even though both members of a prospective pair are expected to contribute to the joint decision to form a (monogamous) pair because both rear the offspring. Song type appears to play no role in the choice of a mate. Nor does beak size, body size, or territory quality. Females pair preferentially with males who have had previous breeding experience. They choose mates on the basis of courtship behavior and black adult plumage. We suggest that females use conspicuous black plumage to identify old males at a distance, then use interactions through courtship to modify initial assessments. Behavioral signs of past breeding experience and black plumage are reliable age- and condition-dependent traits. By mating with experienced black males, they gain a fitness advantage in terms of fledgling production and recruitment of young into the breeding population. The benefits they gain from pairing with experienced males apparently derive mainly from enhanced parental care.

Females change mates between breeding seasons at a frequency of 12 percent to 42 percent. They re-pair with males who are generally old, experienced, and hold territories adjacent to the deserted male or nearby. Females that desert their mates and re-pair within a breeding season may gain a benefit, whereas the deserted males incur a cost through failure to attract another mate. In some years the cost is more than 50 percent of their future potential production for that season. Bigamy is rare, is restricted to old males, and yields no clear reproductive benefits to the participants over monogamous breeding.

Females paired with inferior males may increase the genetic quality of their offspring through extra-pair fertilizations. They do not simultaneously gain enhanced parental care because only the male associated with the nest helps to raise the offspring. Extra-pair copulations have been seen three times. Results of heritability analyses of morphology are consistent with the suggestion that partially black males are more likely to be cuckolded than fully black males. Observations at nests provide additional support. Visits of males from neighboring territories occur most frequently during the female's fertile period. Resident males in black plumage expel all intruders, whereas residents in partially black plumage are unable to drive away intruders in black plumage. Furthermore black males increase the time spent guarding the female during the fertile period, whereas males in partially black plumage do not. Neighboring pairs tend to breed synchronously, which may reduce the chances of cuckoldry.

We conclude that females in choosing males seek reliable indicators of potential parental care, and in addition they may seek indicators of genetic quality.

8

Phenotypic and Genetic Variation

Introduction

In Chapters 3–7 we quantified and discussed many of the distinctive behavioral and ecological features of cactus finches. They show how, and to some extent why, finches vary greatly in the success they achieve. Survival is very low from fledging to breeding, but thereafter it is high. Mortality, largely attributable to starvation, occurs mostly in the dry season and varies annually. Breeding success likewise varies strongly among years. Once having survived a dry season and then bred, finches have a high chance of surviving for several years and breeding many more times, and they have a reproductive advantage over inexperienced birds through acquiring mates and starting to breed early. Overall, whatever determines survival ultimately determines reproductive success. The difference between the most and the least successful individuals is substantial.

We will now examine the question of whether relative success is determined by phenotype; whether having a particular beak shape or body size is important in determining survival and reproductive success. So far our attempts to do this have been restricted to investigating mating success in relation to these morphological traits. An association between them was not found.

In the next two chapters we will describe morphological variation and attempt to understand it in the light of the ecological features of finches and the environmental forces to which they are subjected. The present chapter establishes the pattern of phenotypic variation and the degree to which it reflects an underlying genetic variation. The next chapter explores the processes that are responsible for the variation.

Morphological Variation

Tables 8.1 and 8.2 provide a summary of the population characteristics of beak and body size variation. Males are consistently larger than females to

Table 8.1 Mean Morphological Characteristics of 246 Males and 149 Females

Trait	Males			Females			% Difference	Repeatability
	$\bar{x}$	SE	CV	$\bar{x}$	SE	CV		
Mass, g	25.61	± 0.14	8.47	24.41	± 0.19	9.55	4.9	0.679
Wing length, mm	76.8	± 0.16	3.15	73.7	± 0.20	3.38	4.2	0.829
Tarsus length, mm	22.44	+ 0.05	3.70	21.72	± 0.07	3.68	3.3	0.823
Bill length, mm	14.88	± 0.05	5.17	14.54	± 0.06	5.23	2.3	0.743
Bill depth, mm	10.76	± 0.05	7.43	10.15	± 0.06	7.19	6.0	0.945
Bill width, mm	10.07	± 0.04	6.55	9.56	± 0.05	6.38	5.3	0.948

Note: Resident males (178) were combined with 68 nonresident males because they did not differ in any trait. Differences between the sexes are expressed as percentages of female means. Repeatabilities were calculated from two set of measurements taken on 41 adults from one month to ten years apart. Symbols refer to mean ($\bar{x}$), standard error of the mean (SE), and coefficient of variation (CV). Bill width is the width of the lower mandible.

Table 8.2 Phenotypic Correlations for Males above the Diagonal and for Females below It

		Males				
	Weight	Wing Length	Tarsus Length	Bill Length	Bill Depth	Bill Width
Females						
Weight	—	0.53	0.50	0.42	0.63	0.53
Wing length	0.28	—	0.29	0.32	0.49	0.45
Tarsus length	0.37	0.28	—	0.34	0.41	0.25
Bill length	0.50	0.28	0.51	—	0.28	0.27
Bill depth	0.58	0.41	0.39	0.56	—	0.76
Bill width	0.52	0.37	0.42	0.50	0.82	—

Note: All coefficients are significant at $P < 0.05$ or lower. Sample sizes are the same as in Table 8.1; 246 males and 149 females. Bill width refers to the lower mandible, as in Table 8.1.

a small extent, more so in beak depth and width than in beak length, although in all traits there is considerable overlap of the sexes. These are typical values for the degree of sexual size dimorphism in small passerine birds (Price 1984b).

On the other hand, coefficients of variation in both sexes are atypically large, especially in weight and beak dimensions (Table 8.1). A contributing factor to the high variation in weight and beak length is the change that takes place in the lives of individuals. Wing feather tips become progressively worn, but then the whole feathers are replaced annually by molting. Weight fluctuates diurnally as well as seasonally. Beak length decreases as a result of abrasion at the tip, and increases as a result of regrowth of the horny sheath (rhamphotheca). Beak depth does not undergo similar fluctuations as it does in the great tit (Gosler 1987), and none of the dimensions shrink with age as reported for the song sparrow (Smith et al. 1986).

Despite these extraneous sources of variation, the repeatability of measurements is high. Forty-one adults were measured twice, from less than one to ten years apart in all months from December to August. The repeatability of measurements (Falconer 1981; Lessells and Boag 1987) was calculated as the percent total variance attributable to differences among birds in a one-way analysis of variance.

Repeatabilities vary from 67.9 percent for mass to 94.8 percent for bill width (Table 8.1). In general they are lower for the seasonally varying traits than for the more stable dimensions of tarsus length, bill depth, and bill width. Repeatabilities for males ($N = 27$) and females ($N = 12$) treated separately are close to the values for the combined sample, though usually but not always slightly lower. There is one exception. Repeatability for mass is much lower for females (4.9 percent) than it is for males (72.1 percent). The probable explanation is that several females were first captured and weighed in the breeding season when they may have been carrying eggs or were unusually fat.

Correlations between traits are moderately large and all positive in both sexes (Table 8.2). The highest correlation is between beak depth and beak width. This is understandable because the traits are functionally related. Other traits less closely related in function, such as wing length and bill length, are correlated relatively weakly.

Correlations and variances are inflated by measurement error. Corrections can be made by estimating the error from repeated measurements of the same birds and removing it from the calculated variances (Schluter and Smith 1986). It is useful to do this when comparing parameters of different traits or different species. We have listed unadjusted variances (actually coefficients of variation) and correlations in the tables for two reasons. First, the adjustments have little effect—as low as 2 percent for bill width variance, with a maximum of 16 percent for mass variance. Second, the error involved in measuring the total sample of 246 males and 149 females is likely to have been lower than in the repeated measurement of 27 males and 12 females.

Genetic Variation

We are interested in the degree to which variation in beak and body dimensions reflects an underlying genetic variation. Do large and small birds differ because they have different combinations of genes? If the answer is yes, the variation we observe has an evolutionary significance, because the population is evolutionarily responsive to the forces of selection and to drift. Or do large and small birds differ because they were reared under different environmental conditions? If the differences are nongenetic, we need to know why they arise in development.

Some morphological traits vary continuously on a scale of measurement; others vary discontinuously. We have given greatest attention to beak size and body size, which are examples of continuously varying traits. One discontinuously varying (polymorphic) trait, the beak color of nestling finches, soon became relevant to our study in the context of population subdivision (Chapter 10). We will discuss the two sets of traits separately because different branches of genetics theory apply to each. Continuously varying traits are in the realm of quantitative genetics, whereas polymorphic traits are material for Mendelian genetic analysis. The beak color of nestlings will be considered at the end of the chapter.

Heritability

Phenotypic variation in traits like beak size is caused by genetic and nongenetic factors. The question to be answered is how much of the variation can be attributed to each of these two classes of factors. According to standard quantitative genetics theory, the phenotypic variance of a trait (V_P) can be partitioned into components that represent separate genetic (V_G) and environmental (V_E) effects (Falconer 1981). Thus,

$$V_P = V_G + V_E .$$

The genetic variance, in turn, is made up of separate additive (A), dominance (D), and interaction (I) or epistasis components, so that

$$V_G = V_A + V_D + V_I .$$

We focus on the additive component of the genetic variance because it is by far the largest for polygenic traits (Falconer 1981), and because its magnitude for a given trait determines the evolutionary response (R) to selection acting solely on that trait, according to the formula:

$$R = h^2 s .$$

Here s is the selection differential, which is a measure of the strength of selection, and h^2 is the heritability:

$$h^2 = V_A / V_P ,$$

that is, the proportion of the total phenotypic variance of the trait that can be attributed to the additive effects of genes.

Heritability can be estimated in several ways. In an uncontrolled field study the most satisfactory way is to estimate the degree of resemblance between the offspring and their parents by regressing the average of the values of the offspring, when fully grown, on the average of the values of the two parents. The slope of the regression provides an estimate, with confidence limits, of the heritability as defined above. It is a measure of the degree to which variation among the midparent values predicts variation in the offspring values. It may be viewed as a measure of the strength of the genetic resemblance between offspring and their parents when phenotypic resemblance is not inflated by such factors as parents and offspring growing up in similar, particular, environments.

When Are Offspring Fully Grown?

Before proceeding with analysis we first had to find out the age at which young birds stop growing. The nestlings of song A males and song B males did not differ in their growth (Grant 1981a) and were therefore combined in the following treatments. Composite growth curves for four of the traits are shown in Figure 8.1. They are based on measurements of nestlings of known age taken in 1978 (Grant 1981a) and on measurements over the next five years of birds banded as nestlings and hence of precisely known age (Grant 1983b). A logistic model of growth was used to estimate the age at which asymptotic size is reached. Tarsus length is the first to reach that size, being virtually fully grown at the time of fledging (approximately 2 weeks), whereas beak dimensions take the longest time to reach full size (Table 8.3). By day 70 post-hatching (i.e., after 10 weeks), asymptotic size has been reached in all traits except bill length and bill depth, and for these two traits 99.8 percent of asymptotic size has been attained by then. Therefore measurements of only those offspring of this age or greater were used in the estimation of heritabilities.

Even though asymptotic size is reached by about day 70, further growth remains in body size (mass or weight) and bill length. This is indicated by the fact that mean offspring measurements of these traits are lower than both mean adult male and mean adult female measurements (Table 8.3). A

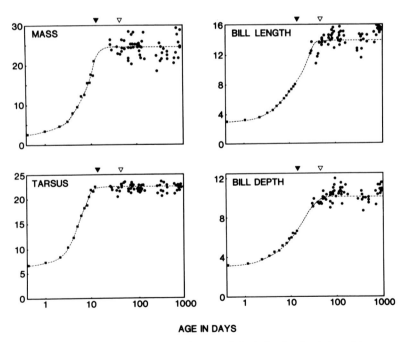

AGE IN DAYS

Figure 8.1 Composite growth curves for four morphological traits. Symbols: solid square, daily means for nestling measurements; solid circle, measurements of individuals, post-fledging; solid triangle, modal day of fledging; open triangle, modal day of departure from the natal territory. From Grant (1983b).

Table 8.3 Morphological Parameters for Parents and Their Offspring

Trait	Male Parent	Female Parent	Offspring	Estimated Asymptote	Age (days)
Mass, g	25.92 ± 1.96	24.54 ± 2.16	24.26 ± 2.11	24.13 ± 0.28	46
Wing length, mm	78.28 ± 2.36	74.00 ± 2.85	74.41 ± 2.47	74.67 ± 1.65	35
Tarsus length, mm	22.59 ± 0.67	21.69 ± 1.04	22.05 ± 0.78	22.64 ± 0.17	33
Bill length, mm	14.83 ± 0.83	14.56 ± 0.85	14.36 ± 0.94	14.17 ± 0.10	75
Bill depth, mm	11.03 ± 0.82	10.01 ± 1.03	10.42 ± 0.68	10.30 ± 0.08	80
Bill width (U), mm	6.71 ± 0.42	6.25 ± 0.35	6.45 ± 0.31	6.29 ± 0.06	70
Bill width (L), mm	10.18 ± 0.62	9.37 ± 0.67	9.61 ± 0.65	—	—
N	28	23	28	383	

Source: Grant (1983b).

Note: Offspring mean sizes with one standard deviation are shown in relation to the means of the parents in the left-hand columns. The right-hand columns give the estimated asymptotic sizes of all traits, and the age in days at which they are reached. U = upper and L = lower mandibles. N = sample size of measurements, one for each bird except for growth data (< 6) that yield the estimated asymptote. Only the eldest offspring from the five families represented by more than one offspring was included in the calculations.

small amount of growth may take place at about one year of age; 17 offspring measured at ages 300–900 days had longer beaks on average than 20 measured at 70–160 days. Strictly, the complication of further growth renders the logistic model inapplicable to the growth of body size and bill length, because the assumption that the parameters do not vary in time is not met. Nevertheless, this complication can be ignored without substantial error because asymptotic sizes are estimated quite closely (Table 8.3), and offspring sizes in these two traits are within one standard deviation of the midparent means.

Estimates of Heritability

Our first attempts to estimate heritabilities were frustrated by small sample sizes. A little more than 100 nestlings were produced by the 14 pairs of adults which had been measured in 1978. Of those nestlings only 4 were recaptured when fully grown. So instead of using a single cohort for the analysis, as was done in a parallel study of *fortis* on Daphne Major (Boag and Grant 1978; Boag 1983), we had to accumulate a sufficient sample of measured offspring of the appropriate age over the next few years.

The results are shown in Table 8.4. Heritabilities calculated from the regression of offspring values on midparent values are uniformly high and significantly different from zero. One of the regressions is illustrated in Figure 8.2.

The table also provides estimates of the heritability of principal component (PC) scores, which are multivariate indices of size and shape. The scores were obtained from an analysis of the mothers, fathers, and offspring of the 52 families with no missing data. Factor loadings given in Table 8.5 help us to interpret the components. Component I, accounting for about 57 percent of the total multivariate morphological variation, is an overall size factor. This can be seen in the uniform and positive loadings of all six traits. Component II is a shape factor representing mainly bill pointedness, since the loading for bill length is very large and positive, while the loadings for the other two bill dimensions are small and negative. Component III is a more complex shape factor, comprising bill shape and body proportion elements. In combination the three components account for about 80 percent of the variation.

All three sets of component scores are heritable. Overall size (PC I) is highly heritable, as is to be expected from the high heritabilities of all of the traits considered one at a time. But even after the effects of body size are statistically removed, the residual, relatively small, subtle variation in bill shape and body proportions (PC II and PC III) is highly heritable.

The same pattern of results is obtained from single-parent regressions. The heritabilities in Table 8.4 are twice the slope coefficient of offspring

Table 8.4 Heritabilities of Morphological Traits

Trait	Mother-Offspring	Father-Offspring	Midparent-Offspring
Mass, g	0.80 ± 0.25** 49	0.37 ± 0.26 52	0.69 ± 0.20*** 55
Wing length, mm	0.78 ± 0.28* 50	0.61 ± 0.27* 53	0.58 ± 0.18** 56
Tarsus length, mm	1.08 ± 0.23*** 48	0.85 ± 0.29** 53	0.82 ± 0.16*** 54
Bill length, mm	0.86 ± 0.21*** 51	0.49 ± 0.25* 54	0.66 ± 0.20** 57
Bill depth, mm	0.79 ± 0.18*** 51	0.62 ± 0.20** 54	0.69 ± 0.11*** 57
Bill width, mm	0.74 ± 0.21** 49	0.60 ± 0.24* 53	0.77 ± 0.16*** 55
PC I	0.84 ± 0.11*** 43	0.48 ± 0.28 45	0.81 ± 0.18*** 52
PC II	0.54 ± 0.34 43	1.64 ± 0.30* 45	0.67 ± 0.17*** 52
PC III	0.94 ± 0.26*** 43	0.83 ± 0.27** 45	0.63 ± 0.16* 52

Note: Estimates are given with one standard error, and sample sizes are below them. U = upper, L = lower, and PC = principal component. Statistical significance is indicated by *($P < 0.05$), **($P < 0.01$), or ***($P < 0.001$). In a few instances where families were represented by more than one offspring, values were averaged. Factor loadings of the principal components are given in Table 8.5, and the components are interpreted with reference to them in the text (p. 181). Single parent and midparent samples differ because in some cases one of the parents was not known, and in other cases a parent raised two or more offspring with different mates.

values regressed on single-parent values. Mother-offspring heritabilities are consistently larger than father-offspring heritabilities, but to a small and nonsignificant extent. There are no demonstrable enduring maternal effects on adult size or effects of incorrect assignment of paternity owing to cuckoldry (Chapter 7). If such effects exist, they are too small to be detected with our small samples. Similarly, there are no detectable maternal effects on the size of *fortis* offspring after their first few days as nestlings (Price and Grant 1985).

Potential Distortions to the Estimates

Heritability estimates for the six morphological traits are uniformly high. Estimates obtained from the midparent regressions vary from 0.58 (wing length) to 0.82 (tarsus length). Compared with other species of wild passerine birds, these are high values (e.g., see Smith and Zach 1979; van Noordwijk et al. 1980; Dhondt 1982; Alatalo and Lundberg 1986; Gustafsson 1986; Schluter and Smith 1986; Mousseau and Roff 1987).

The values are high partly because it is relatively easy to measure the

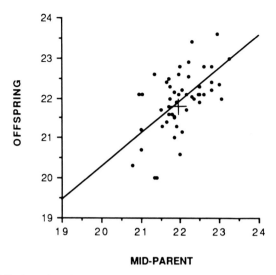

MID-PARENT

Figure 8.2 Offspring values for tarsus length (in millimeters) regressed on midparent values. The heritability is the slope; 0.82. The joint mean is indicated by a cross. See also Table 8.4.

traits reliably. Phenotypic variation is large, and repeatabilities are generally high. Repeatability provides an upper limit to the estimate of the heritability of a trait (Falconer 1981), hence it is a check on the direct estimates. Repeatabilities listed in Table 8.1 are all higher than the heritabilities estimated by midparent regression given in Table 8.4, except for mass to a trivial degree. The samples used for the direct estimation are much larger than those used in the calculation of repeatabilities.

Table 8.5 Results of a Principal Components (PC) Analysis, Based on the Correlation Matrix of Mothers, Fathers, and Their Offspring Averaged ($N = 140$)

	Factors Loadings		
Trait	PC I	PC II	PC III
Mass, g	0.4248	− 0.0068	− 0.2407
Wing length, mm	0.3800	− 0.5027	0.3118
Tarsus length, mm	0.3866	0.0330	0.7563
Bill length, mm	0.3211	0.8498	0.0740
Bill depth, mm	0.4569	− 0.1450	− 0.3055
Bill width, mm	0.4624	− 0.0551	− 0.4171
% variance	57.3	13.1	10.2

Note: The factor loadings show the contribution of each variable to the first three principal components used in the heritability analysis (Table 8.4).

Some obvious sources of bias in the heritability estimates can easily be dismissed. Assortative mating of the parents on the basis of one or more phenotypic traits, which would inflate the heritability estimates from single-parent regressions although not from midparent regressions (Falconer 1981), did not occur. There were no significant parametric correlations between parents for any of the traits at either the beginning or the end of the study (Chapter 7). Variances of the parents were equal. Mean sizes reached by offspring did not vary significantly among years. The offspring were not biased toward one sex: of the 46 of known sex, 20 were males and 26 were females. Separate regressions of the small sample of daughters (18 in 16 families) and sons (16 in 15 families) on midparent values did not differ for any traits in either slope or intercept. All slopes were high (> 0.4), and all but two of the fourteen were significantly different from zero. Thus combining the possibly unequal samples of sons and daughters in the analyses (Table 8.4) is unlikely to introduce a distortion to the estimates of heritability that would arise if the sex ratio of the offspring varied systematically with parental size.

The estimates would be distorted if selection has occurred on the offspring before some of the survivors are captured and measured. For example, the low weights and relatively short beaks of the offspring could be due to selection against large size in early life, as has been demonstrated with *fortis* on Daphne Major (Price and Grant 1984), rather than to or in addition to incomplete growth. This possibility was examined by comparing the nestling weights (at day 8) of individuals subsequently recaptured with the nestling weights of their sibs not recaptured and, separately, those of all other birds not recaptured. In neither case was a difference found. For example, in the first comparison neither large, small, or intermediate individuals were overrepresented among the survivors, either in the total sample or in subsamples from large and small parents treated separately. As we show in the next chapter, there is little evidence of selection on morphological variation in the first year of life.

A more difficult problem to deal with is the distortion that arises when there is a correlation between genotypes and environments.

Genotype-Environment Correlations

The resemblance between offspring and parents could arise, in part, because they experienced the same type of environmental conditions when growing to maturity. For example, if genetically large parents have the best territories with an abundance of food, their offspring will be large partly because they fed well while growing, and they in turn might breed on good territories and produce large offspring; conversely, small parents, because they are small, may secure poor territories, raise small offspring, and so on. Offspring will then strongly resemble their parents, partly because they share genes and partly because they share environments. These common environment effects inflate the estimates of heritability.

When the influence of correlations on the heritability of morphological traits has been sought in other field studies of passerine birds with similar growth patterns by techniques of experimentally randomizing genotypes among environments (Smith and Dhondt 1980; Dhondt 1982), by statistical randomization (van Noordwijk et al. 1980; Alatalo and Lundberg 1986), or by correlation analysis (Boag and Grant 1978; Boag 1983; Alatalo and Lundberg 1986), they have either been found to be small or not found at all. In the *conirostris* population they may have had a small influence on some of the heritability estimates, but we have not been able to detect it.

The majority of growth occurs in the nest (Grant 1981b; also Fig. 8.1), and most of the remainder occurs in the next 10–20 days or so when the fledglings are being fed on their natal territory by their parents. Fledging success was uniformly high from nests that escaped predation in all years of the study, except 1984. Few nestling runts were observed; in contrast, the growth of two congeneric species on Daphne Major was much poorer in a relatively dry year than in a relatively wet year (Price 1985). The probability of surviving to breed was independent of such environmental factors as the experience of the parents or the brood of origin (first, second, etc., in a season), except in one year (Chapter 4). Thus any effects of genotype-environment correlations are likely to have been small at the nestling stage but possibly larger in the initial fledgling stage on the natal territory. This is the time when fledglings are learning to feed by themselves and are becoming independent of their parents. It is a time of increased mortality. They then leave the territory and complete their growth, but even at this time there could be an influence of a genotype-environment correlation if the growth of fledglings off-territory is conditioned by the stage of growth reached on-territory.

If there are common environment effects, we would expect those traits that take the longest time to reach full-size to be the most affected during the initial fledgling stage, because variation in territory quality is likely to be most mirrored in growth performance when the young are beginning to feed themselves. This expectation is not realized. The overall impression from the midparent estimates in Table 8.4 is uniformly high heritabilities among the traits, not a division into traits with low heritabilities and traits with high ones. Contrary to expectation, the trait taking the shortest time to reach full adult size (tarsus length) has ostensibly the highest heritability.

A second line of reasoning brings us to the same conclusion. If favorable and unfavorable patches occur in the environment on a scale greater than the size of a territory, and if patch quality has a determining influence on final size, we would expect offspring born on neighboring territories to be more similar to each other in size than expected by chance. We have tested this possibility by comparing the difference in final weights between offspring born on neighboring territories with the difference in weights between each of these same birds and a randomly chosen contemporary. The two sets of differences are not statistically distinguishable ($t_{24} = 0.75$, $P > 0.1$); neighbors are no more similar than is to be expected by chance. Nor are offspring born on adjacent territories more similar than are their fathers ($t_{14} = 1.00$, $P > 0.1$), and their fathers are no more similar to each other than either one is to a randomly chosen contemporary ($t_{22} = 0.73$, $P > 0.1$). On this scale, therefore, there is no evidence of environmental heterogeneity contributing to either final adult size or to a common environment effect on the size of adults and their fully grown offspring. This fits with the observation that territory size and quality have little influence on fledging success (Chapter 7). We conclude that if common environment effects have inflated the heritability estimates, they are small and undetected.

The Expression of Genetic Variation during Ontogeny

Much of the variation among individuals is due to genetic factors. At what age is the genetic variation manifested? To answer this question we should perform regression analyses of offspring at age X on midparent values when the parents were also at age X. The latter measurements are not available, however, so instead we can only assess the resemblance between offspring at age X and the parents when fully grown.

The age at which offspring detectably resemble their parents differs among traits (Grant 1981a). From as early as the second day after hatching and onwards there is a significant resemblance in bill depth (Fig. 8.3). The slope of the relationship increases from about 0.35 to 0.66 on day 11, which is close to the 0.69 for fully grown offspring (Table 8.4). Bill width (upper mandible) shows a similar pattern, the slope being significantly different from zero from day 6 onwards, increasing to final size by day 11. In contrast, offspring do not resemble their parents in bill length at any stage in the first 12 days of nestling growth. Nor do they resemble them in wing length at any stage. Tarsus length is virtually fully grown by day 11 (Grant 1981a), and at this time the slope of offspring values on midparent values (0.87 ± 0.19) is indistinguishable from the slope of fully grown offspring values on midparent values (0.82 ± 0.16), in other words, from the heritability of the trait (Table 8.4). Maternal effects on size at a particular age are not pronounced for any trait (Grant 1981a).

Thus genetic variation apparently becomes expressed very early in life in three traits, and later in the other three. This is not just a consequence of different rates of maturation. It is true that tarsus length, one of the "early" traits, is almost fully grown by the time birds leave the nest, but the other two early traits, bill depth and bill width, take the longest time to reach asymptotic size (Table 8.3). Whatever the cause of the differences among traits in the pattern of expression of genetic variation, it has an important implication for studies of evolution; the early traits are potentially more responsive to forces of selection acting early in life than are the late traits.

Regressions of offspring values of tarsus length, bill depth, and bill width on midparent values have slopes of ~ 0.7 to 0.8 for both nestlings (< 2 weeks old) and fledglings (> 10 weeks old). Because we believe feeding conditions to have been uniformly good in the nestling phase of

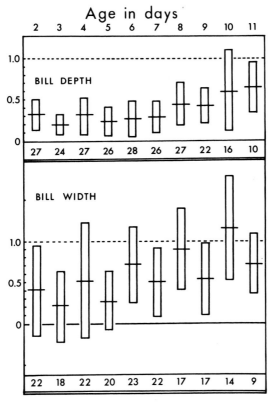

Figure 8.3 Estimates with 95 percent confidence limits of the slope of the regression of nestling measurements on the midparent values. Estimates of the slope approximate the heritabilities. Even-day and odd-day nestlings belonged to different groups, as they were measured every other day, and this is the cause of the apparent but not real oscillation in heritabilities. Sample sizes are shown at the bottom. From Grant (1981a).

growth in all years, we are confident that the estimates of heritabilities of these three early traits are free from bias. Since the estimates for the other three traits are not higher but are the same or lower, they may also be free from bias.

Genetic Correlations among Traits

The genetic variance of a trait determines the potential evolutionary response of that trait to selection. The genetic covariance between two traits is a similar measure that represents the degree to which genes governing the expression of one of the traits also influence the determination of the other trait, through pleiotropy or linkage disequilibrium. In formal terms, the genetic correlation is given by

$$r_A = \frac{COV_{xy}}{\sqrt{(COV_{xx} \, COV_{yy})}} \quad ,$$

where COV_{xx} and COV_{yy} are the offspring-parent (or midparent) covariance of each character, x and y, separately, and the numerator COV_{xy} is the so-called cross-covariance (Falconer 1981) between traits and relatives. The cross-covariance is calculated by taking the average of two covariances; it is the average of the covariance of value X in parents and Y in offspring, and the value Y in parents and X in offspring.

Genetic correlations are important because they determine the evolutionary responses of correlated traits to selection acting on only one (or both) of them. Suppose that there is a strong and positive genetic correlation between body size (mass) and bill depth. Forces of selection that acted solely against small body size, for example, would produce comparable evolutionary responses in both body size and bill size, in direction as well as magnitude; in the next generation, mean bill depth as well as mean body size would be higher. If there was no genetic correlation between the traits there would be no change in the unselected trait, i.e., bill depth. As a third possibility, if the genetic correlation between the traits was strongly negative, an increase in mean body size would be accompanied by a decrease in mean bill depth.

As a general rule, both genetic and phenotypic correlations among *conirostris* traits are fairly high (Tables 8.2 and 8.6). This pattern of similarity is not surprising, given the high heritabilities of each trait. Thus the arithmetic mean of the 15 genetic correlations among the six traits in Table 8.6 is 0.43. The corresponding mean for the phenotypic correlations calculated from the same set of data, sexes combined, is 0.55. All phenotypic and most genetic correlations are significantly different from zero.

The signs of the correlations are as important as their magnitudes. All are positive. Therefore the potential exists for evolution in an unselected trait in the same direction as in a selected trait. The potential for correlated responses to selection appears not to be uniform among the traits, however. Body size (mass) is the most strongly correlated with other traits genetically, especially with those most often used to represent body size in the absence of weight, namely wing length and tarsus length. The functionally related beak depth and beak width are strongly correlated with each other. Bill length, on the other hand, is not genetically correlated with any of the other traits except for tarsus length. So if mass, for example, was subjected to directional selection, there would be positive responses in the next generation in wing, tarsus, bill depth, and bill width but not in bill length.

All of the potential problems of estimating heritabilities accurately afflict the estimation of genetic correlations. In addition, estimates of the correlations are unreliable if the reciprocal elements of the cross-covariance are not approximately equal. This is a serious problem in some studies (van Noordwijk 1984), but it was a problem in our data with only one correlation, that between mass and bill length.

Discontinuous Variation: A Beak Color Polymorphism

Beak color of nestlings and young fledglings is a polymorphic trait; the beak is colored either pink or yellow (Plate 1). A single brood of nestlings comprises all pink morphs, all yellow morphs, or a mixture of pink and yellow morphs. The color is recognizable at hatching and remains unaltered until masked by melanin, which is manufactured by cells in the dermis. Melanin deposition begins basally in both mandibles, spreads rapidly throughout the upper mandible during the second week of nestling life but more slowly in the lower mandible, and is completed after about

Table 8.6 Genetic Correlations between Morphological Traits above the Diagonal and Standard Errors below the Diagonal, Based on Fifty-two Families

	Mass	Wing	Tarsus	Bill length	Bill Depth	Bill Width
Mass	—	0.814***	0.707***	0.195	0.529***	0.510***
Wing	0.072	—	0.400**	0.303	0.489***	0.371*
Tarsus	0.084	0.147	—	0.327*	0.426***	0.299*
Bill length	0.202	0.197	0.154	—	0.207	0.220
Bill depth	0.110	0.120	0.102	0.149	—	0.659***
Bill width	0.129	0.155	0.130	0.169	0.073	—

Note: Bill width refers to the lower mandible. Statistical significance is indicated by *($P < 0.05$), **($P < 0.01$), or ***($P < 0.001$). Note that bill length is genetically correlated only with tarsus length, whereas all other traits are correlated with each other.

two months. In the nonbreeding season the amount of melanin deposited is reduced, and most birds born that year and a few older ones reveal a paler color again. Most, and perhaps all, of the Darwin's Finch species exhibit this polymorphism (Swarth 1929; Grant et al. 1979; Grant 1986a).

In chickens, yellow pigment in body, skin, legs, and beaks is due mainly to xanthophyll, a carotenoid pigment obtained from food (Hutt 1949). The yellow pigment is ingested by all birds and is either deposited in the beaks, legs, and other peripheral tissues, and is visible, or else it is stored solely in the blood and body fat, in which case the external appearance is white or pink where the underlying blood supply can be seen (Hutt 1949). By controlled crosses, Bateson (1902) found that the trait was under genetic control, white skin being dominant to yellow. Dunn (1925) confirmed that yellow was a simple recessive character and showed the inheritance to be autosomal (also Lambert and Knox 1927). Diet can influence the expression of the trait (Hutt 1949), and infections of coccidea can reduce the level of carotenoids circulating in the blood (Ruff et al. 1974), but coccidea were not found in our study of *conirostris* (p. 61).

We take the known fact of simple Mendelian inheritance in chickens as our starting point and examine the *conirostris* data to see if it fits.

AUTOSOMAL INHERITANCE

Our observations are clearly consistent with an autosomal basis of inheritance. The sexes of 75 birds of known nestling color morph were determined by subsequent observation. Five of 33 males (15 percent) were yellow morphs, and 10 of 42 females (24 percent) were also. Frequencies of the two morphs did not differ between the sexes ($\chi^2_1 = 0.41, P > 0.1$).

A ONE-LOCUS MODEL WITH DOMINANCE

Pedigrees are required for a direct test of the hypothesis that yellow is the phenotypic expression of recessive alleles. Unfortunately we know the pedigrees of only three families. Both the male and the female of pair 1 were pink morphs, and they produced six pink and no yellow offspring. Pairs 2 and 3 each comprised a pink father and a yellow mother. Pair 2 produced three yellow offspring and no pinks, whereas pair 3 produced six pink and one yellow offspring.

These meagre data are consistent with the yellow-recessive hypothesis. One or both members of pair 1 may have been homozygous dominants (*WW* in the terminology of Dunn 1925), and the pink fathers of the other two pairs may have been heterozygotes (*Ww*). The hypothesis could only be rejected by yellow parents (*ww*) producing pink offspring, yet none of the pairs were yellow only. We are therefore forced to test the hypothesis indirectly.

MORPH RATIOS

In the population-at-large pink and yellow morph offspring are produced in the approximate ratio of 3:1, i.e., yellow morphs comprise 25 percent on average, although the frequency varies somewhat among years (Table 8.7). The frequency of the gene (w) should be $\sqrt{0.25} = 0.50$, if the population is in Hardy-Weinberg equilibrium. Likewise the frequency of the dominant gene (W) should be $1 - 0.50 = 0.50$. The frequency of genotypes should be 0.25 (WW), 0.50 (Ww) and 0.25 (ww). Unfortunately the heterozygous class cannot be distinguished phenotypically from the homozygous dominant class. Nevertheless, random mating should result in predictable frequencies of families with only pink offspring (WW and Ww), a mixture of pink and yellow offspring (WW, Ww, and ww), and only yellow offspring (ww). We have calculated the frequencies to be 0.44, 0.50, and 0.06, respectively, on the basis of mates being available in the ratio of three pinks to one yellow.

The data to be tested against these expected frequencies are shown graphically in Figure 8.4. As expected from the one-locus model with dominance, there is a strong mode in the 0 percent yellow category and another at a frequency of about 25 percent. A third expected mode at 50 percent appears to be lacking. Grouped into the three categories—pinks only, yellows only, and a mixture—the observed frequencies differ significantly from those expected ($\chi^2_2 = 7.40$, $P < 0.05$). There are too many families with a mixture of pinks and yellows, and too few with just one morph, especially pinks. Sampling bias arising from small numbers should give a deviation in the opposite direction, toward a relative scarcity of mixed families.

Table 8.7 Frequencies of Pink and Yellow Morph Nestlings in Each of the Years of Breeding

| Year | Numbers | | Percentage Yellow |
	Pink	Yellow	
1978	103	35	25.4
1980	153	47	23.5
1981	181	55	23.3
1982	225	62	21.6
1983	328	141	30.1
1984	9	1	(10.0)
1986	181	43	19.2
1987	331	113	25.4
Total	1511	497	24.7

Either the model is wrong or the assumptions of Hardy-Weinberg equilibrium are not met. For example, the character may display incomplete penetrance, mating may not be random with respect to (nestling) beak color, and selection may sometimes occur between hatching and breeding. The proportion of yellow morphs in the sample of 83 recaptured birds of known nestling color (0.20) is lower than in the total sample of nestlings (0.25). By itself this is not statistically significant ($P > 0.1$), but it could reflect either differential survival or an incorrectly estimated proportion of yellow morphs among the breeders. We have recalculated the expected frequencies on the basis of a 4:1 ratio of pinks to yellows among the breeders and found that this does not account for the observed frequencies. In fact the discrepancy between observed and expected frequencies is even greater ($\chi^2_2 = 12.31$ $P < 0.005$). To reduce the discrepancy we would have to make an adjustment in the opposite direction, by increasing the frequency of the yellow genotype to around 0.30. But even this is not sufficient to eliminate the discrepancy, because although the frequency of families producing only pink morph offspring is now well predicted, the frequency of families producing only yellows is predicted to be far higher (0.09) than it actually is (0.01).

A simple one-locus model is inadequate to account for the observations.

TWO-LOCUS MODEL

On Daphne Major the same beak color polymorphism is expressed in the two resident species, *fortis* and *scandens*. The samples of families of known pedigrees are far larger, 138 and 46 respectively (Grant 1986a). They provide no support for single-locus inheritance with dominance, because both pink morphs and yellow morphs are produced by all possible combinations of parental phenotypes. Nevertheless genetic factors do seem

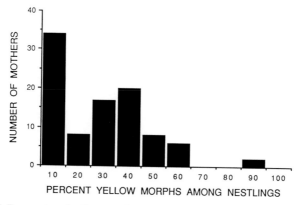

Figure 8.4 Frequencies of yellow morphs among nestlings in 97 families.

Table 8.8 Nestling Beak Color Morphs of Offspring in Relation to Nestling Beak Color of One of Their Parents

| Individual No. | Sex | Morph | No. Broods | No. Offspring | | Percentage Yellow |
				Pink	Yellow	
6129	♀	Pink	7	23	0	0
7631	♀	Pink	7	21	5	19.2
7679	♀	Pink	8	16	4	20.0
7873	♀	Pink	6	17	1	5.4
6137	♂	Pink	17	32	13	28.9
7667	♂	Pink	4	15	0	0
7834	♂	Yellow	6	1	16	94.1

Note: In all cases the other parent was banded, but its nestling beak color was not known.

to be involved in the determination of nestling beak color, since pink parents tend to produce mostly pink offspring and yellow parents tend to produce mostly yellow offspring in both species; the same is true for *conirostris* (Table 8.8). Alleles at two loci may interact epistatically to produce the observed phenotypes (Grant 1986a).

MORPH RATIO CHANGES AFTER A CHANGE OF MATES

We gain further insight into the inheritance of beak color by comparing morph ratios among offspring produced by the same bird with different mates. Pairs that produce pinks only are WW or Ww in the single-locus scheme. If a member of such a pair re-mates at random, it is expected to produce pinks only again with a probability of 0.35, when we take the original observed value of the frequency of pink-only pairs, or 0.44 when we take the theoretically expected value.

Thirty-five re-pairings took place, 17 of which produced only pink offspring (0.49). The estimate is biased because some of the families were small (e.g., only three or four offspring), and the frequency of families producing some yellows would have been underestimated by their inclusion. Even so the observed incidence of families with only pink offspring in these re-matings is quite close to the theoretical expectation from a one-locus model (0.44). We do not have enough information to derive an expectation from a two-locus model.

Re-pairings also show that the predisposition of an individual to produce offspring of a certain beak color is a property of the individual and not of the territory upon which the offspring are raised. Thirty males produced only pink offspring ($N \geq 8$ offspring in each case) with their original mates, then they re-paired on the same territory with new females. Thirteen (0.48) of the new pairings resulted in some yellow offspring being produced. In two additional cases with smaller samples of offspring per family ($\leq$ 7), a male who was simultaneously polygamous on the same territory with two females produced only pink offspring with one and a mixture of pink and yellow offspring with the other.

SELECTION AND DRIFT

Estimation of gene and genotypic frequencies is possibly hampered by two distorting influences. The first is differential survival, arising from natural selection or drift, as indicated above in a comparison of ratios between nestlings and recruits. Another reason for believing selection or drift may occur is that morph frequencies among nestlings vary annually (Table 8.7). The high proportion of yellows produced in 1982–83 differed significantly from the two lowest proportions, in 1986 ($\chi_1^2 = 8.63$, $P < 0.01$) and in 1982 ($\chi_1^2 = 6.07$, $P < 0.05$).

The differences appear to be due largely to the presence of many yellow-producing pairs in 1982–83 rather than to an environmental factor, such as xanthophyll-rich foods being especially common in 1982–83, although this could be a contributing factor. The 19 pairs that bred both in 1982 and in 1982–83, for example, had indistinguishable proportions of yellow morphs in their offspring in the two breeding seasons, yet the proportion of yellow morphs in their offspring in 1982–83 (0.24) is distinctly lower than the proportion (0.30) in the total sample which includes them.

It is possible that yellow morphs survive relatively well in favorable years of high potential recruitment (1982–83) and relatively poorly in years of low per capita recruitment (1982, 1986). This could be tested with data from *fortis* and *scandens* on Daphne Major.

CUCKOLDRY

A second possible distortion arises from cuckoldry and the incorrect assignment of offspring to parents. For example, in the 85 families of eight or more offspring there are no yellow-only producers (there are some among the smaller families). One family produced 16 yellows and 1 pink offspring. This was the closest to an all-yellow family. In this case we strongly suspect mixed paternity, since the mother, paired with the silent male, was one of only three females observed to engage in extra-pair copulations with a neighbor (p. 159). The neighbor in question produced predominantly pinks with his own mate. Consequently, in preceding analyses we considered the focal pair to be an all-yellow producer.

TIME-DEPENDENT GENE EXPRESSION

In the first two months of life, an individual's beak becomes progressively darker as melanin is deposited throughout. After this time, the melanin is partly absorbed in the nonbreeding season, and the pale color now revealed is not necessarily the same as the earlier one.

We have information on 40 birds whose nestling color could be compared with their color at age three months or later. Twenty-six stayed pink, 5 changed from pink to yellow, and 8 changed from yellow to pink (or white); only 1 stayed yellow.

In chickens, when their diet is deficient in carotenoid pigments, the skin color is white in genetically yellow (*ww*) birds, and is hence a phenocopy of skin made white by *W* (Hutt 1949). The color changes in *conirostris* could be explained in terms of environmental factors in a similar way. Diets could be deficient in carotenoids at one time of the year and not at another. Yet there is no seasonal pattern in the production of pink morphs and yellow morphs which could be interpreted as support for this idea. Within a breeding season the proportion of either color morph among nestlings does not change systematically, and those which later changed from one color to another did not hatch at a particular time within the breeding season. These negative statements are based on the results of χ^2 tests.

An alternative explanation for the change in beak color is a temporal variation in the expression of genes. If the gene(s) involved in late expression of color are the same as those controlling early expression of color, we could expect those individuals who change color to come, to a disproportionate extent, from families producing both pink and yellow morphs. This is what we observe. Nine individuals out of the 19 from families producing both morphs changed color, whereas only 1 out of 21 from families producing solely one morph (pink or yellow) changed. The difference in proportions is highly significant ($\chi^2_1 = 9.19$, $P < 0.005$).

Alternatively a different gene may control late expression of color. If so, the above result suggests it is linked to the gene(s) governing early expression. The late gene, if it is different, may be more comparable to the identified gene in chickens than is the early one. Hutt (1949) reports that it is difficult to classify young chickens as white-skinned or yellow-skinned until they are over three months of age. In this respect (and many others) chickens differ from Darwin's Finches.

Summary

In this chapter we examine the genetic basis of variation in two sets of traits: beak and body dimensions which vary continuously, and nestling bill color which varies discretely.

Levels of additive genetic variation underlying the beak and body size variation in the population are high. Heritabilities of seven morphological traits calculated from the regression of offspring values on midparent values are uniformly high, significantly different from zero, and range from 0.58 for wing length to 0.82 for tarsus length. Bill shape is also highly heritable, as determined by a principal components analysis that first statistically removes most of the variation in body size. These facts are important because they demonstrate a strong potential for evolutionary change in these dimensions in response to selection.

Some potential biases in the heritability estimates can be ruled out. There is no evidence of assortative mating by size, annual variation in the growth of offspring, enduring maternal effects, or inflation of the heritability estimates by common environment effects arising from a correlation between genotypes and environments. The degree of resemblance between parents and offspring in three traits—tarsus length, bill depth, and bill width—has reached its maximum value before the offspring leave the nest, hence the scope for common environment effects in these is small. Heritabilities of these traits lie in the range of 0.69 to 0.82.

All traits are correlated with each other. Both phenotypic and genetic correlations are moderately large and all are positive. The weakest genetic correlations appear to be between bill length and other traits. These facts mean that selection on only one trait will have moderately strong effects in the following generation on other traits, except bill length, whereas selection on bill length itself will have little effect on the other traits.

The nestling beak color polymorphism appears to be under fairly simple genetic control, as it is in chickens. In chickens inheritance is autosomal and pink is dominant to yellow. There are too few *conirostris* pedigrees to test a one-locus model, but the data from three families are consistent with it. However, morph ratios in the population do not quite fit those expected from the model. This fact, combined with the much more extensive data from *fortis* and *scandens* on Daphne Major, suggests a two-locus model with epistasis would be more accurate. A time-dependence in gene expression is indicated by changes in beak color, from pink to yellow and vice versa, after about 90 days of age.

9

Hybridization and Selection

Introduction

In Chapter 8 we showed that beak and body size variation is pronounced and highly heritable. We now consider the various agencies that are responsible for the high level of size variation.

As explained in Chapter 1 and summarized in Figure 1.6, two counteracting processes occur against a background of recurring recombination each generation. New alleles enter the population, thereby increasing the genetic variance, and these and old alleles are lost, which decreases the variance. New alleles are introduced through introgression from other populations and mutation, and these and old alleles are lost through selection and drift. The amount of genetic variation in the population at any one time is determined by the resolution of these opposing processes. The amount of phenotypic variation at any one time is a function of those genetic processes and of the degree to which the genes have phenotypic effects. Rates of gain and loss may or may not be equal, hence genetic and phenotypic variance may or may not change from year to year.

In this chapter we will explore how variation is enhanced through introgression of genes from other populations, how forces of natural selection act on this variation, and why natural selection occurs. Mutation and recombination are not known directly and are best considered as background processes. Although both contribute to variation (Brooks 1988; Maynard Smith 1988), their effect on the increase in variation is likely to be small compared to the effect of introgression (Falconer 1981; Grant and Price 1981). We start with introgression.

Introgression

HYBRIDIZATION WITH RESIDENTS

Three unequivocal cases of interbreeding were observed. They constitute about 1 percent of the 333 pairings of banded or unbanded birds, involving at least one *conirostris*, which we followed over a period of 10 years. A *conirostris* (♂) × *difficilis* (♀) pair attempted to breed once, but was unsuccessful. A female *conirostris* bred with a male *magnirostris* three times, twice successfully; a male *conirostris* bred once successfully with a female *magnirostris*. There was no indication of a reduced fitness in the number of eggs laid and hatched, the number of nestlings fledged, or the weights of nestlings. To our great regret none of the fledglings survived in the local area to be recaptured and measured, let alone to breed.

Occasionally birds with strange measurements were captured in the nets (Figs. 9.1 and 9.2). Some were probably F_1 hybrids. One male with an unusually deep beak for *conirostris* even sang a *magnirostris* song (Fig. 9.3). Its territory overlapped two territories of neighboring *conirostris*, but was contiguous with a *magnirostris* territory. Correspondingly, it chased *magnirostris* males and ignored *conirostris*. It paired with a *magnirostris* female in 1981 and 1982, and fledged at least 10 offspring of normal weight and appearance. Unfortunately we never saw them again in subsequent years.

In addition to this almost certain hybrid of *magnirostris* (♂) and *conirostris* (♀) parents, others are probably hybrids by virtue of their intermediate measurements. For example, one *conirostris* had a beak depth exactly at the midpoint between the *conirostris* and *difficilis* averages, and four others had smaller beak depths (Fig. 9.1). There is a clear morphological separation between these five birds and the largest *difficilis*, on the

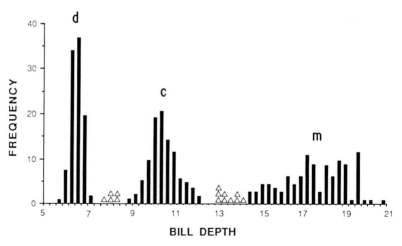

Figure 9.1 Beak depth frequency distributions of adult male and female *difficilis* (d; $N = 123$), *conirostris* (c; $N = 382$), and *magnirostris* (m; $N = 114$), and possible hybrids indicated by open triangle.

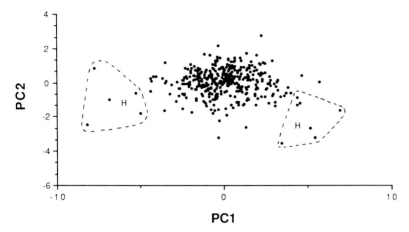

Figure 9.2 Probable hybrids (see Fig. 9.1), enclosed by broken lines and labeled with the letter H, occupy outlying positions on a principal components plot of morphological trait variation (see Table 8.5). PC1 represents body size variation, increasing from left to right. PC2 represents complex shape variation; from bottom to top beaks become relatively more pointed and wings become shorter. Probable hybrids with *difficilis* are on the left, probable hybrids with *magnirostris* are on the right.

one hand, and the next smallest *conirostris,* on the other. Other dimensions were approximately intermediate too.

All five of these birds were females. Either they had never grown to their full genetically determined potential, or they were hybrids (F_1 or F_2) of *conirostris* and *difficilis*. All of them bred successfully with *conirostris* males, and five of their offspring were measured when fully grown. The measurements of the offspring tend to support the hybrid explanation by being lower than the *conirostris* mean, on average. One resembled its father's (larger) measurements, two were close to the midparent value, and two (one a male) were well below the midparent value and slightly larger than the mother's measurements.

A small sample of measurements can only be suggestive. This sample does not suggest that the mothers were stunted *conirostris*. It is consistent with the highly heritable nature of beak depth variation.

None of the birds with strange measurements were at the midpoint between *magnirostris* and *conirostris* beak depth means; all were slightly smaller, although others may have been hybrids but indistinguishable from *magnirostris* (Fig. 9.1). Nine had beak depths equal to or exceeding 13.0 mm, including the one who sang a *magnirostris* song (13.6 mm). They ranged from 2.9 to 4.2 standard deviations above the *conirostris* mean. The next largest *conirostris,* three males, had beak depths of 12.3 mm, which is 1.9 standard deviations above the mean. Therefore this group may have been F_1 or F_2 hybrids, or the product of backcross matings to *conirostris*. The fact that they were all below the interspecific midpoint in mean measurements does not rule out the possibility that they were F_1 hybrids. On Daphne Major one offspring of a *magnirostris* × *fuliginosa* pair has been measured and found to be much closer to the *fuliginosa* mother's size than to the *magnirostris* father's size. A lasting maternal influence on size is possible; the size of a mother constrains egg size, which in turn could influence the adult size of the offspring.

Three of the group disappeared before breeding, two bred with *magnirostris*, and four bred with *conirostris*. If these birds were really stunted *magnirostris*, some of them should have produced offspring larger than themselves when they bred with *conirostris*. In fact the three measured offspring were all smaller than these birds and smaller than the midparent values in all dimensions. A hybrid origin seems probable.

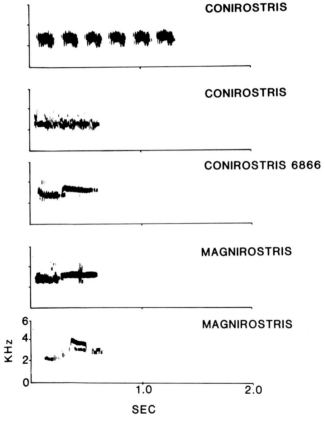

Figure 9.3 Songs of *conirostris*, *magnirostris*, and a *magnirostris* × *conirostris* hybrid (No. 6866) on Genovesa.

Putting all this information together, we can now estimate the frequency of hybrids in the breeding population. If interspecific pairs form at a frequency of 1 percent as our observations suggest, and if breeding occurs without fitness loss, hybrids should also be at a frequency of 1 percent. Our estimate is larger, at 3.0 percent; that is, four *conirostris* × *magnirostris* hybrids and five *conirostris* × *difficilis* hybrids out of a total of 304 individuals that were banded, measured, and breeding with *conirostris*. The discrepancy might be due to chance, or hybrids might sometimes survive at a higher rate than nonhybrids. A third possibility is that hybridization occurred more frequently before our study began than during it for unknown reasons. Five out of the 14 hybrids were born before the study began.

RELATIVE FITNESS OF HYBRIDS

At each breeding episode to which they contributed, the supposed hybrids laid as many eggs and fledged as many young as did nonhybrids. They did not survive for as long, and they did not breed as many times as did the nonhybrids, as can be seen from the following.

Their numbers rose from five in 1978 to a maximum of seven in 1980, then fell to one in 1985; none were alive in 1986–88. All populations of finches suffered after 1983, and the hybrid group may have been eliminated in the two study areas because they were rare at the outset, or maybe because they were at a selective disadvantage. (Reasons are given later in this chapter for believing they were at a selective disadvantage.)

While they were present in the years 1978–83, the presumed hybrids constituted 4.1 percent of the banded *conirostris* breeders. This is a higher frequency than for the study period as a whole and gives further reason for believing that hybridization occurred more frequently before than during the study. Given this frequency (4.1 percent), they should have contributed to 8.2 percent of the pairs. Their actual frequency among pairs in the period 1978–83 was only 4.4 percent, and their involvement in breeding attempts (nests with eggs) was at a lower frequency of 4.1 percent.

These estimates, while not precise because of sample size limitations, indicate that hybrids are at a relative disadvantage.

HYBRIDIZATION WITH IMMIGRANTS

Immigration of *conirostris* from the only other populations on Española and Gardner is not known to occur. If it did occur, could we detect it? If a male immigrant sang we could identify its extraneous origin by the unique features of its song (Fig. 9.4; see also Bowman 1983). No such song was heard on Genovesa throughout the study, even though we have heard and recorded four *magnirostris* on Genovesa singing songs very similar to one sung at Santa Cruz and quite different from the typical ones

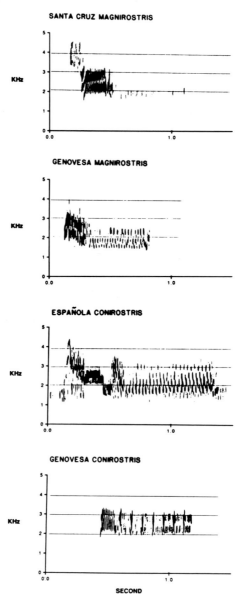

Figure 9.4 Songs of *conirostris* on Genovesa and Española, and of *magnirostris* on Genovesa (a probable immigrant which was measured) and Santa Cruz. For typical songs of *magnirostris* on Genovesa see Figure 6.7. The main part of the songs of *magnirostris* and Española *conirostris* has a lower frequency than the Genovesa *conirostris* songs, and sounds quite different. The Santa Cruz and Española sonagrams were supplied by L. M. Ratcliffe.

sung on Genovesa (Fig. 9.4). In the absence of song, we would have to rely on morphological cues to identify an immigrant from Española or Gardner, cues that include the distinctive features of bill shape, body form, and dark female plumage (Fig. 9.5; see also Grant et al. 1985; Grant 1986a). We have not had our suspicions raised by a single individual on Genovesa.

Immigration of other *Geospiza* species is known. In addition to the *magnirostris* individuals, we have captured one *scandens* and four *fortis*, all in the early dry season months of June–August 1979–83 (Fig. 9.6). Two specimens of *fortis* were collected on the island in earlier years (Lack 1945; Yang and Patton 1981). We have observed but not captured seven other *fortis* individuals, two of them present in February 1988, and one *Camarhynchus parvulus* (small tree finch).

Despite a very slow trickle of immigrants, we have no evidence that they stay to breed. Searching for immigrants in the breeding season is, for most of the island, little better than looking for a needle in a haystack. The search would certainly be successful in 13 percent of the island area where *conirostris* breed, but unlikely to be successful outside the two study areas. A one in five chance of detection seems reasonable. To judge from the study on Daphne Major (Grant et al. 1975; Boag and Grant 1984), most immigrants are likely to die or emigrate without breeding. Therefore interbreeding between a resident species and an immigrant species probably does not occur any more frequently than once a decade. As a source of alien genes added to the *conirostris* population, it is insignificant in comparison with interbreeding with another resident species. Nevertheless, if it occurs it could have important genetic effects.

Effects of Interbreeding on Morphological Variance

Hybridization increases both phenotypic and genetic variance. If we knew with certainty which individuals were hybrids and which were not we could estimate the magnitude of the increase in variance by simply recalculating the variance after first removing the hybrids. In the absence of certain knowledge we can perform the exercise with individuals tentatively identified as hybrids, the results being useful as an illustration of the potential importance of hybridization. They show the actual effects of hybridization only to the extent that we have correctly identified them as hybrids.

Phenotypic results are shown in Table 9.1. For both sexes, presumed hybrids have the greatest effect on beak depth variation. Elimination of five hybrids from the female sample reduces the coefficient of variation by 22 percent, from 7.19 to 5.59, and removal of four hybrids from the male sample reduces the coefficient by 11 percent, from 7.43 to 6.62. Effects on the variance are approximately twice as large.

Figure 9.5 Sexual dimorphism of *conirostris*. Upper: Española, female left and male right. Genovesa, female (middle) and male (lower). Beaks are deeper on Española, and females have a relatively dark plumage there. Photos by P. R. Grant.

Figure 9.6 Upper: *G. conirostris* female above and an immigrant *fortis* below. Lower: *G. conirostris* male on the right and an immigrant *scandens* on the left. Photos by P. R. Grant.

Table 9.1 Effects of Suspected Hybrids on Phenotypic Variation

	Females Reductions in		Males Reductions in	
Trait	Variance	Coefficient of Variation	Variance	Coefficient of Variation
Weight	19.2	9.9	6.8	3.0
Wing length	9.9	5.3	3.1	1.6
Tarsus length	27.0	15.2	0.0	0.0
Bill length	19.3	10.9	1.7	0.0
Bill depth	38.9	22.3	21.9	10.9
Bill width	24.6	13.4	13.6	5.8

Note: Entries in the table are percentage reductions in variance and coefficients of variation resulting from exclusion of six suspected hybrids from the sample of 146 females and four suspected hybrids from the sample of 249 males with complete measurements.

The revised sample lacking the hybrids includes three offspring of the hybrids. Removal of the two females reduces coefficients for the several traits by only 1–2 percent. There are probably additional birds in the sample that are hybrids of varying degrees of relationship to the other two *Geospiza* species. If we could identify and remove them, the coefficient would come down further, perhaps to less than 5.0. The full effects of hybridization are impossible to measure, but an elevation of the coefficient of variation for beak traits by at least 20 percent and probably more is a safe conclusion to draw.

Hybridization has different effects on the variance (and coefficient of variation) of different traits because some traits differ between the parental species more than others (Table 9.2). Interspecific differences in wing and tarsus length are small. Differences are much larger in bill dimensions, especially in bill depth. Likewise, the effects of hybridization on variances

Table 9.2 Morphological Comparison of *G. conirostris* with Its Two Sympatric Congeners

Trait	*difficilis*	*conirostris*	*magnirostris*	Percentage Larger Size of *magnirostris*	Percentage Smaller Size of *difficilis*
Mass, g	11.6	24.7	35.5	43.7	53.0
Wing length, mm	62.7	76.6	84.7	10.6	18.2
Tarsus length, mm	17.8	22.1	23.6	6.8	19.5
Bill length, mm	9.5	14.7	16.4	11.6	35.4
Bill depth, mm	6.3	10.6	17.6	66.1	40.6
Bill width, mm	4.3	6.4	8.6	34.4	32.8

Source: Mean values for adults, sexes combined, are taken from Grant (1981a).
Note: Bill width refers to the upper mandible.

are greater for bill dimensions than for wing and tarsus length, again especially for bill depth (Table 9.1).

There are two exceptions to these general statements: tarsus length variance of females was greatly increased by hybridization, and bill length variance of males was scarcely affected by hybridization. Both exceptions are explicable. Female hybrids are mainly the product of *conirostris* × *difficilis* matings, and mean tarsus length differs more between these species than between *conirostris* and *magnirostris*. The absence of an effect of hybridization on the variance of male bill length is understandable because all male hybrids were the product of *conirostris* × *magnirostris* matings, and these two species differ very little in mean bill length. The one feature not explained is the relatively small effect of hybridization on the variance in mass (Table 9.1) despite a large interspecific difference in means (Table 9.2).

The effects of hybrids on genetic variance are less well estimated because only three families of measured parents and offspring include a suspected hybrid. When they are excluded, the calculated heritabilities of all traits fall by a few percent, from 3 percent for mass to 13 percent for bill width. A decrease in heritability results from a decrease in genetic variance relative to phenotypic variance, since $h^2 = V_a/V_p$. All phenotypic variances decreased when we excluded the hybrids (Table 9.1), therefore all genetic variances must have decreased as well.

To conclude as we began, these estimates are meaningful only insofar as we have correctly identified the hybrids.

Selection

Under most regimes of selection, genetic variation is reduced. Stabilizing and directional selection have the effect of decreasing the frequency of alleles possessed by individuals of extreme size. To the extent that extreme individuals differ from the rest of the population as a result of the alleles they possess, those particular alleles will decline in frequency towards zero. On the other hand, disruptive selection, if strong enough, can maintain high levels of variation by favoring the extreme individuals (Maynard Smith 1979; Felsenstein 1981).

Selection can occur at any stage of the life cycle. It follows that traits like beak size or shape can have effects on fitness at any stage of the cycle. Fitness of an individual, as measured by its contribution of offspring to future breeding generations, is the outcome of all processes governing survival, obtaining a mate, and rearing offspring. Overall fitness can therefore be divided into its component parts (Fig. 9.7; adapted from Falconer 1981). There could be a relationship between one or more of these components and beak size or shape.

A desirable procedure is to partition the total selection on a trait over the lifetime of a cohort into individual events which correspond to particular fitness components (Howard 1979; Brown 1988), because by doing so we would be able to compare them quantitatively on a common scale and thereby reveal the most important selection forces acting on a trait, their direction, and their time of action (Arnold and Wade 1984a, b). Unfortunately we cannot adopt this approach as our samples are too small, they are not nested, and variances and covariances between characters do not remain constant among groups, largely for reasons of sampling error. Instead we have analyzed selection events one at a time, relying heavily on Lande and Arnold's (1983) statistical technique for separating the direct effects of selection at each episode from indirect effects arising from character correlations. This is a powerful method for analyzing directional selection. For stabilizing and disruptive selection, however, it requires larger samples than we possess, so we have resorted to univariate analysis.

We have already considered mating success in Chapter 7 and found this

FITNESS

COMPONENTS	SUB-COMPONENTS
SURVIVAL	SURVIVAL IN FIRST YEAR
	LONGEVITY
MATING SUCCESS	OBTAINING A TERRITORY
	OBTAINING A MATE
BREEDING SUCCESS	AGE AT FIRST REPRODUCTION
	NUMBER OF CLUTCHES
	NUMBER OF OFFSPRING
PARENTAL CARE	PREDATOR AVOIDANCE
	FEEDING RATE

Figure 9.7 Components of fitness. From B. R. Grant (1985).

component of fitness to be uninfluenced by bill traits. This leaves survival and reproduction. Although they are obviously interconnected, survival is the more important of the two (Chapters 3 and 4). To reiterate a conclusion reached at the end of Chapter 4, whatever determines survival indirectly determines reproductive success. We concentrate on survival over the dry season in particular because this is the season when food supply declines, and when beak size or shape is most likely to be important in determining success or failure. Adult mortality during a breeding season is almost negligible.

PROCEDURE

For a given group of birds and period of time, we gave an absolute fitness value of 1 to survivors and a value of 0 to those that disappeared, following standard protocol (Lande and Arnold 1983; Endler 1986; Price and Boag 1987). These values were converted to relative fitnesses by dividing by the mean absolute fitness. Relative fitness was regressed on morphological measurements which had been log-transformed, standardized to have zero mean and unit variance, and checked for multivariate normality. The coefficients estimated by multiple regression analysis are standardized directional selection differentials and gradients. The selection differential (s) is the difference in a character mean before and after selection and is measured in standard deviation units. It is produced by both the direct effects of selection on the character and the indirect effects of selection on other characters correlated with it. The selection gradient (β) is a measure of the direct effects alone and is obtained by statistical removal of the indirect effects. Standardization makes it possible to compare selective effects on different characters (e.g., see Lande and Arnold 1983; Manly 1985; Endler 1986; Price and Boag 1987).

Statistical and interpretational problems arise both when potentially important variables are excluded from an analysis and when highly intercorrelated variables are included (Arnold 1983b; Lande and Arnold 1983; Endler 1986; Mitchell-Olds and Shaw 1987; B. R. Grant and P. R. Grant 1989). Our full set of morphological data comprises three beak characters and three other traits that represent overall body size. This set does not exclude any morphological trait that we have reason to believe is important in determining survival. Among the three body-size traits, weight and wing length were less repeatable than tarsus length (Table 8.1) and were therefore deleted. Beak width was deleted because in the sample of adults in 1983 it was strongly correlated with beak depth ($r = 0.87$). In effect the two dimensions can be considered a single trait, at least in this sample. The next highest correlation, between tarsus length and wing length, was much lower ($r = 0.59$). For consistency we used the remaining three traits in all selection analyses.

Dry-Season Survival

FIRST YEAR OF LIFE

If extreme variants are culled from the population because they are poorly suited to exploiting the food resources, we might expect the winnowing process to occur in the first year of life because survival of young birds is generally so low.

Selection may occur in the first two months post-fledging, but if it does it escapes detection because morphological traits are not fully grown and sizes are not known. Contrary to our expectation, we have no evidence that selection occurs after the traits are fully grown and before the birds are one year old.

The closest we have come to detecting such an effect of selection on traits affecting survival was in 1980–81, when 16 survivors were compared with 8 nonsurvivors (B. R. Grant 1985). Variances were actually higher in the group of survivors than in the remainder, indicating a possibly disruptive rather than stabilizing effect of selection. The difference was not quite statistically significant however (two-tailed Levene's test). At this time there were no directional selection effects on mean bill size, nor were there any detected selection effects during the long dry period after the 1983 El Niño (B. R. Grant and P. R. Grant 1989). In this last case the sample size of 87 birds was surely large enough for us to detect directional selection if it had occurred.

These negative results suggest that factors other than beak size or shape determine survival in the first dry season of a bird's life. Chance factors may play an important, perhaps dominant, role in governing which few birds will survive. Another, probably important, class of factors is behavioral. It would seem crucially important for a bird to learn to find new food items and to perfect techniques for extracting and dealing with the foods, as the characteristic breeding-season foods of small seeds and caterpillars dwindle. New skills for finding and extracting dry-season foods are probably acquired by imitation of adults and by trial-and-error learning (Grant and Grant 1980).

The importance of learning these skills is well illustrated by events in the long dry period after the rains stopped in 1983. During 1984, 87 banded and measured members of the 1983 cohort were observed feeding on as many as 20 different days and as few as 3. Forty-five of them fed only in the typical manner of the wet season, gleaning arthropods from the vegetation and eating small seeds, despite the scarcity of these resources (Table 2.4). The remaining 42 displayed the typical dry-season foraging behaviors of bark stripping, pad ripping, and so on (Fig. 9.8). Those who had developed dry-season feeding skills had the higher fitness, for 14 of them survived well into 1985 whereas none of the others did ($\chi^2_1 = 15.49$, $P < 0.001$).

Figure 9.8 *G. conirostris* stripping bark from a *Croton* bush in search of hidden arthropods (upper), tearing open an *Opuntia* pad to reach fly and beetle larvae and pupae (middle), and cracking small seeds on the ground (lower). Photos by O. Jennersten.

Table 9.3 Selection on Young Birds Feeding in a Dry-Season Manner, 1984–86

Character	Coefficients		
	β ± SE	s	N
Beak length	0.19 ± 0.26	0.29(*)	42
Beak depth	0.08 ± 0.30	0.30(*)	(0.33)
Tarsus length	0.47 ± 0.26*	0.38*	

Note: Coefficients are standardized directional selection gradients (β ± standard error), and differentials (s), and their statistical significance is indicated by *(P < 0.05) or (*) (P ≈ 0.06). The number of individuals present in 1984 is given under N, and the proportion surviving follows in parentheses.

The two groups of young birds characterized by their feeding did not differ in any of the bill or body-size dimensions. Therefore beak size and shape had no bearing on the development of dry-season feeding skills. Behavioral factors, uncorrelated with age, were presumably the determining influence (B. R. Grant and P. R. Grant 1989). However, after the dry-season feeding skills had been acquired, beak size (or body size) did influence survival. Large birds with deep beaks were at a selective advantage (Table 9.3). Entries for individual bill dimensions in the selection gradient (β) were not significant, and it appears that overall size was favored.

To understand why large birds with large beaks were favored, we need to look more closely at the dry-season feeding activities, their relationship to body size and to beak size and shape, the survival of adults over the same period, and the food supply.

DRY-SEASON FORAGING

Apart from the rare consumption of small seeds and arthropods picked up from the ground, arthropods gleaned from the vegetation, and nectar taken from the flowers of *Cordia lutea* and *Waltheria ovata,* there are only six foraging activities of cactus finches in the dry season. Five of them involve *Opuntia* cactus. The activities are: (a) extracting seeds from cactus fruits and eating the surrounding fleshy aril, (b) cracking the moderately large and hard seeds of *Opuntia,* (c) ripping open fallen cactus pads that are rotting beneath the bushes to extract fly and beetle larvae and pupae, (d) probing *Opuntia* flowers for nectar and pollen, (e) pecking at cactus buds to obtain the developing anthers, and (f) stripping bark from dead trees and shrubs to obtain hidden arthropods (see Figs. 9.8–9.12).

The middle third of the dry season, from August or September to about November, is probably the time of least food. New foods become available afterwards. Buds on cactus plants develop in large numbers usually during December in a typical year, and cactus flowers begin to open late in that

Figure 9.9 Cactus exploited by *conirostris*. Upper: Seeds have been removed from the fruit, their surrounding arils have been consumed, and they have been discarded. When the seeds harden they are cracked open by *G. conirostris* occasionally, but more often by *G. magnirostris*. Lower: a fallen pad excavated for fly and beetle larvae and pupae. Photos by P. R. Grant.

month or in January. From August to December there are essentially only four feeding activities (see a, b, c, and f above).

FORAGING EFFICIENCY AS A FUNCTION OF BEAK SIZE AND SHAPE

Different skills are required in the four activities, and different beak sizes and shapes are best suited to performing those activities. Mechanical analyses of beak shapes of Darwin's Finches have been carried out by Bowman (1961). These show that a relatively short and deep beak allows a powerful crushing force to be applied at the base of the bill for the purpose of crushing seeds or tearing at bark (Fig. 9.13). A relatively long and shallow bill is better suited to hammering, probing, and pecking at the tip of the bill. Biting at the tip is enhanced by a strong curvature there.

Field observations made prior to 1983 showed that dry-season diets do reflect beak size and shape to a certain degree (B. R. Grant 1985). For example, 5 birds that opened *Opuntia* fruits to reach the enclosed seeds had significantly longer beaks on average than the 26 birds observed at the same time and in the same places but never seen opening fruits ($t = 2.54$, $P < 0.02$). They did not differ significantly in the other two bill dimensions ($P > 0.1$). In this case the bill dimension and proportions associated with a feeding skill—hammering, pecking at the tip, and probing—are clearly identified and consistent with our reasoning from simple mechanical considerations.

Individuals observed cracking *Opuntia* seeds had significantly longer and deeper bills than those not observed to crack *Opuntia* seeds (two-tailed t tests, $P < 0.05$ in each case). The same was true for those stripping bark off trees compared with those who did not. It appears that overall beak

Figure 9.10 *G. conirostris* cracking an *Opuntia* seed. Photo by P. R. Grant.

Figure 9.11 *G. conirostris* exploiting *Opuntia helleri* cactus for food. Upper: attacking a bud for pollen (see also Fig. 4.4). Middle: on a flower. Lower: probing a flower for nectar at the base (see also Fig. 4.4). Photos by O. Jennersten.

Figure 9.12 *G. conirostris* removes the bark from dead *Bursera graveolens* branches (upper) in order to reach termites (lower) and other cryptic arthropods. Photos by P. R. Grant (upper) and O. Jennersten (lower).

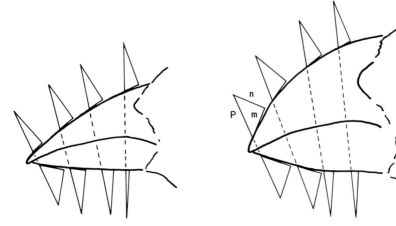

Figure 9.13 Forces (*P*) on the beak during biting in two extreme forms of *conirostris*. They are shown as passing into the bill matrix at right angles to the surface of contact between upper and lower mandibles (the commissure), and are resolved into pressure components (*m*) and fracture-risk components (*n*). The fracture-risk components represent those parts of the forces that must be resisted by the strength of the bill matrix. Note that in the less pointed beak (right) more of the forces are taken up as pressure components (*m*) in the bones and the fracture risk (*n*) in biting at the tip of the bill is lower. Adapted from Bowman (1961).

size, as opposed to one beak dimension or shape, makes the difference between those who do and those who do not feed in these two ways. The problem of character correlation clouds the interpretation, however. It is possible that only one bill dimension is important, and the others appear to be as a result of the correlations.

To overcome this problem, we used multiple regression analysis to separate direct and indirect effects of bill dimensions on diet. In order to calculate "trophic" differentials (*s*) and gradients (β), we gave a value of 1 to individuals engaged in a particular feeding activity and 0 to individuals never observed feeding in this way. One analysis succeeded in separating correlated effects. The 22 individuals (out of 42) seen to strip bark from trees were apparently suited to the task by virtue of deep beaks, because beak depth entered the trophic gradient significantly and positively (0.50 ± 0.25 SE, $P < 0.05$), whereas the coefficient was negative and not significant for beak length ($P > 0.1$). In contrast to this result, none of the trophic gradient coefficients were significant in an analysis of *Opuntia* seed cracking. Selection differentials were positive and significant for bill depth ($s = 1.03$, $P < 0.05$) and bill length ($s = 1.04$, $P < 0.05$). We interpret the differentials and gradients to indicate the importance of bill size as a whole as opposed to any single bill dimension for this particular feeding activity.

Opuntia pad ripping requires the same behavioral techniques as does bark-stripping. Most birds were observed to rip pads. They did not differ from the few not observed to forage in this manner in any bill dimension. At the time we made these observations rotting pads were relatively soft and easy to excavate.

ADULT FITNESS UNDER TYPICAL CONDITIONS

During the dry seasons in the first half of the study, 1978–82, adult survival was generally unrelated to phenotype. Survivors and nonsurvivors did not differ detectably in variances. Males that disappeared did not differ in mean bill dimensions from those who survived. The one significant selection event occurred among the females and was not restricted to a single dry season. Twenty-six females survived for three or more, two, or just one breeding season. Beak length was a target of selection, as identified by a significant entry in the selection gradient (0.17 ± 0.08 SE, $P < 0.05$). Tendencies in the opposite direction on other bill dimensions probably counteracted the direct effect of selection on beak length, because none of the selection differentials was significant.

Thus under usual dry-season conditions, selection on the morphological features we have studied is either weak or absent among adults. Birds died presumably because they were weakened by disease, were eaten by owls, or failed to find enough food and not, apparently, because they had beaks of a particular size or shape.

CHANGES IN DRY-SEASON FOOD SUPPLY

The fact that beak length was a target of selection suggests that the foods best exploited with a relatively long (pointed) bill, namely fruits and flowers of *Opuntia,* may have been relatively plentiful in the early years as they certainly were in November 1978 (Grant and Grant 1980). Unfortunately we have few estimates of the supply of the different dry-season food resources to examine this possibility. Nevertheless, if we are correct, a change in resource composition should lead to predictable changes in the relative fitness of phenotypes of different beak sizes and shapes. Here we were fortunate, because a change in the food supply gave us an opportunity to construct and test some specific predictions.

Opuntia flowers and fruits declined drastically in numbers during and after the heavy rains of 1983. This natural "experiment" occurred as a result of three factors. *Opuntia* bushes were smothered by a rapidly growing vine, *Ipomoea habeliana* (Fig. 2.16). Many fell over in high winds because they had absorbed too much water; they had become top-heavy, and their shallow roots were incapable of supporting them. In coastal regions, especially on the northeastern cliffs, salt spray may have contributed to their destruction (Fig. 2.17). Even though most bushes remained standing, having lost several pads, their subsequent production of

flowers and fruits was seriously depressed throughout the ensuing drought (Fig. 9.14).

Recovery began very slowly in 1985. A factor contributing to the slowness of the recovery was the behavior of the finches. We had shown previously that cactus finches snip the stigmas with their beaks in up to 78 percent of the flowers when alternative food items are scarce (Grant and Grant 1981). They also destroy developing flower buds at this time. This behavior represents a short-term gain, in that it facilitates removal of pollen and nectar from deep within the flower, at the expense of a long-term reward, namely production of seeds (Fig. 9.15). We showed experimentally that stigma snipping prevents fertilization of the ovules (Grant and Grant 1981). In the dry conditions following the heavy rains of 1983, the incidence of stigma destruction rose to nearly 100 percent and fluctuated at a high level until full breeding activity resumed in 1986 (Table 9.4).

The decline in cactus flower and fruit production was accompanied by

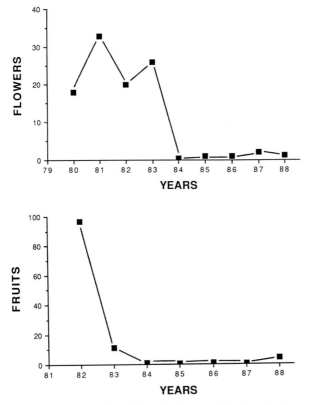

Figure 9.14 Decline in the production of *Opuntia* flowers and fruits in the dry period after 1983. Based on weekly counts of 10 bushes throughout the wet season.

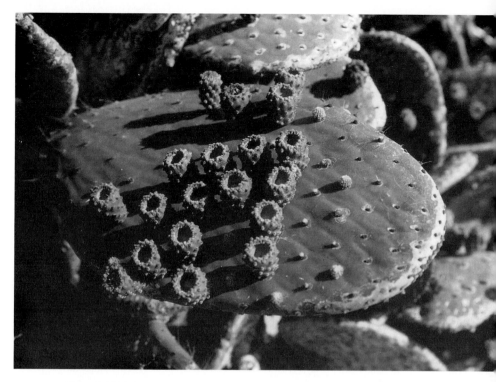

Figure 9.15 Fruits on an *Opuntia* pad. Many of these will not develop further because the stigmas in the flowers were damaged or destroyed by *conirostris*. Photo by P. R. Grant.

a large increase in the number of cactus pads rotting on the ground and an increase in the arthropod fauna they contained. The rotting pads dried out as the dry season progressed, making it increasingly difficult for birds to extract the arthropods. The problem for the finches was exacerbated by the near disappearance of small seeds, which had been produced in very large numbers in 1983 and were plentiful at the beginning of the dry period.

Thus most of the dry-season feeding niches of cactus finches were drastically reduced by this natural experiment. One niche, arthropods in rotting cactus pads, increased; another, arthropods beneath bark, appears to have been relatively unaffected.

ADULT FITNESS UNDER ALTERED CONDITIONS

In the dry conditions following the 1983 El Niño, relatively long-billed birds should have been at a selective disadvantage because the foods they are best able to exploit, the fruits and flowers of *Opuntia,* became scarce. Relatively deep-beaked birds should have been at a selective advantage because the resources best exploited with a deep-beak, arthropods beneath bark and in the tissues of hardened, dead *Opuntia* pads, became relatively common.

One hundred and thirty-seven banded and measured adults (sexes combined) bred in 1983. By the middle of 1984 half of them had

Table 9.4 Percentage of *Opuntia* Flowers with Stigmas Damaged by Cactus Finches in Relation to the Onset of Heavy Rain

Year	Jan	Feb	Mar	Apr	July	Month (Week) of Rain
1978	78.1	0.8	0.0	0.0	X	Jan (2)
1980	72.0	32.8	3.5	0.0	0.0	Jan (4)
1981	50.3	43.3	37.5	0.8	X	Jan (3), Mar (2)
1982	X	25.3	3.3	7.3	0.5	Feb (1)
1983	0.0	1.5	0.3	4.5	0.0	Before Jan (1)
1984	79.8	92.5	72.5	47.3	95.0	Mar (1)
1985	72.2	70.7	88.0	84.8	88.9	No rain
1986	61.0	19.5	8.4	3.5	X	Jan (4)
1987	70.8	0.0	0.0	4.3	0.0	Jan (2)

Note: 100 flowers were observed at weekly intervals. After the first rain other foods appear (e.g., caterpillars), and stigma damage declines. X = missing data.

disappeared. Disappearance (presumed mortality) was nonrandom with respect to phenotype (Table 9.5). As predicted, relatively long-billed birds were at a selective disadvantage. Beak length was a target of selection (significant β) and displayed a net change (*s*) as a result of selection. There was no selection on bill depth nor on body size (as indexed by tarsus length), which is positively correlated with the beak traits.

Deep-beaked birds were not at a selective advantage in the latter part of 1983 because the rotting pads that harbored a large supply of arthropods were moist and soft, and hence relatively easy to excavate. By the middle of 1984, however, most of the fallen pads had become dry and hard. At this time birds were excavating the tough *Opuntia* trunks and pads in a manner similar to stripping bark off trees. Less than half of the adult survivors to

Table 9.5 Selection over the Dry Period, 1983–85, among Adults

		Coefficients		
Period	Character	β ± SE	*s*	*N*
1983–84	Bill length	−0.22 ± 0.08**	−0.24**	137
	Bill depth	−0.05 ± 0.08	−0.07	(0.54)
	Tarsus length	−0.02 ± 0.07	−0.10	
1984–85	Bill length	−0.05 ± 0.18	0.14	74
	Bill depth	0.37 ± 0.19*	0.31**	(0.39)
	Tarsus length	0.14 ± 0.19	0.22**	
1983–85	Bill length	−0.28 ± 0.22	−0.01	137
	Bill depth	0.27 ± 0.21	0.15	(0.19)
	Tarsus length	0.24 ± 0.17	0.17	

Note: Coefficients are standardized directional selection gradients (β ± standard error) and differentials (*s*), and their statistical significance is indicated by *($P < 0.05$) or **($P < 0.01$). The number of individuals present at the beginning of each interval is given under *N*, and the proportion surviving follows in parentheses.

1984 survived to 1985, and large birds with deep beaks were selectively favored (Table 9.5). Large overall size was favored, as shown by a significant, positive, selection differential for tarsus length. It was not a target of selection, however, but increased presumably as a consequence of the positive correlation with beak depth, which was a target of selection. At the same time, the 1983 cohort experienced the same regime of selection in favor of large individuals with deep beaks (p. 212).

We attribute the differential survival at least partly to unequal skills at excavating hidden arthropods in hardened *Opuntia* pads and beneath bark. The association between survival and beak depth arises from the relationship between beak depth and bark stripping (p. 217). We conjecture that the same relationship holds between beak depth and the behaviorally similar foraging by pad ripping.

An additional factor affecting survival in 1984 was the presence of a large crop of the cherrylike stones on *Bursera* trees (Table 9.6; illustrated in Plate 4). These are the largest and hardest seeds regularly cracked by cactus finches; the even harder *Opuntia* seeds are cracked by only a few individuals. *Bursera* seeds are not likely to have been an important factor in survival as the supply is only available for about one month, and that month in 1984 happened to be July when we made our feeding observations (Table 9.6). For the rest of 1984 the observations made in 1985 are more applicable. As much as 85 percent of foraging time was spent on substrate tearing and ripping activities.

Table 9.6 Diets in July, Expressed as Proportions of the Total Feeding Observations Made in Each of Three Years

	Years		
Food and Foraging	1983	1984	1985
Seeds and fruits			
Small seeds	**0.45**	0.09	0.01
Bursera seeds	0	**0.77**	0.06
Opuntia seeds	0.03	0	0
Opuntia arils	0.09	0.01	0.08
Pollen and nectar			
Opuntia flowers	0.03	0.02	0
Arthropods			
Opuntia pad ripping	0.03	0.06	**0.71**
Bark stripping	**0.21**	0.03	**0.14**
Leaf gleaning	**0.18**	0.01	0.02
Minimum number of birds	103	75	46
Number of observations	161	114	77

Note: Principal foraging activities (> 10 percent) are shown in **boldface**. Small seeds have a size-hardness value (geometric mean of depth in mm and hardness in kgf) of 2 or less. *Bursera* seeds have an average value of 5 and *Opuntia* seeds have an average value of 7 (Abbott et al. 1977).

From the middle of 1985 to the beginning of 1986 there was little mortality—only four more birds died—and it was not size selective.

These selection episodes following the El Niño event of 1983 were not repeated after the 1987 El Niño event. This shows, first, that very wet years can have very different consequences, and second, that preceding as well as succeeding conditions are important in determining those consequences. Adult densities differed markedly in the two years (Fig. 3.6), and *Opuntia* cactus was not as drastically affected in 1987 as it had been in 1983. So adult mortality, which had been pronounced and size-selective after the breeding season of 1983, was almost negligible and nonselective after the breeding season of 1987, at least until March 1988.

We have combined males and females in all these analyses because they showed the same trends at all phases of the long dry period. There was no sign of stabilizing or disruptive selection at any time (Levene's tests, all $P > 0.1$).

SELECTION AGAINST HYBRIDS

Three putative hybrids were present at the start of this study in 1978. Their numbers rose to a maximum of seven in 1980 and fell to four in 1983. The last one, a female, was seen in early 1985. In view of what we know about selection after the breeding season of 1983, it is not surprising that hybrids disappeared completely. In the early phase of the following dry period, hybrids with a *magnirostris* parent may have been at a selective disadvantage if, with their long and deep beaks, they had specialized to any extent on *Opuntia* fruits, flowers, and seeds. In the middle phase, hybrids with a *difficilis* parent would have been at a selective disadvantage under conditions that favored large body size and deep beaks.

Reproduction

MATING SUCCESS

It was shown in Chapter 7 that the ability of a male to attract a mate was not affected by its beak morphology. However, there is one exception.

In 1984 the small amount of rain was sufficient to stimulate a breeding attempt by 44 adult males, but it was not sufficient to bring an equal number of females into reproductive condition. These are just the circumstances where previous nutritional and physiological states could have an influence on potential reproductive success. And to the extent that previous nutritional state is in part determined by beak size or shape, we would expect breeding activity of females and mating success of males to be nonrandomly associated with their morphology.

Our analysis of this situation gave mixed results. Females that attempted to breed did not differ morphologically from those who did not attempt to

Table 9.7 Sexual Selection in 1984

Character	Coefficients		
	β ± SE	s	N
Bill length	−0.01 ± 0.14	0.13	44
Bill depth	0.32 ± 0.13*	0.36*	(0.63)
Tarsus length	−0.08 ± 0.11	−0.01	

Note: The number of territorial males is given under N, and the proportion acquiring mates follows in parentheses. Coefficients are standardized directional selection gradients (β ± standard error) and differentials (s), and their statistical significance is indicated by *(P < 0.05).

breed. Male mating success, on the other hand, was associated with beak depth (Table 9.7). Potential fitness gain was not translated into actual fitness gain as all but one of the nesting attempts failed to yield a single fledgling. Nevertheless, the differential mating success indicates that the advantage to birds possessing a deep beak under natural selection (Table 9.5) had already emerged in the first two months of 1984.

A similar phenomenon has been documented in the population of *fortis* on Daphne Major. Medium ground finches with deep beaks had a strong selective advantage during a drought in 1977 (Boag and Grant 1981; Price, Grant, Gibbs, and Boag 1984). At the resumption of breeding in early 1978, males with the deepest beaks had a mating advantage. Price (1984a) interpreted this result as showing a female preference for males with a particular morphology but was unable to resolve the question of where the advantage of such a preference might lie. He argued against a female preference for male morphology itself and in favor of a preference for reliable signs of good parental care. Our interpretation is similar. According to our reasoning, beak size influences nutritional state, and this in turn determines the amount of energy that is available for, and devoted to, courtship; good parental care is likely to follow. Males with deep beaks were at a mating advantage after an episode of natural selection because their courtship was superior. In this way natural selection and sexual selection on a trait were serially linked.

If this reasoning is correct, *conirostris* females would not be expected to show a mating preference with respect to male morphology after a dry period without natural selection, because the courtship quality of the males would not vary systematically with their morphology. As expected, they did not show a preference for males with deep beaks or any other beak trait when pairing in 1986 after a period (late 1985) in which beak traits were not subjected to natural selection. Both the mating success of males and the order in which they bred (Tables 7.3 and 7.4) were independent of beak variation.

BREEDING SUCCESS

Morphological traits have little or no influence on breeding success. The reproductive fitness of males—measured by the production of eggs, nestlings, and fledglings—was not associated with beak morphology at any time during the study (Table 7.5; B. R. Grant 1985; B. R. Grant and P. R. Grant 1987). Reproductive fitness of females was also independent of beak morphology, although we identified through selection analysis a tendency for long-billed individuals to reproduce best in the early part of the study (Table 9.8).

Variation in the number of clutches produced by females in 1978 showed signs of selection in favor of individuals with long bills (Table 9.8). A significant entry in the selection gradient was not translated into a significant net effect of selection (s), presumably because of counteracting effects arising from character correlations. These trends persisted in 1980, 1981, and 1982. Coefficients in the selection gradient for bill length were significant and positive ($P < 0.05$), and those for bill depth were not significant and negative. None of the selection differentials was significant. The consistency among years is due partly to certain individuals breeding more successfully than others in all years.

Table 9.8 Components of Adult Female Fitness in Relation to Bill Morphology in the First Half of the Study, 1978–82

Components and Characters	Coefficients	
	$\beta \pm SE$	s
Longevity		
$N = 26$		
Bill length	$0.17 \pm 0.08*$	0.12
Bill depth	-0.11 ± 0.16	-0.07
Bill width	-0.03 ± 0.16	-0.07
Number of clutches		
$N = 13$		
Bill length	$0.21 \pm 0.09*$	0.09
Bill depth	-0.31 ± 0.14	0.02
Bill width	0.19 ± 0.13	0.05
Recruits		
$N = 36$		
Bill length	$0.43 \pm 0.19*$	0.17
Bill depth	-0.57 ± 0.32	-0.21
Bill width	0.13 ± 0.29	-0.16

Source: B. R. Grant (1985).

Note: Number of clutches applies to 1978 only. The number of recruits per female was divided by the number of clutches she produced to correct for large variations in longevity and production. Coefficients are standardized directional selection gradients ($\beta \pm SE$) and differentials (s), and their statistical significance is indicated by *($P < 0.05$).

In their contribution of offspring to future generations, females showed an almost identical tendency. That is, increased bill length was again identified as a target of selection, but the effect was nullified by character correlations and selective tendencies in opposite directions, even after the effects of varying longevity had been removed (Table 9.8). All of these results may reflect the combination of a relatively poor performance of the hybrids produced by *conirostris* × *difficilis* pairs and a relative abundance of cactus flowers, since the early part of the study was the time when both hybrids and cactus flowers were most frequent.

LIFETIME REPRODUCTIVE SUCCESS

The overall fitness of an individual can be assessed by the number of breeding individuals, that is, recruits, it produces. We know from Chapters 3 and 4 that fitness varies enormously among individuals. Does it vary in relation to morphological traits? A positive answer would imply a tendency for morphological traits to change systematically over the decade. If this were the case we would need to know how the change arises and why it occurs; whether it is a transitory process, perhaps oscillating in direction about some long-term average phenotype; or whether it is a slow, gradual, unidirectional, long-term process. The analyses performed so far point to a negative answer, but these are not definitive because they have not been comprehensive. Some stages of the life cycle had to be omitted.

Individuals that became recruits in 1980–87 were born in the years 1978–86, years that included known selection events and possibly unknown ones. Effects of all selection events are integrated in a simple analysis of recruitment in relation to the beak morphology of male parents and female parents, treated separately. Parents who bred for the first time in 1986 and 1987 are excluded because their lifetime recruitment will not be completed until sometime in the 1990s. Parents who bred in 1978 are included, even though their recruitment in the following year and earlier years is not known. Nine males and two females who had bred in 1983 were still alive in 1988 at the end of the study, so their lifetime reproductive success is underestimated. Inclusion of parents with partial reproductive data is not likely to distort the analysis to a significant degree because they are a minor fraction of the total samples. Two other distortions are likely to be minor. The first arises from the awkward fact that fitness of a parent is a complex function of the number of recruits it produces and the time it produces them (Charlesworth 1980), either early or late in its life, when the population is changing in size. The second arises from the fact that selection occurred in the 1983 cohort prior to recruitment, and strictly, this should not be allowed to affect the measurement of the fitness of the parents because the ability to survive the first year or two of life is a property of the individual and not of its parents.

For the male sample ($N = 110$) and the female sample $N = 85$), only one of the selection coefficients was significant; beak length had a direct effect on female fitness ($\beta = 0.66 \pm 0.29$, $P < 0.05$). This result is not surprising in view of three facts. First, the breeding period was necessarily restricted to a portion (1978–83) of the total study period. Second, during most of that period (1978–82) long-billed females were the most successful in producing eggs, nestlings, and fledglings, as described in the previous section. Third, the number of recruits produced by a parent was positively correlated with the number of offspring fledged. Part of this result might be explained by selection against small hybrids. However, when we deleted the five suspected hybrids from the list of females and reanalyzed the data, we found, as before, a significant association between beak length and fitness. This implies that the hybrids had no effect upon the relationship.

Phenotypic traits have a detectable influence upon fitness over short periods of time and under unusual conditions, such as droughts. In the absence of sufficient data we are left to speculate that over a long period spanning contrasting environmental conditions, all but perhaps the most extreme phenotypes have approximately equal fitness. To place that speculation on a more solid foundation we would have to study the population for two decades or longer.

Directional and Stabilizing Selection

Directional selection on bill dimensions was detected at different stages of the life cycle, but not in the first year of life. To summarize, beak length of females was subject to selection, with several components of fitness being affected: longevity, and the production of clutches and recruits in the early part of the study. In contrast, birds of both sexes with short bills survived best in the early phase of a long dry period when *Opuntia* floral resources declined. Birds with deep beaks were favored at a later stage when there was little food other than arthropods beneath bark and in *Opuntia* pads. A carryover of this effect of natural selection into a reproductive context occurred in 1984 when male mating success was positively correlated with beak depth. Natural and sexual selection occurred in sequence and in the same direction, as found also by Price (1984a) in his study of *fortis* on Daphne Major.

Stabilizing selection appears to have been rare or nonexistent, but our powers to detect it are limited, and we suspect it occurred, perhaps repeatedly, but weakly. The problem is a statistical one with small sample sizes. The Levene's test is more sensitive than the Lande and Arnold (1983) selection analysis, but even this lacks the sensitivity we need.

As an example, consider the reduction in numbers of adult finches over the first phase of the dry period after the 1983 rains. Half of the starting sample of 137 birds survived (Table 9.5). Given these sample sizes of 74 surviving and 63 dying, stabilizing selection would only be detected if the variance of those who died was about 67 percent greater than the variance of the survivors; the critical F value for a two-tailed test, with degrees of freedom 60 and 60, is 1.67. This is equivalent to a reduction in the variance as a result of selection of about 25 percent. In contrast, small directional shifts in the mean as a result of selection can be detected easily. At this same time short-billed birds survived best. The (significant) difference in mean beak length between those who survived and those who did not was 2.5 percent.

Thus stabilizing selection has to be sufficiently intense to result in a reduction in phenotypic variation of at least 25 percent for us to detect it with our largest samples. It has to be much higher for our typically smaller samples.

EFFECTS OF SELECTION ON MORPHOLOGICAL VARIATION

Hybridization occurred in the first half of the study, and directional selection on bill dimensions and body size was strong in the second half. The influence of selection predominated, with the net result that variation in each of the three beak dimensions was lower at the end of the study than at the beginning. In two cases variation was significantly reduced; females varied less at the end in beak depth and beak width (Table 9.9). Coefficients of variation were compared in two-tailed F tests (Van Valen 1978) to demonstrate the reduction because in one instance the mean had changed as well. The average beak width of females was larger at the end of the study than at the beginning (($t_{34} = 2.26$, $P < 0.05$, with logarithmically transformed data), while the variation decreased. Variation among

Table 9.9 Changes in Bill Variation between the Beginning and End of the Study

Character and Year	Males		Females	
	N	CV	N	CV
Bill length				
1978	21	5.44	13	6.26
1987	29	4.35	24	5.08
Bill depth				
1978	21	8.06	13	9.50
1987	29	5.65(*)	24	4.73***
Bill width				
1978	21	5.86	13	6.27
1987	28	5.17	23	3.71*

Note: Coefficients of variation (CV) are shown with sample sizes (N). Statistical significance is indicated by *($P < 0.05$), ***($P < 0.001$), and (*)($P \approx 0.08$).

the males was similarly reduced in all bill dimensions but to a lesser extent (Table 9.9).

We previously found, by removing hybrids from the sample of all breeding birds and recalculating coefficients of variation and variances, that the effect of hybrids is to elevate the coefficient by a minimum of 20 percent and the variance by 35 percent (Table 9.1). The disappearance of all hybrids by the end of the study undoubtedly contributed to the reduction in variation and provides us with another means of estimating the effect of hybridization upon phenotypic variation. In the most extreme case, hybridization doubled the coefficient of variation; the initial coefficient for female beak depth was twice as large as the final one (Table 9.9). The effect on the phenotypic variance was even stronger. A final beak depth variance of 0.235 contrasts with an initial one of 0.893. These are much stronger effects of interbreeding and recruitment of hybrids than those calculated earlier, presumably because they are inflated by the consequences of selection on nonhybrids as well.

The variance in beak depth of females had declined significantly by 1982 (B. R. Grant 1985). We interpreted this to indirectly signify a regime of disruptive selection during the drought of 1977, the argument being that predrought variation was probably similar to the low 1982 value. It fits with the demonstration of multiple-niche occupancy in the dry season. With more information now available, and a better though not certain characterization of the hybrids, we should complement the disruptive selection argument with a hybridization hypothesis to explain the high level of variation observed in 1978. Hybridization, unlike disruptive selection, has been observed. Disruptive selection has been inferred from the pattern of survival of birds in their first year of life (p. 210). Why hybridization occurred more frequently before our study began is a question that none of our observations helps to answer.

EVOLUTIONARY RESPONSE TO SELECTION

Selection occurs within a generation; evolution occurs between generations. Selection occurred on different bill traits in different directions in different phases of the dry period from 1983 to 1986. Did this result in an evolutionary change in bill dimensions? We cannot answer this question directly, because only four offspring of measured parents survived to breed in 1986. As an alternative we can estimate the potential evolutionary change by integrating effects of the different selection episodes between breeding in 1983 and its resumption in 1986, a period in which less than 20 percent of the adults survived. Entries in the selection gradient over the whole period, together with the genetic variance-covariance matrix (G) for the set of morphological traits (z), determine the evolutionary response of those traits ($\Delta \bar{z}$) to selection in the next generation (Lande 1979).

As shown in Table 9.5, there was no selection over this period. The effects of strong selection for short beaks were apparently nullified subsequently by indirect effects of selection in the opposite direction on the positively correlated trait of beak depth.

Evolution of these traits did not occur because all β values were effectively zero (Table 9.5); there were no direct effects of selection on a single dimension independent of other correlated dimensions. Thus short-term changes tended to cancel.

The Maintenance of Variation

The model depicted in Figure 1.6 has served as a framework for investigating the maintenance of variation. We have ignored the invisible factors of mutation and recombination and have concentrated on those measurable ones identified as probably the most important. We have demonstrated enhancing effects of introgression and a reduction in variation as a result of differential survival. Selection was mainly directional and fluctuating. Its effects over a decade were perhaps equivalent to weak overall stabilizing selection. Directional selection may have been augmented by stabilizing selection itself, which was nonetheless too weak for us to detect. Comparable observations on introgression and selection have been made on *fortis* on Daphne Major. Although the details differ, the points of similarity are introgression of genes through rare hybridization (Grant 1986a), fluctuating directional selection (Price and Grant 1984; Gibbs and Grant 1987c), and either the absence of stabilizing selection or its occurrence at too low a level to be detected.

A COMPARISON OF SPECIES

The two *Geospiza* congeners on the island, *difficilis* and *magnirostris,* differ in their variation. The smaller species, *difficilis,* displays markedly lower variation. Its coefficient of beak depth variation, for example, has been estimated from samples of both sexes combined ($N = 105$) at the beginning of the study (Grant 1981a) to be 4.85 ± 0.33 SE. The coefficient for *magnirostris* ($N = 44$) is 9.83 ± 1.05, and that for *conirostris* ($N = 45$) is intermediate (8.88 ± 0.94). In terms of the model, the differences between species can be accounted for by different rates of introgression or different rates of selective loss or both. This applies also to the differences between Darwin's Finches and related species in other continental and insular locations (Tables 1.1–1.4).

Our few data give no reason for believing that hybridization rates differ among the ground finch species on Genovesa. *G. difficilis* and *magnirostris* are not known to breed with each other, whereas each breeds with *conirostris* at approximately equal frequency. It is possible that hybrids of *conirostris* × *difficilis* parents do not breed with *difficilis*, whereas hybrids of *conirostris* × *magnirostris* do breed with *magnirostris*. We observed no instances of the former and one instance of the latter. It would be strange, however, if *difficilis* bred with *conirostris* but failed to breed with the more similar *conirostris* × *difficilis* hybrids. A barrier to gene exchange is not a likely explanation for the low variation in *difficilis*. However, inasmuch as *difficilis* are consistently more numerous than *magnirostris* (Fig. 9.16), the same rate of interbreeding with *conirostris* would have given a lower per capita rate of addition of *conirostris* alleles to the *difficilis* population.

We favor instead the idea that selection has a more constraining influence on *difficilis* bill variation than it does on *magnirostris* bill variation. The principal cause is the narrowness of the dry-season niche of *difficilis*. It is defined by small seeds, small arthropods on the surface of the ground and vegetation, and nectar in small flowers (Schluter and Grant 1982, 1984a, b). It is almost discretely different from the niche occupied by *conirostris* in the dry season. In contrast, the dry-season niche of *magnirostris* almost entirely comprises just two food items, the large and hard seeds of *Opuntia helleri* and *Cordia lutea,* and this is not discretely different from the *conirostris* niche, since *conirostris* also eat *Opuntia* seeds.

This last remark must be qualified. In the long dry period after the breeding season of 1983, cactus seeds all but disappeared, and the low overlap in dry-season niches between *conirostris* and *magnirostris* was

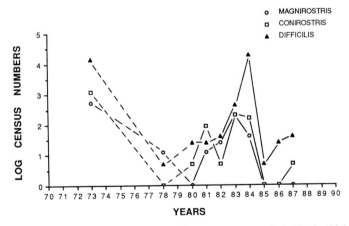

Figure 9.16 Numbers of finches captured in mist net censuses conducted in the third week of March in each of several years. 1973 data, obtained in a year following an El Niño event, are from Abbott et al. (1977). Note that abundances covary, and that *difficilis* is generally the commonest species. Two nets were used at the same six locations on six consecutive days for a total of 200 meter-net-hours.

Table 9.10   Selection during the Dry Period, 1984–85, among *G. magnirostris* Adult Males

Character	Coefficients		
	β ± SE	s	N
Bill length	0.33 ± 0.28	0.45*	23
Bill depth	0.62 ± 0.27*	0.58**	(0.39)
Tarsus length	−0.14 ± 0.27	0.27	

Note: Coefficients are standardized directional selection gradients (β ± standard error) and differentials (s), and their statistical significance is indicated by *($P < 0.05$) or **($P < 0.01$). The number of individuals present in 1984 is given under N, and the proportion surviving follows in parentheses. Females are not included because very few were measured.

almost completely eliminated (P. R. Grant and B. R. Grant 1987). In the absence of the smaller of its two food sources, cactus seeds, *magnirostris* was restricted to feeding on *Cordia lutea* seeds (B. R. Grant and P. R. Grant 1989), and selection favored those individuals with large beaks (Table 9.10).

Quantitative data in support of the reasoning from niches to bill morphology are given in Figure 11.10 in Chapter 11, where the role of *conirostris* in the community of finches is discussed more fully.

EFFECTS OF POPULATION SIZE

We conclude this chapter with a discussion of morphological variation in light of the population structure described in Chapter 7 (additional implications are discussed in Chapter 12 and Appendix IV). There are two important features: first, the population fluctuates in size in response to environmental fluctuations; second, it is not panmictic and local neighborhood sizes are small (see Chapters 10 and 12). Conditions are right for occasional, extreme reductions in local neighborhood size. Under these conditions random effects on allele frequencies may be very important. A population or subpopulation passes through a bottleneck in which alleles are lost through drift (Wright 1932; Frankham 1980; Lande 1988). Levels of inbreeding are increased, and linkage groups may be created, modified, or eliminated (Carson and Templeton 1984).

Phenotypic variation should decrease in these circumstances, but it is not always observed to do so; in fact, sometimes it increases (Lints and Bourgois 1982; Bryant et al. 1986). For example, Bryant et al. (1986) observed a decrease in additive genetic variance for a series of three traits largely independent of body size in housefly populations that had been experimentally forced to pass through bottlenecks, and an increase in additive genetic variance for a group of five body-size-related traits. They attributed the increase to nonadditive genetic variation (dominance, epistasis) affecting these traits. Soulé and Stewart (1970) have argued that sustained directional selection, which was not observed in this study, could have the same effect of increasing phenotypic variation transiently and for similar reasons. Thus the expressed (phenotypic) variation in the scheme depicted in Figure 1.6 can fluctuate as a result of fluctuation in the contributions from recombination which exposes hidden genetic variation (see also Lande 1976).

If the population of *conirostris* has been present on Genovesa for hundreds of generations (Chapter 2), it will have experienced these conditions many times, so severe reductions in population size observed now are not likely to have more than minor effects on variation. It is possible that a bottleneck effect of the type observed by Bryant et al. (1986) could have contributed to the high level of variation observed at the beginning of our study, when population size was low as a result of the preceding drought. An effect in this direction was not observed after a drought towards the end of the study. Variation decreased during the study, and selection is a sufficient explanation to account for it.

Summary

The model depicted in Figure 1.6 provides a framework for investigating the maintenance of variation in traits such as beak dimensions. We have ignored the invisible factors of mutation and recombination and have concentrated on those measurable ones identified as probably the most important. In this chapter we demonstrate enhancing effects of introgression and a reduction in phenotypic variation associated with differential survival.

Interbreeding with the two resident congeners, *difficilis* and *magnirostris*, occurs at a frequency of about 1 percent. Two other sources of introgression, conspecific and heterospecific immigrants, are not known to make any contribution, although their effects are difficult to detect because our sample sizes are small. *G. scandens* and *fortis* immigrate in small numbers, but they have never been known to stay to breed. Conspecifics may immigrate from Española and Gardner very rarely, but they have never been identified on Genovesa.

Nine birds were identified as probable hybrids by their measurements. Their effect on variation in bill depth was calculated to be an increase of 20 percent in the coefficient of variation and an increase of 35 percent in the variance. Hybrids reproduced successfully, but they did not live long and contributed less than average to total breeding.

Stabilizing selection either did not occur or occurred weakly and was not detected. Directional selection was detected on several occasions. It occurred mainly in the second half of the study, when a change in food supply arising from the El Niño conditions of 1983 favored birds with beaks of the appropriate shape to exploit the food resources that were relatively common during an ensuing drought. The net effects of selection reflect a combination of direct and indirect effects on each beak dimension. In the first phase of the drought, when *Opuntia* flowers and fruits declined drastically, relatively long-billed birds were at a selective disadvantage. In the second phase, when the major feeding activities required tearing and ripping of the substrate, large birds with deep beaks were favored. Hybrids were at a selective disadvantage during this long dry period.

Variation was maintained by the opposing processes of introgression and selection, but not in a state of balance. The effects over a decade of changing directional selection were perhaps equivalent to overall stabilizing selection. These effects were stronger than the enhancing effects of hybridization. By the end of the study a significant reduction in the variance of beak depth and beak width had taken place among females, and nonsignificant differences in the same direction were apparent among the males.

G. magnirostris is highly variable for the same reasons, we suggest, that apply to *conirostris*. *G. difficilis* is less variable. One probable reason is that selection strongly counteracts any tendency to drift away from an adaptive peak that is narrowly specified by the dry-season feeding niche. Another possible reason is that effects of introgression are weaker in this species than the others because the population size always remains relatively large.

10

Population Subdivision and Sympatric Speciation

Introduction

The allopatric model of speciation is widely accepted as the most probable mode of speciation for vertebrate organisms (Mayr 1963; Futuyma 1986). Naturalists are not entirely convinced that it is sufficient to account for all the taxonomic diversity they see, principally because they would have to invoke an improbable number of range splittings and rejoinings to explain the existence of many closely related species in local assemblages. They are led, therefore, to consider other possibilities, including sympatric speciation. We have been led to consider the possibility of sympatric speciation for different reasons (B. R. Grant and P. R. Grant 1979, P. R. Grant and B. R. Grant 1989).

When we began this study in 1978, the population appeared to be ecologically subdivided into two groups, and there were hints of a reproductive subdivision too. Subdivision is incipient speciation, or at least it provides the potential for speciation. Subdivision can arise in two ways. The population could be in the secondary contact phase of allopatric speciation, in which the descendants of immigrants from a partly differentiated conspecific population are in the process of diverging, under selection, from residents both ecologically and reproductively (e.g., Grant 1986a). Arguing against an immigration hypothesis is the fact that the nearest conspecific population is situated very far away at the south of the archipelago on Isla Española and its satellite Gardner (Fig. 1.2). Moreover, there is no evidence among recently discovered fossils that the species has been present on the intervening island of Santa Cruz in the past two thousand years (D. W. Steadman 1985, pers. comm.). This forces us to confront the sympatric alternative: *in situ* origin of the differences, with a stage having been reached on the way to full sympatric speciation.

Maynard Smith (1966) developed the first genetical model to show how speciation could, in theory, occur sympatrically. He demonstrated, with a simple one-locus, two-allele model, how a stable polymorphism can exist in a heterogeneous environment with two niches, even when the adults form a single randomly mating population. This is the first step towards sympatric speciation. The second step is the subsequent evolution of reproductive isolation between the populations in the two niches. The conditions for this to happen are stringent. The least likely to be realized involve pleiotropic effects of the niche-adapting alleles, effects which directly govern mate choice. A more likely condition is habitat (or niche) selection, with individuals breeding only in the habitat (or niche) to which they are adapted. Alternatively, assortative mating could be under the control of alleles at another locus with either modifying effects on the niche-adapting alleles or independent effects. Maynard Smith (1966) concluded, "Whether this paper is regarded as an argument for or against sympatric speciation will depend on how likely such a polymorphism is thought to be, and this in turn depends on whether a single gene difference can produce selective coefficients large enough to satisfy the necessary conditions" (p. 649).

The selective coefficients can be made large in models (Dickinson and Antonovics 1973; Pimm 1979; Rice 1984) by setting up conditions that result in low fitness of the heterozygotes at or near polymorphic equilibria (Pimm 1979; Loeschcke and Christiansen 1984; Wilson and Turelli 1986). Thus the first step of developing a stable polymorphism, which Maynard Smith (1966) regarded as crucial, poses no special theoretical difficulties.

The second step, the evolution of reproductive isolation, is equally crucial yet less well investigated theoretically (Futuyma and Mayer 1980; Felsenstein 1981). It is generally thought to require a genetic association between traits conferring adaptation to a niche and traits causing mating preferences. The association may be created by strong linkage disequilibrium, but this is opposed by reassortment of alleles at different loci through recombination (Maynard Smith 1966; Felsenstein 1981; Rice 1984). Alternatively, pleiotropy may be involved. Either individuals breed only in the habitat to which they are adapted (Rice 1984, 1987), a passive form of reproductive segregation, or they actively choose to breed with similar types who are adapted to the same niche as themselves (Seger 1985). A different class of models was introduced by O'Donald (1960) to consider assortative mating of phenotypes (and genotypes) without explicit attention being given to ecological factors such as resources or habitats. Since a gene for assortative mating can increase to fixation sympatrically (Endler 1977), speciation could be the outcome of sexual selection alone (Lande 1981; West-Eberhard 1983; Lande and Kirkpatrick 1988).

In nature the evidence for sympatric speciation is ambiguous. On the one hand, events leading to speciation in sympatry can be plausibly reconstructed by using ecological, behavioral, and genetic data from a few, unusually suitable organisms, mainly insects (Bush 1975; Tauber and Tauber 1977a, b; Tauber et al. 1977; Gibbons 1979). On the other hand, even these well-investigated examples can be explained alternatively in terms of the classical model of allopatric speciation (Hendrickson 1978; Futuyma and Mayer 1980), and it is nearly impossible to clearly reject one mode of speciation as an explanation in favor of the other. In response to this dilemma, some biologists stress genetic rather than geographic criteria for classifying modes of speciation (Templeton 1981), although this does not eliminate the problem of identifying and understanding the causes of sympatric divergence.

Those causes are best investigated with polymorphic species that show current signs of splitting into two, which is just what was indicated by our initial observations of *conirostris* in 1978. In this chapter we will review the relevant evidence that has been presented in several preceding chapters for different purposes and integrate it with new information in order to reach conclusions about the nature of the population subdivision and the likelihood of it leading to speciation.

The Initial Observations

ECOLOGICAL ISOLATION

We made the surprising discovery in 1978 that males of the two song groups differed in bill length (Table 10.1). Had the difference been statistically significant at the 5 percent level, we would have regarded it with skepticism, because the comparison is slightly biased by a lack of complete independence of the samples. Given the cultural inheritance of song, local settlement, long life of many of the breeders, and general mate fidelity, members of each group of males are likely to be more related to each other by descent than to members of the other group. Nevertheless the

Table 10.1 Morphological Characteristics in 1978 of Males That Were Separated into Two Groups on the Basis of Song Type (*A* or *B*)

Trait	Type *A* (*N* = 11)	Type *B* (*N* = 10)	Statistical Test of Difference in Means	
			t	*P*
Mass, g	25.85 ± 0.52	24.96 ± 0.73	1.01	> 0.1
Wing length, mm	78.3 ± 0.7	77.7 ± 0.6	0.62	> 0.1
Tarsus length, mm	22.67 ± 0.23	22.33 ± 0.21	1.08	> 0.1
Bill length, mm	15.47 ± 0.14	14.55 ± 0.20	3.87	< 0.002
Bill depth, mm	10.79 ± 0.18	11.02 ± 0.36	0.58	> 0.1
Bill width, mm	9.99 ± 0.45	10.23 ± 0.23	0.89	> 0.1

Source: B. R. Grant and P. R. Grant (1979).

Note: Means and one standard error are shown. Abbreviations refer to grams (g), millimeters (mm), sample size (*N*), Students *t* value (*t*) and probability that the difference in means between song groups is due to chance (*P*).

unadjusted probability of obtaining a difference of the observed magnitude by chance was 1 in 500. The average bill length of song *B* males was 6.3 percent smaller than the average for song *A* males. Other bill dimensions did not differ significantly, although if we forget the sampling errors for the moment it appears that song *B* males were not only shorter in the bill but deeper and wider. Sampling errors cannot seriously be dismissed, but the opposite trends in the bill dimensions suggest a difference in bill shape between the two groups of males (Fig. 10.1). Females that were mated to the two sets of males showed no parallel variation. Their measurements, separately or combined, were statistically indistinguishable from those of the song *B* males.

The difference in bill length between the two groups of males was found to be correlated with a difference in diets. Song *A* males spent relatively more time feeding on *Opuntia* cactus flowers, whereas song *B* males spent relatively more time feeding on rotting *Opuntia* pads on the ground where they obtained the larvae and pupae of flies and beetles. The frequencies of the two groups of males feeding in these two ways differed significantly (Fisher's Exact test, $P = 0.006$).

On a return visit to the island in the dry season (November) of that year, we had difficulty in finding banded males whose song type had previously been registered and who were also seen feeding; they were almost entirely

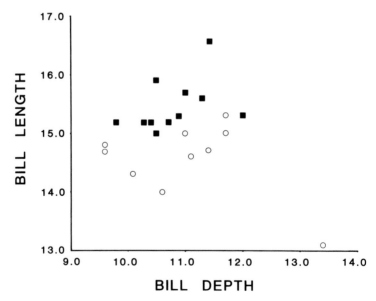

Figure 10.1 Beak characteristics of song *A* males (solid square) and song *B* males (open circle) in 1978. From Grant (1983b).

silent then and often cryptic in their feeding. Eight males from each song group were observed to feed on *Opuntia* flowers (Fig. 9.11) and buds. The two groups differed strikingly in two other respects. Five males, all song *A* type, were observed to feed from fruits on cactus bushes by hammering a hole in the fruit with the bill and eating the pulp from around each seed (Fig. 9.9). Six males, all song *B* type, were observed ripping open cactus pads in just the manner they exhibited earlier in the year (Figs. 9.8 and 9.9). It seems extraordinary that no song *A* males ripped open pads and no song *B* males opened fruits, since the difference in bill dimensions, though statistically significant, is not profound. Nevertheless the observations reject the hypothesis that the two groups of males used the three cactus-feeding activities with equal frequency ($\chi^2_2 = 11.48, P < 0.005$). The associated bill difference between the two groups of males is functionally related to the foraging differences; a long bill is advantageous in probing, and a short and relatively stout bill is advantageous in crushing or tearing large and hard food items (see Chapter 9, specifically Fig. 9.13).

Females were observed feeding less often. In both seasons there were no significant differences in foraging between the two groups of females mated with the two groups of males, or between either female group and their respective male groups.

GENETIC BASIS OF THE SUBDIVISION

The high heritability of beak length implies that the two groups of males, differing in mean phenotype, differed also in mean genotype.

A second indication of a genetic difference was provided by their nestlings. The frequency of the yellow morph (Plate 1) among the 59 nestlings produced by the *A* males was 36 percent, whereas it was only 18 percent among 79 nestlings produced by *B* males. The proportions are different ($\chi^2_1 = 4.79, P < 0.05$). The mean proportion of yellow morphs is significantly higher for the seven *A* males than for the nine *B* males (data transformed to arcsin $\sqrt{\text{percentages}}$, $t_{14} = 2.14, P = 0.05$). As discussed in Chapter 8, there is a fairly simple genetic basis to the nestling beak color polymorphism, so a difference in morph frequency between the families of *A* males and *B* males is likely to reflect a genetic difference between the two sets of fathers or mothers or both.

REPRODUCTIVE SUBDIVISION

We did not witness pair formation in 1978 and had no direct evidence of a reproductive separation between the two song groups of males. However, the pattern of territory arrangement (Chapter 6) suggested the possibility of some degree of reproductive isolation.

No two mated males of the same song type had adjacent territories (Fig. 6.6). In contrast, the few unmated males showed no such regularity. Their territories were next to males that sang either the same song type as themselves (homotypic) or the opposite one (heterotypic) or both. Among the mated males the ratio of 11 boundaries between heterotypic neighbors to 0 boundaries between homotypic neighbors cannot be attributed to chance. The territories of unmated males were randomly distributed with respect to song type of neighbors.

Most importantly, the difference between the territory patterns of the mated males and the unmated males is statistically significant. The difference could have arisen from female choice of particular males, as elsewhere in the archipelago our observations of pair formation in Darwin's Finches had indicated a large element of female choice of males (Ratcliffe and Grant 1983a, b, 1985). Females could have actively chosen to pair with males who had heterotypic neighbors, as we discussed in Chapters 6 and 7.

Thus the spatial nonrandomness implies that females are responsive to the differences in song types as well as to the particular song type of neighbors. We gained no insights that year into why females should pay attention to the neighbor of a potential mate, but an influence of song type on mate choice opened up the possibility of assortative mating with respect to paternal song type. And if assortative mating was taking place, the population was at least partly subdivided reproductively.

Subsequent Observations

Several questions raised by our initial observations could only be answered by following the fates of offspring of known parents. To take one example, in order to answer the question: "Is mating assortative with respect to song type?", we had to find out if sons sing the same song as their fathers and if daughters pair with males who sing the same song as their own fathers.

SONG AND MATING

Some facts are consistent with the idea that the population is structured along the lines of song. As outlined in Chapter 6, sons do sing their fathers' song type and even reproduce much of the fine structure in their own songs (B. R. Grant 1984). Thus there is a continuity between generations in the structure of the population determined by song, although the continuity is not perfect because the acquisition of song characteristics through imprinting is subject to error. Females are responsive to the difference between song types. This was revealed by experiments involving the playback of tape-recorded songs, which encouraged us to expect that when females first pair they choose males who sing the same song type as their fathers.

The expectation was not realized, however. Daughters do not initially pair assortatively with regard to paternal song type (Chapter 6). No temporal pattern of regularity occurs either, such as the first few females all pairing assortatively and the later ones all pairing disassortatively.

It is hard to escape the conclusion that the choice of a male by a female is independent of any conditioning influence of her father's song type. Moreover there is no evidence of assortative mating by bill size. In none of the years was there a significant parametric correlation between males in bill size or any other measured morphological trait. In none of the years did the females mated to song *A* males differ significantly in any of the measured morphological traits from those mated to song *B* males. Hence there is no direct evidence of reproductive subdivision in the population.

A less likely possibility is that the offspring fit enough to survive and breed come predominantly from the assortative matings with respect to song type. If this did occur, it would effect a reproductive subdivision of the population in the face of random mating. If there is any trend, it is in the opposite direction. Out of eight recruits from known parents and grandparents, six were produced by mothers who had paired disassortatively with respect to paternal song type.

MORPHOLOGY AND ECOLOGY

The random mating of females implies that the difference in bill size between the two song groups should have disappeared in subsequent generations, the rate depending on the degree to which bill size is heritable. In line with this reasoning, the difference in average bill length between the *A* males and the *B* males did disappear after 1978 (Fig. 10.2) as a result of recruitment to both groups of young males. If their mothers had mated in the same way as the group of birds whose initial pairings were known, that is, randomly, half of them would have had maternal grandfathers of the opposite song type from their fathers' song type.

With the breakdown of the morphological subdivision of the population, the ecological subdivision disappeared too. In the dry season of 1981, for example, the different methods of feeding on cactus resources were still associated with different beak sizes and shapes in the same manner as observed in 1978, but the association of each of them with song type was no longer present. Finally, the alternating pattern of territories of mated males disappeared in 1979, reappearing weakly at the beginning of the breeding seasons of 1980 and 1982 (Chapter 6). The regular pattern disappeared because new recruits established new territories next to previously established ones (Fig. 6.6), and some females chose as mates males with homotypic neighbors. It did not disappear because males reestablished their territories in different places; once having bred, males hold their territories for life. This last fact, combined with the long life of these birds, was responsible for the brief reappearance of the nonrandomness in the territory boundaries of mated males.

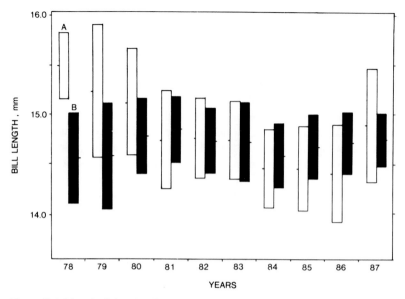

Figure 10.2 Mean beak lengths of song *A* males (open bars) and song *B* males (solid bars). The vertical bars show 95 percent confidence limits, and the means are indicated by short horizontal lines.

Note from Figure 6.6 that territories are interchangeable between males of the two song types: the territory of an *A* male one year may be occupied by a *B* male in another year, after the *A* male has died. There is no systematic association between males of a song type and habitat features, and therefore there is no spatial isolation of the two groups other than that provided by territoriality itself. This applies to the nonbreeding (dry) season as well as to the breeding (wet) season.

    In short, the signs of subdivision, both ecological and reproductive, which we had observed in 1978, subsequently disappeared.

How the Population Became Subdivided

The change in population structure gave us the same disadvantage we experience when attempting to reconstruct the events leading to speciation, namely the processes of interest had already occurred. In our case the subdivision of the population occurred prior to our study. In the year immediately preceding the study (1977), very dry conditions prevailed (D. Day 1978, pers. comm.; Grant and Grant 1983). To judge from events on Isla Daphne Major at this time (Boag and Grant 1981, 1984) and the absence of males in black 0–2 plumage on Genovesa, there was probably no (successful) breeding and substantial mortality (Grant and Grant 1983).

The stressful conditions of 1977 may have created the structure in 1978 in some way.

The important question to answer is how the beak size difference came to be associated with the song difference in males.

RANDOM EFFECTS

One possibility is that it was simply a matter of chance. The original study area constitutes 10 percent of the island area that is occupied by breeding cactus finches. Events in the study area may not be typical of the island as a whole, and chance is likely to play a greater role in small samples than in the total population. By chance, heavy mortality in 1977 may have been biased with respect to beak size in the two groups of males.

Two methodological checks failed to turn up evidence of bias of this type. In 1987 we did what we should have done in 1978; we walked around the island and censused birds to see if the proportions of *A* and *B* males in the study area differed from those outside. They did not (Table 10.2). This does not directly address the question of whether the beak size difference between the song groups in 1978 was representative of a population-wide difference, but at least it gives no indication that the study area and the finches in it are atypical. In vegetation also the study area is representative of the whole island (Grant and Grant 1980; Hamann 1981).

The second check was on the beak difference itself. To ascertain that the difference was real and not the result of measurement error or of different regimes of growth and abrasion experienced by the two groups of males, we recaptured and remeasured as many as possible of the original 21 males in the years 1979 to 1981 and compared the original and second set of measurements. No differences were found between them (Grant and Grant 1983) or between sets of measurements of five recaptured females. The recaptured *A* males (4) and *B* males (4) differed significantly in bill length at second measurement ($P < 0.01$; Grant and Grant 1983), just as the same individuals did at first measurement ($P < 0.01$). Repeatability of

Table 10.2 Frequencies of Males That Sang Song *A*, Song *B*, or Both *A and B* in the Main Study Area (Area 1) and Outside in March 1987

Males	Study Area	Outside
Song *A*	0.32	0.33
Song *B*	0.66	0.64
Song *A* and *B*	0.02	0.03
Total individuals	38	124

Note: The main study area constitutes 10 percent of the island area occupied by breeding cactus finches. The censused area outside, on the west, north, and east coasts and in the center, including area 2, amounts to an additional 23 percent.

these measurements is fairly high (Table 8.1), thus the bill differences were real and not an artifact of sampling.

While random effects cannot be dismissed (see also Appendix IV), we have sought an explanation for the association between song type and bill size in terms of the properties of the birds revealed in the first few years of the study.

SYSTEMATIC EFFECTS AND THE DEVELOPMENT OF A HYPOTHESIS

In 1983 we reasoned that if song is used principally in the breeding season, then an association between bill size and song type is likely to arise at that time (Grant and Grant 1983). It could arise, for example, if females of a particular morphology choose to pair with males of a particular song type, for then the male offspring would inherit some of the alleles which govern that particular morphology from their mother and acquire the song from their father. From this reasoning we developed the following hypothesis to account for the bill length–song type association observed in 1978.

Large females are postulated to come into breeding condition earlier than small females, and they are the first to choose mates. They pair to a disproportionate extent with males that have heterotypic neighbors. Males which sing the A song have heterotypic neighbors more frequently than do the B males simply as a result of their relative scarcity; hence not only do they have a frequency-dependent mating advantage, they tend to pair with the largest females. The slight tendency for males of the two groups in the next generation to differ in size is enhanced by disruptive selection in the dry season. Disruptive selection occurs on beak size; the longest-billed and the shortest-billed males are favored over intermediates as a result of the feeding advantage each experiences in the two cactus niches of fruit and pad exploitation. Long-billed birds come predominantly from the A males, and short-billed birds come predominantly from the B males, since bill size is strongly correlated with body size (Table 8.2). In a drought, such as occurred in 1977, disruptive selection is stronger, with the result that the association between bill length and song type is stronger in the following year (e.g., 1978).

Retrospective hypotheses are satisfactory to the extent that they are consistent with the known facts. They are useful to the extent that they predict events which can be tested against future observations. We will discuss these two features in turn.

CONSISTENT FACTS

The three most important facts consistent with the hypothesis are as follows. First, large females paired with new males significantly earlier than small females in 1980 and 1981 (Grant and Grant 1983). Nonsignificant tendencies in the same direction were observed in the other three breeding seasons for which we have sufficient data: 1982, 1986, and 1987. Second, males with heterotypic neighbors had a mating advantage over those without heterotypic neighbors in each of the years 1980–82 (B. R. Grant 1985). This may have come about largely through female preference for experienced males (Chapter 7), with a direct preference for males with heterotypic neighbors being a possible, but secondary, additional factor (B. R. Grant and P. R. Grant 1987).

Third, there is evidence of disruptive selection, although it is indirect. We first note that the frequency distribution of bill lengths in Figure 10.1, with males of the two song types combined, gives no hint of the bimodality we might expect if a normally distributed character had been subjected to disruptive selection. On the other hand, for small samples such as ours bimodality would be likely only if selection had been extremely intense. If disruptive selection had occurred in 1977, the variance in bill size of the survivors in 1978 would have been greater than the variance of the total sample in 1976 before the drought. We lack information from 1976 to make the direct comparison, but, as explained in Chapter 9, if we assume that adults in 1982 had the same characteristics as their predecessors in 1976, we can compare the 1978 adults with the 1982 adults. The comparison showed that the 1978 survivors were indeed exceptionally variable in bill length.

Although these facts are consistent with the hypothesis, they have alternative interpretations. For example, the change in variance may have been brought about by hybridization, or it may have been produced by both hybridization and disruptive selection. Tests of the hypothesis are needed to distinguish between these alternatives.

PREDICTIONS FROM THE HYPOTHESIS

Droughts occur in the Galápagos archipelago on average about once a decade (Grant and Boag 1980). We were fortunate to witness a complete drought in 1985; no rain whatever was recorded in the normal wet season months of January to May, and birds did not breed. This should have been the ideal circumstance for observing the events which previously had only been hypothesized to occur, because we knew the sizes and breeding histories of many individuals in the preceding year (1984). The resulting characteristics of the population in 1986 should have been the same as those we encountered in 1978.

In fact, disruptive selection did not occur in 1985. Variances did not change, and song *A* and song *B* males did not differ in bill length or in any other measurement in the following year (Fig. 10.2).

The reason was apparent even before the drought began. Its origin was in the El Niño event two years earlier. As described above, one of the consequences of the exceptional rainfall was an extreme growth of vines, principally *Ipomoea habeliana*. They smothered the cactus, and many bushes died (Figs. 2.16 and 2.17). The poor condition of the surviving cactus continued through the dry year of 1984 (57 mm total rainfall) and into 1985. Very little flowering occurred, and to a rough approximation it may be said that one of the two cactus niches had disappeared. So although the other niche, in the form of insect larvae and pupae in rotting pads, was abundantly present, the conditions for disruptive selection to occur were not. And, in fact, directional selection occurred at this time (Chapter 9), whereas when two food niches were present in the drought of 1977 disruptive selection may have occurred. This shows that different droughts do not necessarily cause the same patterns of mortality (cf. Chapter 3).

A CONTRIBUTING FACTOR: THE FAMILY EFFECT

Since sons sing the same song as their fathers, resemble them in beak characteristics, and tend to settle close to where they were born, males of a given song type that survive a drought and then breed in the local area may be close relatives. We refer to this as the family effect because it seems likely that close relatives would survive well under the same conditions, when beak size and shape and associated feeding skills are important determinants of survival. The family effect, while not restricted to fathers and sons, may have contributed to the difference in beak lengths between *A* males and *B* males in 1978. It constitutes a subtle form of pseudoreplication that biases the statistical comparison of the groups (D. Schluter 1988, pers. comm.; see also p. 237).

The evidence for this effect is mainly circumstantial because relatedness of the 1977 drought survivors is not known. It amounts to the following: One of the four survivors of the 1977 drought depicted in Figure 4.10, No. 6980, survived the next drought of 1984–85, as did the son (No. 6137) of another with whom she bred. After the drought of 1984–85, when relatedness of some of the survivors was known, one son bred in a territory separated from his father's by one intervening territory. In addition, two daughters of another father who also survived settled to breed four and six territories away respectively. At another time (1987), we have known one son (No. 8057) to breed next to his father (No. 6137). Thus the argument is supported by these observations, but they are too few to be expressed quantitatively.

Sympatric Speciation?

The population was partly subdivided ecologically for a brief period and only once in the decade. We understand how the subdivision collapsed through the combined effects of random mating with respect to song and the transmission to the offspring of parental features of song and beak morphology. We understand less well how the subdivision was created. A hypothesis to explain it in terms of a frequency-dependent mating advantage of the rarer males and disruptive selection in the dry season could not be tested because the environment changed.

Thus the population is not currently undergoing a split into two species. There are two interrelated reasons for this. First, despite an occasional subdivision of the population, the dry-season niches are not different enough to support two very different feeding types. Second, there is neither a detectable deviation from random mating nor an identifiable factor that would favor it. Mate choice is not genetically correlated with the use of a particular niche.

In our decade-long study we witnessed a fraction, of unknown magnitude, of the range of conditions experienced by the finches. Perhaps over a time span of many decades, centuries, or even millennia, the right combination of conditions for promoting speciation could occur. As a prelude to a discussion of what that combination might be, we will first summarize the main features of coexisting species and how they are believed to have typically evolved.

All sympatric species of ground finches differ by at least 15 percent in at least one bill dimension (Grant 1981b, 1986a). Since their niche differences are directly proportional to their bill differences (Abbott et al. 1977), there is a minimum ecological difference between them too. These differences are believed to have evolved under directional selection, entirely in allopatry in some cases and partly in allopatry and partly in sympatry in others (Grant 1981b, 1986a). Reproductive isolation evolved as a correlated effect of the ecological and morphological divergence (Grant 1986a). Sympatric species do not normally interbreed because they discriminate between conspecific and heterospecific potential mates at least partly on the basis of their bill size and shape (Ratcliffe and Grant 1983a).

The two groups of *conirostris* males differed at most by only 6 percent in our study. This is probably too small to foster any discrimination in a mating context. Admittedly in the secondary contact phase of allopatric speciation, incipient species are likely to differ initially in some cases by less than the final amount of 15 percent or more, but 6 percent falls so far short of the minimum apparently required for coexistence that something else would have to cause further divergence to make speciation possible. Furthermore, one of the dry-season food niches was observed to decline catastrophically, reminding us that a degree of temporal stability would be needed for further divergence to occur.

The ''something else'' required for further divergence would have to be another feeding niche not presently exploited, or seldom exploited, and differing markedly from the other two. Two candidates are easy to identify.

One comprises the many types of small seeds produced by grasses and other herbs, and the small quantities of nectar in the small but plentiful flowers of such species as *Waltheria ovata* (Fig. 2.9). The other comprises the large and hard seeds of *Opuntia* cactus and the even larger and harder seeds of *Cordia lutea* (Plate 5). Mature *Opuntia* seeds are exploited rarely by the cactus finch individuals with the largest bills, whereas *Cordia* seeds appear to be too hard for any of them to crack. These two niches are easy to identify because they are currently occupied, the former (small seeds) by *difficilis* and the latter (large seeds) by *magnirostris*.

Thus one factor of great importance suggested by our study, as well as by some theoretical investigations of the speciation problem (e.g., Maynard Smith and Hoekstra 1980; Wilson and Turelli 1986), is the way in which the environment is heterogeneous. Only when the heterogeneity is strong, the range of available food resources is broad, the difference between the extremes is large and persistent, and population variation is also large, will disruptive selection be strong enough to produce the divergence that can lead to speciation (P. R. Grant and B. R. Grant 1989). These conditions could arise if one of the congeneric species on Genovesa becomes extinct. The conditions may have been present in the past if the arrival of the ancestral cactus finches had preceded one of their congeners. Variation among cactus finches would have been large initially only if there was interbreeding with another congener at that time.

A second factor of potential importance is a genetic one. Genetic drift may do what selection does not or cannot do. Chance association of alleles under sustained conditions of low population size (Wright 1980) may result in a coupling of niche adaptation and selective mating in a heterogeneous environment from which, under selection, speciation could proceed. The population may have remained small for a long time more than three thousand years ago when the climate was drier than now (Chapter 2). Even under present conditions of strongly and erratically oscillating climate, a long series of drought years and sustained low population size are not out of the question.

Thus the right combination of conditions for sympatric speciation to occur may be a very rare sequence of dry years and the absence of congeners. These conditions are most likely to have been present in the early stages of the adaptive radiation of Darwin's Finches.

Population Subdivision

We conclude this chapter with a return to the secondary theme of a subdivided population. There may be many ways in which the population is structured. We can easily recognize a clear subdivision along the lines of song. The two groups of males differed not just in song but to a small degree in gene frequencies as well.

FREQUENCIES OF COLOR MORPHS AMONG NESTLINGS

In 1978 yellow morph nestlings were proportionately more common among offspring produced by type *A* males than among those produced by type *B* males. In 1980 the difference was even stronger ($\chi^2_1 = 9.72$, $P < 0.01$). The relatively frequent occurrence of yellow morphs among broods of the *A* males continued without exception for the remainder of the study. Proportions fluctuated among years (Fig. 10.3), but in the last year the difference was statistically significant ($\chi^2_1 = 5.35$, $P < 0.05$) and almost the same as in the first year of study.

These observations were made in study area 1. In 1980 we expanded the study to include area 2 (Fig. 3.1) and there found a nonsignificant difference in the same direction. Thereafter opposite trends developed in the two study areas (see Table A8 in Appendix V, and Fig. 10.3). On area 2 the *B* males became the most frequent producers of yellow morph offspring, and that trend persisted without exception from 1981 until the

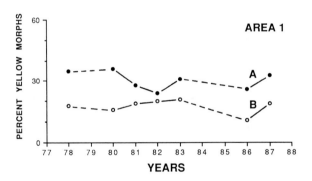

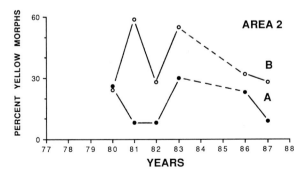

Figure 10.3 Fluctuations in the proportions of yellow morphs among nestlings of *A* and *B* males on study area 1 (above) and study area 2 (below). Broken lines span years when breeding did not occur or when the nestling colors were not recorded.

end of the study (Fig. 10.3). In two years, 1981 and 1983, the difference in proportions produced by the two groups of males was significant ($P < 0.05$).

The striking feature is not so much the significant difference in some years as the persistence of the trends in each study area and the consistent difference between them after 1980. Thus we have now identified two factors associated with the production of offspring with different beak colors: song type of father and study area. These factors interact.

DIFFERENCES AMONG BREEDING MALES

The persistent trends in the two study areas were determined by two factors: consistent average differences between the A males and the B males in their tendency to produce yellow morph offspring, and the breeding of some of those males in successive years. The differences among males are presumed to be a reflection of genetic differences (Chapter 8). Those differences are shown by characterizing each male by the proportion of its nestlings that were yellow morphs, and then testing the difference between the two groups of males in each area with a t test, having first transformed the proportions to arcsin $\sqrt{}$ percentage values. To make tests in different years independent, we considered each male only once, in the year in which it was first recorded breeding.

In area 1 the higher mean proportion of yellow morphs produced by A males in 1978 ($t_{14} = 2.14$, $P = 0.05$) was repeated among the new breeding males in 1980 ($t_{17} = 3.03$, $P < 0.01$) and again in 1986 ($t_{13} = 2.21$, $P < 0.05$). In area 2 the difference between A and B males in the opposite direction was significant in only one year, 1983 ($t_{13} = 2.76$, $P < 0.02$), but sample sizes were always much smaller there.

The same pattern is manifested when we classify males into those which did produce yellow morph offspring and those which did not. We assume that the two groups of males so classified are genetically different. In all seven years the proportion of yellow producers was higher among the A males than among the B males in study area 1 (Table A8). Considering once again just the new breeding males in each year, we find the difference between the two groups of males to be significant ($P < 0.05$) in three years (1980, 1981, and 1987) and close to being significant ($0.1 > P > 0.05$) in two others (1978 and 1983; Fisher's Exact tests). In study area 2 the proportion of yellow producers was significantly higher among the B males than among the A males ($P < 0.05$) in one year (1983).

Thus both types of analysis indicate genotypic frequency differences between the A males and the B males in each of the study areas in some years, and differences between study areas. We have discussed the polymorphism without reference to the breeding females, even though the genetic control of offspring beak color is as much a function of the mother's as of the father's genotype (Chapter 8). It is conceivable that the two groups of males were genetically the same, and that their offspring differed genetically because their mothers differed genetically. This does not seem likely in view of the persistence of the difference despite frequent mate changes which are random with respect to male song type. Nevertheless female differences would certainly contribute to offspring differences.

SPATIAL HETEROGENEITY

These results indicate a genetic heterogeneity in the population on the spatial scale of the study areas. Echoing the discussion of the bill size differences, we have little idea how the differences in morph frequencies within and between areas arose, even though we witnessed one of them. Possibly it was simply a matter of chance (Appendix IV). The number of birds breeding in the two areas was small, especially in area 2.

We have a better idea of why the differences persisted. With random mating prevailing, the differences might be expected to disappear quickly in the absence of selection. However, three factors retard the rate of homogenization. First is the cultural inheritance of song by sons from their fathers as opposed to neighbors or others. Genes transmitted from father to son stay in the song group. Second is the generally slow rate of turnover of breeders, especially the males (see Chapter 3 and Appendix III). Thus genetic characteristics of a song group remain similar from one year to the next in part because many individuals remain alive from one year to the next.

The third factor is the most relevant to the maintenance of the difference between study areas. It is the tendency for recruits to breed fairly close to their birth place (Chapter 6). This results in a predominant settlement in the local area. An outstanding example of such a family effect occurred in 1983 when six close relatives bred at the same time in the eastern one-third of study area 1: three of No. 6108's five recruits (see Fig. 4.10), the mother of two of them, one grandson, and his mother. Altogether, of 46 recruits raised in area 1, 34 (73.9 percent) settled to breed in area 1. The figures for area 2 are five out of nine (55.5 percent). To this should be added the very strong tendency for females to stay within the local area when changing mates (Chapter 7); only five were known to breed in both areas. Thus initial local settlement and local movement of females in successive matings combine to create an inertia to forces of change induced by breeding patterns and acting on the difference in gene frequencies in the two areas.

Generalizing from the local study area to the whole population, we see a small degree of heterogeneity in space. Cactus is distributed more or less continuously in some areas and sparsely, if at all, in other areas. Patchiness in this vital habitat feature imposes a patchiness on finch distribution. No patch of finches is completely isolated, but some are connected more strongly than others. As a result of this structure and limited dispersal, local groups of finches show semi-independent fluctuations in frequencies of neutral, or near-neutral, alleles. In contrast, adaptive shifts induced by environmental change are likely to be island wide (Chapter 9).

Summary

Models of sympatric speciation, with varying degrees of complexity and realism, show the conditions under which a single interbreeding population can split sympatrically into two. They describe mechanisms for the evolution of niche differences between two segments of a population and for the subsequent evolution of reproductive isolation between them.

In the first year of our study, the population of cactus finches appeared to be ecologically subdivided into two groups, and there were hints of a reproductive subdivision. However the ecological subdivision disappeared subsequently through the effects of random mating. The subdivision may have been created by disruptive selection under the stressful conditions of a drought in the year immediately preceding our study. A second drought occurred in the eighth year of study, but the hypothesis of disruptive selection could not be tested because the environment had changed; one of the two principal dry season cactus foods had been almost eliminated by the smothering effects of extensive vine growth in the exceptionally long and wet growing season two years earlier.

The major lesson we draw from this study is that environmental heterogeneity has to be of a special kind to foster sympatric divergence leading to speciation. The niches or habitats to which different members of the population adapt should be markedly different and display a long-term persistence, although not necessarily a constancy. The right combination of conditions for sympatric speciation to occur may be a very rare sequence of dry years and the absence of congeners.

Regardless of the likelihood of sympatric speciation, the population is not homogeneous and structureless. The environment is somewhat patchy for cactus finches, and this imposes patchiness on their distribution. On the scale of the study areas, the population is genetically heterogeneous, as indicated by differences between study areas in the frequencies of nestling beak colors among the offspring of the two song groups of males. Nestling beak color is probably a neutral trait under simple genetic control (Chapter 8). As a result of this structure and relatively short dispersal distances, local groups of finches show semi-independent fluctuations in frequencies of neutral alleles.

11

The Place of *G. conirostris* in Its Community

Introduction

The dynamic nature of the *conirostris* population variation, both phenotypic and genetic, has interesting implications which we will explore in the final two chapters. The large amount of additive genetic variation implies that the population could shift relatively rapidly under selection to a new mean phenotype. The moderately strong, positive, genetic correlations imply constraints on some shifts. Thus overall size could undergo an evolutionary change in either direction relatively quickly, unimpeded by genetic constraints because all genetic correlations are positive, but an evolutionary change in shape would be much slower owing to the retarding effects of genetic correlations with other traits. These considerations raise the question of why the population on Genovesa is morphologically where it is. For example, why is the average weight about 25 g? Why is the beak nearly 15 mm long and almost half as long again as it is deep? In these characteristics the population is unique on Genovesa, indeed in the whole archipelago.

These questions are both phylogenetic and ecological. To answer them we need to delve into the evolutionary history of the species, as well as inquire into its place in the contemporary ecological community on Genovesa; for either the Genovesa population has reached an environmentally determined optimal phenotype under past regimes of natural selection, or else it has been constrained from reaching it by its genetic architecture, in which case it remains relatively poorly adapted.

Evolutionary History

Little is known about this subject. Fossils of Darwin's Finches have been found on Floreana and Santa Cruz (Steadman 1986), but *conirostris* bones have not been identified among them so far (D. W. Steadman 1986, pers.

comm.). Fossils have not been found on the islands currently occupied by the species.

A comparative analysis of protein polymorphisms by electrophoresis (Yang and Patton 1981) has shown that the six species of ground finches are so closely related to each other it would be impossible at present to reliably infer ancestor-descendant relationships among them. The biochemical characterization of *conirostris* from Genovesa was based on tissues from only four specimens. Taken at face value, the data suggest *conirostris* became a distinct species about 200,000 to 250,000 years ago, having been preceded by the formation of only one other ground finch species, *scandens* (Yang and Patton 1981).

Populations on Española and its satellite Gardner are morphologically similar to each other, but not identical (Lack 1945, 1947; Abbott et al. 1977; Grant et al. 1985). Birds are known to fly between the islands (Gifford 1919; personal observations); the distance is less than one kilometer (see Fig. 11.1). The population on far-distant Genovesa differs markedly from both (Figs. 9.5, 11.2 and 11.3), enough to raise doubts about whether the populations are truly conspecific. These doubts have not been completely laid to rest. The distinctive feature of all three populations, and hence of the species as currently recognized, is a lateral flattening of the bill, which makes it less convex in cross section than in all other *Geospiza* species (Grant 1986a).

The Gardner population is almost certainly derived from the much larger population on adjacent Española, but whether the Genovesa population is ancestral to both or derived from one of them is open to conjecture. It is also possible to argue that all contemporary populations were derived from one or two ancestral populations on central islands which have since become extinct. If *scandens* existed earlier than *conirostris*, the *conirostris* population on Genovesa, which most closely resembles *scandens*, possibly gave rise to the population on Española. Alternatively, the morphologically more distinctive population on Española may be old, and the population on Genovesa may be new and derived, in which case its similarity to *scandens*, particularly on the neighboring islands of Marchena and Pinta, could be explained as being the result of occasional gene exchange or of convergence. There is no independent evidence that would help us to choose between these alternatives. The island of Española is geologically much older than Genovesa (Cox 1983; Hall et al. 1983), but these relative ages are irrelevant if the radiation of the finches began after all major islands had been formed (Grant 1986a). A much better biochemical characterization of nuclear and mitochondrial DNA would help to clarify questions of phylogenetic relationships at both population and species levels.

RECONSTRUCTING THE FORCES OF SELECTION

Despite these uncertainties, we can make some progress toward understanding how one species or subspecies was evolutionarily transformed into another by applying the techniques of quantitative genetics to field data (Price, Grant, and Boag 1984; Schluter 1984b). The minimum forces of directional selection required to transform the Genovesa phenotype into the Española phenotype, or vice versa, are given by the equation (from Lande 1979):

$$\beta = \mathbf{G}^{-1}\Delta\bar{z},$$

where β is the selection gradient we wish to estimate. It is a vector whose entries represent the forces of directional selection acting directly on each character. The influence of character correlations on the effects of selection on each character enter the equation in the term $\mathbf{G}$, which is the genetic variance-covariance matrix and is here inverted. $\Delta\bar{z}$ represents the morphological distance to be traversed between Española and Genovesa phenotypes. It is the vector of mean differences between the characters of the two populations.

For the purpose of calculation it does not matter which population gave rise to which. Let us suppose the Española population gave rise, through the dispersal of a few individuals, to a population on Genovesa. Let us further suppose that variances and covariances, estimated from field measurements on Genovesa, are accurate and have remained unaltered throughout the evolutionary transition. The assumption of constancy is not likely to be strictly correct (Turelli 1988), but small departures from constancy would not alter the main conclusions reached below. The calculated minimum forces of selection under these conditions are given in Table 11.1. The transition requires a vector length (B) of 1.071. This may be compared with the maximum observed, not in the present study but in the companion study of *fortis* on Daphne Major, to give some idea of how many selection events would be required for the full transition to take place. The maximum observed in the Daphne Major study was 0.12 (Price, Grant, and Boag 1984). Therefore nine such events would be required.

It is a matter of guesswork as to how long the process of transformation would have taken. For estimating it, one strong selection event every decade seems too rapid, whereas one each century seems quite likely, yielding estimates of approximately one hundred and one thousand years respectively. These might be considered to bracket the target, although inclusion of other traits in the analysis would make the upper estimate the more realistic one.

But even if selection episodes had occurred at constant intervals, the

Figure 11.1 Española with its satellite island, Gardner. Goat damage to the vegetation has been extensive, causing erosion and the loss of much of the cactus. Cactus has the growth form of a tree, unlike on Genovesa. Photos by P. R. Grant.

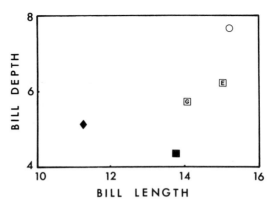

Figure 11.2 Average beak sizes of *conirostris* (open square) on Genovesa (G) and Española (E), and of three other ground finch species, *magnirostris* (open circle) on Genovesa, and *fortis* (solid diamond) and *scandens* (solid square), on Daphne Major. From Grant and Grant (1982).

approach to the new phenotypic mean is unlikely to have been constant. Initially it was fast and then became progressively slower as the retarding effects of genetic correlations among characters became increasingly effective obstacles to further progress toward the new phenotype, unless the correlations themselves changed. Because the approach to the new phenotypic mean was probably initially fast, we used the strong selection event on Daphne Major in the calculations rather than the weak events on Genovesa.

Table 11.1 Selection Gradient (divided by 100) for the Transition between Española and Genovesa Phenotypes, Based on the Genetic Covariance Matrix of the Population on Genovesa

Traits	Forces	$\Delta \bar{z}$
Weight	0.597	0.213
Bill length	−0.519	0.003
Bill depth	0.545	0.203
Bill width	0.473	0.135
Net force	1.071	

Note: Entries in the gradient show the minimum forces of selection on each of four morphological traits involved in the transition. The net force of selection is estimated by the vector length, which is calculated by summing the squared values for the six traits and then taking the square root (Schluter 1984b). The $\Delta \bar{z}$ values give the differences between the logarithmically transformed means of the two populations. Notice that beak depth differs substantially whereas beak length differs hardly at all. Nevertheless, a strong downward force of selection on beak length is needed to overcome the correlated effects on beak length of an upward force of selection on beak depth. It is assumed that beak length is genetically correlated with other traits (values in Table 8.6), even though for some traits correlations have not been demonstrated statistically. If beak length is truly not correlated genetically with other traits, the net force of selection involved in the transition would be smaller.

Figure 11.3 *G. conirostris* from Genovesa (lower) and Española (upper right) compared with *magnirostris* on Genovesa (upper left). Photos by M. P. Harris (upper right), D. Schluter (upper left) and O. Jennersten (lower).

If we assume from the results of biochemical analysis (Yang and Patton 1981) that *conirostris* evolved as a species more than two hundred thousand years ago, and that transformation of one population phenotype (Española) to another (Genovesa) required about one thousand years, or less than 1 percent, then enduring effects of genetic constraints would appear to be minor. Despite inevitable uncertainties at almost every step in our estimation procedures, we can draw the following conclusion. There has probably been ample time, environmental pressure, and additive genetic variation for an approximate alignment of the mean phenotype on Genovesa to be reached with an environmentally determined optimum. Ecological studies illuminate that optimum.

Community Ecology

COMMUNITY COMPOSITION

Before discussing the optimum, we need to make a few remarks about the finch community on Genovesa, because the other finch species have a bearing on the niche that *conirostris* can occupy. The finch community

Table 11.2 Dry-Season Diets of Four Species of Darwin's Finches on Genovesa in January 1988, Expressed as Proportions of the Total Feeding Observations Made for Each Species

Food Category	G. magnirostris	G. conirostris	G. difficilis	C. olivacea
Seeds and fruits				
Small seeds	0.04	0.09	**0.40**	—
Cacabus seeds in capsule	0.04	**0.19**	—	—
Cordia seeds	**0.85**	—	—	—
Cordia arils	—	0.05	—	—
Pollen and nectar				
Waltheria flowers	—	—	**0.39**	**0.33**
Cordia flowers	—	—	0.03	—
Ipomoea flowers	—	0.05	0.01	—
Bursera flowers	—	—	0.01	—
Cryptocarpus flowers	—	—	0.03	—
Chamaesyce flowers	—	—	0.03	—
Arthropods				
In wood	—	**0.21**	—	—
In cactus	—	**0.19**	—	—
On bark and leaves	—	0.07	0.07	**0.62**
On ground	—	0.07	0.03	0.05
Opuntia cactus				
Buds	—	0.05	—	—
Flowers	—	0.02	—	—
Fruit	—	0.02	—	—
Pulp	0.07	—	0.02	—
Minimum number of birds	22	35	100	18
Number of observations	27	43	111	24

Note: Principal diet categories are shown in **boldface**. Observations were made by G. Rosenqvist and P. T. Boag (1988, pers. comm.). Compare with Table 11.3.

Figure 11.4 *G. magnirostris* cracking the hard stone of *Cordia lutea*, an important food in the dry season. Photo by O. Jennersten.

comprises three ground finch species and the warbler finch (Plate 10; Figs. 2.18 and 2.19). The ground finch species are clearly distinct in morphology (Fig. 9.1). There are small (*difficilis*), medium (*conirostris*), and large (*magnirostris*) species. They are ecologically distinct in a corresponding manner (Table 11.2). *G. magnirostris* is the only species on the island able to crack the large and hard stones of *Cordia lutea* and consume the kernels (Fig. 11.4). *G. conirostris* exploits *Opuntia* cactus comprehensively, as well as removing arthropods from beneath bark. *G. difficilis* feeds on small seeds and on the nectar and pollen of small flowers (*Waltheria ovata*; Fig. 11.5). The warbler finch shares this last resource with *difficilis*, but is principally a gleaner of small arthropods (termites, flies, spiders, etc.) from the surfaces of leaves, branches, and the trunks of trees. Thus ecological segregation of the four species is pronounced, especially in the dry season (see also Grant and Grant 1980).

The niche of *conirostris* is illuminated by a comparison with finches on the neighboring island of Marchena (see Fig. 1.2). Marchena has seven finch species. An important factor here is its greater size, even after allowing for the fact that large areas of this island are covered by recently formed, bare lava (see Fig. 13 in Grant 1986a). Floristically it is similar to Genovesa (Fig. 11.6). It is more than four times as high, but this is not

Figure 11.5 Nectar in the flowers of *Waltheria ovata* is an important food in the dry season for *difficilis* (upper) and *Certhidea olivacea* (lower). Photos by O. Jennersten.

Figure 11.6 Deciduous forest habitat on Marchena, similar to the habitat on Genovesa (see Fig. 2.8), except for extensive areas of bare lava. Photos by K. T. Grant (middle) and P. R. Grant (upper, lower).

Table 11.3 Dry-Season Diets of Seven Species of Darwin's Finches on Marchena in January 1988, Expressed as Proportions of the Total Feeding Observations Made for Each Species

Food Category	Geospiza magnirostris	Geospiza fortis	Geospiza fulginosa	Geospiza scandens	Camarhynchus psittacula	Platyspiza crassirostris	Certhidea olivacea
Seeds and fruits							
Small seeds	**0.27**	**0.45**	**0.76**	—	—		—
Medium seeds	0.14	0.02	—	—	—	0.03	—
Cordia seeds	**0.18**	—	—	—	—	—	—
Rynchosia fruits	**0.32**	**0.49**	**0.18**	0.25	—	0.17	—
Croton fruits	—	—	—	—	—	**0.50**	—
Cambium							
Croton	—	—	—	—	—	0.15	—
Pollen and nectar							
Waltheria	—	—	0.05	—	—	—	**0.52**
Bursera	—	0.01	0.01	—	—	0.15	—
Arthropods							
In wood	—	—	—	—	**1.00**	—	—
In cactus	—	—	—	**0.75**	—	—	—
On bark and leaves	—	—	—	—	—	—	**0.48**
Opuntia cactus							
Pulp	0.09	0.03	—	—	—	—	—
Minimum number of birds	20	82	90	4	21	40	50
Number of observations	22	87	95	4	21	41	52

Note: Principal diet categories are shown in **boldface**. Compare with Table 11.2.

sufficient to cause altitudinal variation in plant composition to a significant degree.

The two species common to Genovesa and Marchena, *magnirostris* and *C. olivacea*, feed in the same distinctive ways on the two islands (Tables 11.2 and 11.3). *G. fuliginosa* on Marchena is the counterpart of *difficilis* on Genovesa both ecologically and morphologically (Schluter and Grant 1982, 1984b). Like *difficilis*, *fuliginosa* feeds mainly on small seeds and the nectar and pollen of *Waltheria ovata*, to an even greater extent than is shown in Table 11.3 (Schluter and Grant 1982). This leaves *conirostris* as the odd one out on Genovesa, for it has no single counterpart on Marchena or elsewhere in the archipelago (Lack 1947, 1969). In fact, it combines the niches of three species on Marchena: it feeds on cactus products, as does *scandens* (Fig. 11.7); it feeds on small and medium-sized seeds such as those of *Cacabus* and *Bursera* (Chapter 9), as does *G. fortis*; and it feeds on cryptic arthropods beneath bark, as does *Camarhynchus psittacula*, the large tree finch (Fig. 11.8).

The odd one out on Marchena is *Platyspiza crassirostris*, the vegetarian finch (Fig. 11.9). It has no counterpart on Genovesa, and the reason for its absence on that island is not obvious. It feeds often on buds and leaves as well as seeds (Bowman 1961; Downhower and Racine 1976), and it has the unique habit of stripping the bark off developing twigs of *Croton scouleri* to extract the sugar-rich phloem and cambium (Table 11.3). When we made our observations on Marchena in January 1988, its major dietary component was fruits from this plant. At this time on Genovesa *Croton* plants lacked fruits, as did those of another important plant on Marchena, *Rhynchosia minima*. This provides the only clue about how food supply might determine the distribution of *Platyspiza*; that is, how *Platyspiza* might survive the dry-season food shortage on Marchena but not on Genovesa. It is also possible that *Platyspiza* has never reached Genovesa.

In summary, the four species of finches on Genovesa are morphologically and ecologically distinct. They exploit approximately the same resources as seven species do on Marchena. *G. conirostris* on Genovesa combines the feeding niches of three of the Marchena species, *scandens*, *fortis*, and *Camarhynchus psittacula*. It is thus ecologically different from the three sympatric species and more similar to three allopatric ones. These facts are relevant to the niche on Genovesa and the mean phenotype for exploiting it. They show where constraints on the niche lie, arising from competitive interactions with the three sympatric species. They also show where competitive constraints are lacking. Absent species may have been competitively excluded in the past (Lack 1947; Grant and Grant 1982).

PREDICTING THE MEAN PHENOTYPE

Food supply governs the abundances of finches and their beak sizes (Abbott et al. 1977; Smith et al. 1978). To understand why a species on a

Figure 11.7 *G. scandens* on Marchena: Upper: male. Middle: female. This species has a more slender bill than *conirostris* on Genovesa (lower) and is slightly smaller but is otherwise very similar. Photos by P. R. Grant.

Figure 11.8 *Camarhynchus psittacula* male on Marchena. It has a shorter, blunter bill than *conirostris*. Both species tear bark off trees. Photo by P. R. Grant.

particular island has a particular beak size and shape, it is necessary to examine the food supply available on that island.

Schluter and Grant (1984a) developed a procedure for doing this quantitatively with generalist granivore populations of ground finches, that is those populations with generalized seed diets, but without habitat specializations. *G. conirostris* on Genovesa was not included. Expected population densities were calculated as functions of beak depth on each island with known frequency distributions of seed sizes. On Genovesa the density curve was found to be bimodal (Fig. 11.10). A small peak in expected population density corresponded with a beak depth of about 6 mm, suitable for dealing with a variety of small seed types. A second, more pronounced peak corresponded with a depth of about 16 mm, suitable for cracking large and hard seeds such as those of *Cordia lutea*. The two granivorous species on Genovesa, *difficilis* and *magnirostris,* have mean beak depths of 6.3 mm and 17.5 mm, respectively (Grant 1981a), which are very close to those associated with the peaks (Fig. 11.10). We may speak of the curve as being an adaptive landscape, and the two peaks as being occupied by *difficilis* and *magnirostris.* The close correspondence between actual and predicted mean beak depths is evidence that the two species are at or close to their environmentally determined optima.

In principle the same procedure could be used to predict the mean phenotype of *conirostris* on Genovesa, but in practice the complexities of their food supply make this extremely difficult to do. We were not able to estimate the availability of nectar and pollen in cactus flowers, or of arthropods in cactus pads and beneath bark. Were those problems solved, we would still have to combine these very different food classes with seeds on some common scale in order to develop an expected population density in relation to beak size.

Figure 11.9 Upper: *Platyspiza crassirostris*, the vegetarian finch, stripping the bark off a twig of *Croton scouleri* on Marchena to get at the cambium and phloem. Lower: Close-up of a stripped *Croton* twig. Photos by D. Day (upper) and D. Schluter (lower).

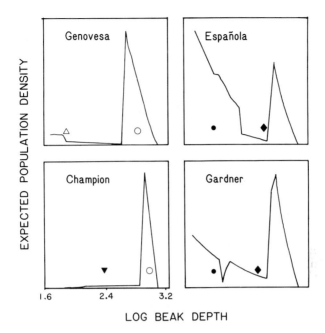

Figure 11.10 Expected population density of finches in relation to beak depth on four islands. Actual mean beak depths of granivorous species on those islands are shown by symbols: *magnirostris* (open circle), *conirostris* (solid diamond), *fortis* (solid triangle), *fuliginosa* (solid circle), and *difficilis* (open triangle). See text for further details. Adapted from Schluter and Grant (1984a).

Lacking the data for employing this predictive technique with *conirostris,* we have to rely on the comparative method as second best. Beak forms of Genovesa and Española populations are compared with food availabilities and diets, in an attempt to account for the difference in beak form in terms of the difference in food. This complements the comparison made in the preceding section between *conirostris* on Genovesa and finches on Marchena.

FOOD SUPPLY ON GENOVESA AND ESPAÑOLA

Both islands are medium-sized, low, arid, and well isolated. They have many plant species in common (Wiggins and Porter 1971; Connor and Simberloff 1978), consequently the distributions of seed sizes are fairly similar (Table 11.4; see also Abbott et al. 1977). At the plant species or genus level there are few conspicuous differences between the islands. *Prosopis juliflora,* a tree that produces large and hard seeds, is abundant on Española but absent from Genovesa. In contrast, *Opuntia, Bursera,* and *Croton* are all much rarer on Española than on Genovesa. The cactus is even rarer on Española than is indicated in Table 11.4 (see also Schluter and Grant 1984b) because the study site on Española was deliberately chosen to include cactus. The relative scarcity can be attributed to a century or more of destruction by introduced goats, which was only stopped by their elimination in 1978 (Hoeck 1984).

Seed sizes of individual plant genera are the same on the two islands, with one mild and one strong exception. *Cordia lutea* stones are slightly smaller and easier to crack on Española than on Genovesa (Grant and Grant 1982). The strong exception is *Opuntia*. *Opuntia megasperma* on Española is aptly named. The seeds are much larger and harder (Table 11.4) and their arils are more fibrous and less succulent than are the seeds of *Opuntia helleri* on Genovesa (see also Racine and Downhower 1974). The two

Table 11.4 Plant Species Whose Seeds Are Eaten by *G. conirostris* on Genovesa and Española

Species	Size-Hardness Index of Seeds	Percentage of Coverage		Eaten on	
		Española	Genovesa	Española	Genovesa
Cordia lutea	13–14	34	36	**E**	
Opuntia megasperma	10–11	X		**E**	
Prosopis juliflora	10–11	50		**E**	
Tribulus cistoides	9–10	X	X	**E**	
Opuntia helleri	7–8		8		G
Bursera graveolens	4–6	X	58	E	**G**
Ipomoea spp.	3–4	X	2	E	G
Neptuna plena	3–4	X		E	
Cenchrus platyacanthus	3–4	2	X	**E**	G
Lantana peduncularis	2–3	14	14	**E**	**G**
Vallesia glabra	2–3	X		E	
Desmanthus virgatus	2–3	8		E	
Trianthema portulacastrum	1–2	16	X	E	
Croton scouleri	1–2	X	44		**G**
Chamaesyce spp.	1–2	X	X		G
Heliotropium spp.	1–2	?	X		G
Boerhaavia erecta	0–1	2	X	E	G
Sida spp.	0–1	18	X		G
Cacabus miersii	0–1	?	X		G
Portulaca howelli	0–1	?	X		G
Aristida subspicata	0–1	X	X	E	G
Panicum fasciculatum	0–1	22	X	**E**	G
Eragrostis spp.	0–1	X	16		G
Cyperus spp.	0–1	X	X		G

Source: Modified from Grant and Grant (1982).

Note: Percentage of coverage is given on the basis of 50 random quadrats at each site. Letters in **boldface** indicate ≥ 5 percent of feeding time was spent on this food item at any one time of the year. The size-hardness index is $(DH)^{\frac{1}{2}}$ where D is seed depth in mm, and H is hardness in kilograms-force (1 kgf = 9.8 Newtons). X = present on island.

Table 11.5 Proportions of Time Spent Foraging for Different Food Items in the Dry Season of 1979: July on Genovesa, August on Española

Food Category	Genovesa	Española
Small seeds on ground	**0.20**	**0.29**
Identified seeds, $(DH)^{\frac{1}{2}}$		
0–1	0.01	—
1–2	0.03	0.01
2–3		0.06
4–6		
7–14	0.01	**0.29**
Tribulus mericarp fragments	—	X
Opuntia cactus		
Flower	0.02	—
Aril	**0.49**	0.04
Pads	0.04	**0.28**
Spine base	—	0.01
Bark ripping	**0.16**	0.03
Leaf gleaning	0.04	—
Total seconds	19,416	26,135

Source: Grant and Grant (1982).

Note: $(DH)^{\frac{1}{2}}$ is the size-hardness index (see Table 11.4). Principal foraging activities are shown in **boldface**. X signifies that a food was known to be eaten but was not recorded in these timed observations.

species of cactus may have been derived from independent mainland stocks (Dawson 1962, 1965; Anderson and Walkington 1967). This does not preclude the possibility of evolutionary change in seed size and coat thickness on the two islands, at least partly in response to the finches that exploit them (Grant and Grant 1981).

The supply of arthropods on the two islands appears to be similar (Schluter and Grant 1984b). On both islands insect larvae and pupae are found in *Opuntia* pads, in the stones of *Cordia lutea,* and under the bark and inside the wood of various trees and shrubs.

DIETS ON THE TWO ISLANDS

Diets are most likely to reflect beak adaptations in the dry season, a time of diminishing food (Smith et al. 1978). Dietary characteristics of the two populations were therefore quantified in the dry season of 1979 under the closely comparable conditions prevailing in successive months (Table 11.5).

There are proportional differences in resource use between the populations. These can be attributed partly to relative availabilities and partly to sampling errors, but aside from these one major difference stands out; Española birds crack *Cordia lutea* stones and eat the kernels, Genovesa birds do not. To a lesser extent the Española birds feed on *Prosopis juliflora* seeds, which are missing from Genovesa. They also feed frequently on the large and hard seeds of *Opuntia megasperma*, whereas on Genovesa feeding on the smaller and softer seeds of *Opuntia helleri* is a rare event; in this particular feeding study only two individuals were observed to do so, apart from the large presumed hybrid which sang a *magnirostris* song (Chapter 9).

Other differences are more matters of degree. Much more time was devoted to exploiting the arils surrounding cactus seeds on Genovesa than on Española. This is understandable because on Genovesa they are succulent whereas on Española they are not. Birds on Genovesa also spent more time stripping bark off trees to reach insects and termites than did their Española counterparts. The tip-biting, twisting, and tearing movements to remove bark are part of the repertoire of the large tree finch, *Camarhynchus psittacula*, on other islands such as Pinta, Marchena (Table 11.3), and Santa Cruz; it is not present on either Genovesa or Española. These two differences between *conirostris* populations are real and not the result of observations being made at slightly different times. On the other hand, the difference in frequency of exploiting *Opuntia* pads for arthropods is more apparent than real. It was scarce on Genovesa in July 1979 (Table 11.5) but not in the preceding November (Grant and Grant 1980) nor in July 1985 (Table 9.6).

MORPHOLOGICAL DIFFERENCES IN RELATION TO DIET

Food supplies on Genovesa and Española are similar, whereas diets and beak morphologies differ. Therefore food availability alone does not explain the Genovesa phenotype.

On the other hand, the morphological differences between the island populations are interpretable in adaptive terms. The relatively long bill of the Genovesa birds is excellently suited to the task, among others, of probing flowers for pollen and nectar (Chapter 9), and the relatively stout beak of Española birds is clearly functionally related to the task of cracking large and hard seeds. Given the similar food availabilities on the two islands, the question becomes: Why, on Genovesa, has *conirostris* adopted a somewhat specialized exploitation of the cactus niche at the cost of forgoing the exploitation of large and hard seeds?

Part of the answer can be seen in Figure 11.10. Genovesa has two seed niches, both occupied by other species. There is no intermediate niche for *conirostris* to occupy on this particular landscape. On Española there are also two niches, but here, in the absence of *magnirostris, conirostris* occupies one of them. So the crucial difference between the islands is the presence or absence of *magnirostris*.

The role of *magnirostris* can best be illustrated by comparing its diet with that of *conirostris*. Dietary relationships between these two species cannot be represented by the formulation in Figure 11.10, but they can be expressed in terms of the similarity in the time spent in various foraging activities. We have used the Renkonen index to compare diets (Grant and Grant 1982). Similarity is calculated as $1 - 0.5 (\Sigma|p_{1i} - p_{2i}|)$, where p_{1i} and p_{2i} are the proportions of foraging time spent by the two species, 1 and 2, on each discrete class of food (i) listed in Table 11.5. On a scale of 0 (diets completely different) to 1 (diets identical), the similarity of *conirostris* and *magnirostris* diets on Genovesa is 0.14; in other words, their niches differ substantially. The niche of *conirostris* on Española is much more similar to the niche of *magnirostris* on Genovesa (0.42), as expected from Figure 11.10, than to the niche of its conspecific population on Genovesa (0.31). On the other hand, the niche of Genovesa *conirostris* is most similar to the niche of *scandens* (0.42).

Thus the presence of *magnirostris* on Genovesa has a constraining influence on *conirostris,* effectively preventing any evolutionary increase in beak size. The selective advantage of large beak size in the *magnirostris* population, under dry conditions and scarcity of *Cordia* and *Opuntia* seeds (Chapter 9), suggests that larger *conirostris* which could crack *Cordia* stones would be at a strong disadvantage under these conditions in the presence of *magnirostris.* On Española, on the other hand, the evolution of beak size towards a *magnirostris* phenotype in the past was presumably unimpeded by a competitive constraint from *magnirostris.*

Some Speculations on Evolution and Competition

The reasoning we have used to account for the mean phenotype of *conirostris* on Genovesa implies that if *magnirostris* were to disappear from the island, perhaps killed off by a pathogen, and its food supply was left intact, *conirostris* would respond evolutionarily with an increase in beak size. No other species would invade the island and take over the vacant niche, because no other species elsewhere in the archipelago cracks *Cordia* stones. While an increase in *conirostris* beak size would be expected, we doubt if the end product would be as large as *magnirostris* for reasons discussed below, unless another species invaded the island and subjected it to competitive pressure.

The adaptive peaks occupied by *conirostris* and *magnirostris* on Genovesa are not separated by a trough so deep and so broad as to prevent evolutionary movement from one to the other. This can be inferred from the broad niche of *conirostris* on Española (Grant and Grant 1982). By way of comparison, the niches on Isla Champion of *fortis* and *scandens* on the one hand and *magnirostris* on the other are separated by an uncrossable gap (e.g., see Fig. 11.10). The comparison with Genovesa is apt because the *conirostris* niche there combines elements of *scandens* and *fortis* niches elsewhere (p. 265).

Evidence for the gap being too large to cross comes from an actual removal "experiment" carried out unintentionally. About a hundred years ago an exceptionally large form of *magnirostris* was driven extinct through human activity on Floreana, and probably as a consequence on Champion as well. Its food supply was drastically depleted on Floreana but left intact on Champion. Measurements of *fortis* and *scandens* on Champion, taken from live birds in 1980 and from museum specimens collected early in the century, give no indication of an evolutionary change. The failure to change is probably not for lack of genetic variation; as on Daphne Major, the morphological traits of these two species are heritable, highly so in the case of *fortis* (Boag and Grant 1978; Boag 1983; Price, Grant, and Boag 1984). The food supply of the missing *magnirostris,* seeds of *Opuntia megasperma,* remain unexploited. As a result the peak in expected population density of a large granivore is very high, and it dwarfs a second peak for a medium granivore (Fig. 11.10).

The niches of *conirostris* and *magnirostris* on Genovesa, although evolutionarily within the reach of a single species, are different enough that they could not be exploited with equal efficiency by a single species, or so we suppose. On Española the two niches are exploited by a single species, *conirostris,* but it is mainly a granivore with a large bill because there is a much greater supply of seeds than of cactus products. So even in the absence of *magnirostris* on Genovesa, *conirostris* would remain largely an exploiter of the cactus niche, at least initially.

However, any shift in the morphology and ecology of *conirostris* permitted by the absence of *magnirostris* would make it more likely that *fortis* or *scandens* would colonize Genovesa. These two species are smaller versions of *magnirostris* and *conirostris,* respectively. Neither of them coexists with *conirostris,* a fact which can be attributed to competitive exclusion (Lack 1947; Grant and Grant 1982). Of the two, *scandens* would appear to have the better chance of coexisting with a larger-than-at-present *conirostris,* by virtue of its specialization on cactus. On the other hand, *fortis* is more likely to colonize the island by virtue of its higher rate of immigration (Chapter 9). The immigrant species, let us call it *scandens,* would be at a competitive disadvantage to a degree dependent upon how much evolutionary change in *conirostris* had occurred prior to the immigration. Some interbreeding would be expected too (Grant 1986a). But if *scandens* succeeded in becoming established, the larger *conirostris* would become yet larger under the competitive pressure from it; large size in *conirostris* would be more advantageous in the presence of a cactus-resource-depleting species than in its absence. Under these circumstances, *conirostris* could evolve into a *magnirostris*-like species. Perhaps this is what happened on Española, with *scandens* later becoming extinct when cactus declined.

So far we have performed a thought experiment with the imaginary removal of *magnirostris* from Genovesa. The reciprocal experiment is to remove *conirostris*. If it became extinct, the ecological and evolutionary result would be different. An evolutionary response from *magnirostris* would probably occur, but before it had got very far *scandens, fortis,* or both, would probably invade the island, in view of their fairly high immigration rates (Chapter 9), and occupy the niche of the missing cactus finch. *G. magnirostris* would be expected to remain a large-seed finch. In other words, ecological replacement would occur much more rapidly than evolutionary replacement (cf. Roughgarden 1972, 1986).

Evolutionary events on the islands of Wolf and Darwin may be the natural analogue to this particular thought experiment, except that *fortis* and *scandens* have apparently failed to establish themselves on these extremely isolated islands. Here, in the absence of a breeding population of *conirostris, magnirostris* has converged toward the *conirostris* phenotype (Grant et al. 1985). Diets are not well known. Some feeding on cactus occurs, but the large-seed niche has not been relinquished (Schluter and Grant 1984a). *G. conirostris* may immigrate rarely (Curio and Kramer 1965) and interbreed with *magnirostris.*

These speculations reflect our view that the number of species of a taxonomic group on an island, such as ground finches on Genovesa, is determined largely by local conditions. Those conditions may be represented as a landscape with peaks. Two have been quantified on Genovesa (Fig. 11.10). A third is known to exist, because it is occupied (by *conirostris*). Only one species occupies a peak (Schluter and Grant 1984a). Which particular species occupies a given peak on a particular island is largely a matter of chance and history. On Genovesa the small-seed niche is occupied by *difficilis,* but on most islands in the archipelago, including Marchena and Española, it is occupied by *fuliginosa* (Fig. 11.10). Likewise the cactus niche is occupied by *conirostris* on Genovesa but by *scandens* on other islands. In fact, *conirostris* occupies two close peaks: one determined by cactus flowers, fruits, and buds, the other determined by cryptic arthropods.

It is remarkable how few species of plants determine the adaptive peaks. On Genovesa just two plant species are vital to one and possibly two of the three ground finches. These key species are *Cordia lutea* and *Opuntia helleri.* Their removal, by pathogens for example, would almost certainly cause the extinction of *magnirostris. G. conirostris* could be similarly affected, although a few individuals might survive in dry seasons by feeding nearly exclusively on arthropods beneath bark. Then, with strong directional selection pressure and large genetic variation in bill dimensions, an evolutionary shift towards a tree finch phenotype might be expected to occur. A third key plant species is *Waltheria ovata,* which, as a source of pollen and nectar, is important to *difficilis* in years of seed scarcity (Table 2.4; see also Table 11.2 and Fig. 11.5). It may not play a vital role, but its disappearance would cause a catastrophic drop in *difficilis* population size. Thus three of the sixty plant species on the island largely control the fate of the three ground finch species. In their dependence on one or two key plant species, the finches resemble primates in a Neotropical rain forest (Terborgh 1983, 1986).

For Darwin's Finches as a whole, a fifth peak exists in the Genovesa landscape, occupied by the warbler finch (*Certhidea olivacea*). The population of this species would also decline if *Waltheria ovata* disappeared, as can be inferred from the frequent feeding of warbler finches on its nectar and pollen (Table 11.2 and Fig. 11.5).

The Populations on Española and Gardner

The comparative method illuminates the reasons for the mean phenotype on Genovesa being where it is, but at the same time it highlights our main areas of ignorance. We conclude this chapter with a few additional comments on the two southern populations of *conirostris* on Española and neighboring Gardner where our ignorance is greatest.

Morphologically these populations are as variable as their relatives on Genovesa (Table 1.1). Aside from their taxonomic status and the unique female plumage (Fig. 9.5), this is the main problem they pose, for in the absence of *magnirostris* as a source of genetic input, they might be expected to vary less, rather than more, than their Genovesa relatives.

In terms of the theoretical scheme in Figure 1.6, one population varies more than another because its genetic variance is enhanced by mutation and introgression at higher rates, or it is subjected less to the depleting effects of selection, or both. It is a pity we were not able to be on the three islands simultaneously to evaluate these alternatives. Our knowledge of the southern populations is fragmentary.

If higher rates of introgression are at least part of the answer, the source must be either the resident *fuliginosa* or an immigrant species. The few, short-term studies and observations of breeding on this island have not detected hybridization (Downhower 1976, 1978; D. J. Anderson and R. L. Curry 1984, pers. comm.). On the other hand, forces of stabilizing selection are likely to be weak. The niche on Española is exceptionally broad (Grant and Grant 1982). This is relevant to the maintenance of variation if individuals differ in their diets in a manner related to their beak differences (Van Valen 1965; Van Valen and Grant 1970; Grant et al. 1976; Grant 1981c; Price 1987). Alternatively, all individuals may be identical generalists, but if so they would differ fundamentally from conspecifics on

Genovesa. A detailed study of individually marked birds is needed to clarify this issue.

Another possible explanation for the high variation in Española and Gardner populations is that it results from the mixing of two populations with different means. The means do differ (Grant et al. 1985); we do not know the reason. So far as available data show (Fig. 11.10, and Abbott et al. 1977), the food supplies are similar on the two islands.

Clearly, intensive field studies on these islands are needed. However, a factor complicating the interpretation of results is the change in vegetation on Española caused by goats (Gardner has never had goats). Exactly when they were introduced is not known. By 1905–6, when members of the California Academy of Sciences spent a year in the archipelago, damage to the vegetation was far advanced and remains so today one decade after the removal of the goats (Fig. 11.1).

Goat damage on Española could have altered selection pressures on *conirostris*. For example, *Opuntia* certainly decreased and *Prosopis* probably increased under the influence of goats. The overall food supply could have changed in a significant way, as the size-hardness values of seeds is lower for *Prosopis* than it is for *Opuntia* (Abbott et al. 1977). Yet we can find no evidence of a recent change in beak morphology through a comparison of measurements of specimens collected in 1905–6 and measurements of live birds made in 1973 (Abbott et al. 1977), in 1979, and again in 1988. The real comparison, however, should have been made with birds in 1805–6! It is therefore still possible that the difference between Española and Gardner *conirostris* is caused by different selection pressures arising from different food supplies that are caused, in turn, by the effects of goats on one island (Española) but not on the other (Gardner).

Summary

This chapter is mainly concerned with explaining the mean phenotype of Genovesa *conirostris*. Either the population has reached an environmentally determined optimal phenotype under past regimes of natural selection, or else it has been constrained from reaching it by its genetic architecture, in which case it remains relatively poorly adapted.

The historical roots of the population are unknown. There are no fossils of the species which could be used to reconstruct those roots, and biochemical information is inadequate. The population on Genovesa may have been derived from the population on Española, or it may have given rise to that population, or both may have been derived from one or two populations which have since become extinct.

The genetic covariance structure of the Genovesa population has been used to estimate the minimum forces of selection necessary to transform one of the populations into the other, regardless of the direction. The result has been combined with estimates of the strength of directional selection on *fortis* on Daphne Major to give a rough value for the number of such episodes to effect the transformation. If selection was as intense as it has been on Daphne Major only once a century, no more than a thousand years would have been required for the whole process. We conclude that there has probably been ample time, environmental pressure, and additive genetic variation for an approximate alignment to be reached between the mean phenotype and an environmentally determined optimum.

The optimum on Genovesa is determined by the availability of different types of food (the adaptive landscape) and the exploitation of part of the food supply by other ground finch species. A comparison with the food niches of related populations and species elsewhere in the archipelago helps us to understand why *conirostris* on Genovesa is largely, but not entirely, a cactus feeder, and hence why it has a relatively pointed beak. In the absence of *scandens,* it exploits *Opuntia* products. In the absence of *fortis,* it eats small and medium-sized seeds, including *Cacabus* and *Bursera.* In the absence of the large tree finch *Camarhynchus psittacula,* it exploits the arthropod fauna beneath the bark of trees and shrubs. And in the presence of *magnirostris,* it does not feed on large and hard seeds; whereas in the absence of *magnirostris* on Española and Gardner, it does.

We speculate that the absence of *magnirostris* from Genovesa, but not its food supply, could lead to an evolutionary replacement by *conirostris* if *scandens* invaded the island and subjected it to competitive pressure. On the other hand, the absence of *conirostris* from the island, but not its food supply, would lead to an ecological replacement by immigrant *scandens,* or *fortis,* or both. Evolutionary replacement by *magnirostris* would be too slow.

Morphologically the *conirostris* populations on Española and Gardner are more variable than the population on Genovesa. The difference in levels of variation between the island populations can be interpreted in terms of the theoretical scheme (selection-introgression balance) in Figure 1.6, although field data from Española and Gardner are insufficient to examine it in detail. Goats have damaged the vegetation on Española. This is a complicating factor in the search for an understanding of the causes of variation, but it is also a potentially illuminating factor, in that goats were never introduced to Gardner and the vegetation there is intact. Intensive, long-term studies on these two islands would be very valuable as a complement to the one we have conducted on Genovesa.

12

Summary, Synthesis, and Some Implications for Conservation

Introduction

Darwin's Finches are unique in being the only group of closely related species of birds that has remained intact in the island setting in which it evolved. Several populations are biologically interesting because they are unusually variable. Genetic variation in quantitative characters is essential for evolutionary adaptations to a changing environment and ultimately for speciation. In this book we describe an eleven-year investigation into the causes and significance of morphological variation in a population of the large cactus finch, *Geospiza conirostris*. Our principal concern is with the unusually large variation in bill size and shape, because these are the traits identified as having been important in the evolution and speciation of the whole group of Darwin's Finches (Grant 1986a), and because by studying variation we can more fully understand the process of evolutionary change.

We chose *conirostris* on Isla Genovesa for three reasons. First, the population is small and exceptionally variable in bill dimensions. Second, Genovesa is well isolated, being a low flat island in the extreme northeast of the Galápagos archipelago. Morphological variation is not likely to be influenced by gene flow from other populations of the same species, rather it should be governed by factors intrinsic to the island. Third, the island has no introduced animals or plants, nor has it ever had human settlements, and as a consequence it possesses a full complement of naturally occurring species.

Studies of isolated, natural populations that maintain sufficient genetic variation in quantitative characters to survive at low density and adapt to changing environments are valuable. Not only are there few of them left to contribute to our understanding of evolution, but they provide us with information for the management of populations of continental species that are isolated in "island patches" through the fragmentation of their habitat (Diamond 1985; Lande and Barrowclough 1987). Such small populations are likely to become extinct when numbers fluctuate erratically for either demographic or environmental reasons (Lande 1988). In addition, it is believed that a major factor in recent extinctions has been the lack of sufficient genetic variation in quantitative characters for populations to respond to changes in the environment, such as climatic changes, deteriorating habitats, and introductions of diseases, predators, and competitors (Futuyma 1983; Diamond 1985; Lande and Barrowclough 1987; Ehrlich 1988).

Synthesis

We set ourselves the task of identifying the factors responsible for some birds succeeding much better than others in living long and reproducing often. Is it a matter of chance who survives to breed and who does not, who reproduces once and who produces many times? Or do some birds succeed because they are better equipped than others to exploit the environment and avoid its hazards? Does success vary in relation to beak size and shape, or in relation to other morphological traits? As the study progressed we learned that the difference in success between the most and least successful individuals was extremely large. Here is a summary of the pertinent facts.

From 100 eggs, about 60 nestlings fledge, on average. About 1 in every 10 fledglings survives its first year, and closer to 1 in 20 survives to breed. Thereafter survival is high (~80 percent per year), as is typical of tropical rain-forest birds. Like them also, the maximum longevity of individuals is high, and high in relation to their size; the oldest birds reach at least 12 years, and 15 years is estimated to be a realistic maximum. Nevertheless, reproductive success varies greatly. Approximately 15 percent of breeders produce no fledglings, and almost 80 percent contribute no recruits to the next generation. In contrast to the failures, the most successful male in our study produced five recruits, and the most successful female produced three. She had to work hard for this reward. She lived for at least 12 years, and at a minimum laid 110 eggs in 31 clutches that yielded 58 fledglings. But the chances of achieving any success are low. In terms of the starting number of 100 eggs, only one is likely to yield a breeding adult which in turn produces a breeding adult.

The key environmental factor for understanding these population features is rainfall and its variation. Sporadic and occasionally heavy rain falls in the first four months of the year. The predominantly drought-deciduous vegetation responds by producing leaves, flowers, fruits, and seeds. Arthropod populations rise dramatically and birds breed. In the remainder of the year, the dry season, there is almost no plant production, and many birds die of starvation.

Superimposed on this seasonal pattern of production is an annual variation caused by the variable and unpredictable rainfall regime. The

extremes are abundant rainfall, associated with an oceanographic disturbance known as El Niño, and droughts. In El Niño years the wet season is
prolonged. For example, the season lasted for eight months in 1983, when
more than 2400 mm of rain fell. Plants, arthropods, and finches reproduced continuously throughout this period. In 1985, by way of maximal
contrast, no rain fell and no production or reproduction occurred. Despite
the unpredictability of rainfall from month to month and from year to year,
there is a tendency for climatic extremes to recur at intervals of four years
or more.

What makes Genovesa (and other Galápagos islands) especially interesting is that it has a mixture of tropical and temperate characteristics. Day
length is almost constant throughout the year, and temperature variation is
not large, as is typical of other tropical locations at similar altitude.
Rainfall variation within and among years is very strong, so the environment is seasonal in the same way that semiarid regions are at temperate
latitudes. Finches correspondingly display a mixture of the features of
tropical and temperate relatives. Adults survive at a high average rate as do
tropical (rain-forest) species, yet they vary strongly in their annual survival
rate as do temperate zone species. Reproductive rates and success are high,
as in temperate zone species. We interpret this mixture as the product of
natural selection in an uncertain environment. Feeding conditions are often
unfavorable, and then survival is at a premium. Yet sometimes food is
abundant, and breeding rapidly and maximally is clearly advantageous.

Finches are ready to breed when the first rains fall, and they breed
rapidly and repeatedly if conditions for breeding remain suitable for a long
time. The first eggs are laid usually one to two weeks after the first heavy
rain. Typically a minimum of 15–20 mm of rain is needed for a sustained
leafing response from the plants, for caterpillars to be produced, and for
finches to breed successfully. The length of the season varies directly with
the amount of rain. It lasted for eight months in 1983. Females adjust their
breeding effort according to feeding conditions. They lay smaller clutches
when food is not so plentiful, and they vary the interval between successive
clutches. Intervals are shortest when food is most plentiful in relation to
demand. Breeding ceases when caterpillars and spiders, which are most
important nestling foods, become scarce.

Morphological traits have little bearing upon whether a finch breeds
successfully or not. A factor of much greater importance is experience.
Experience (or age) plays a role in determining several components of
reproductive success, but most strongly determines the number of clutches
of eggs. Experienced birds gain this advantage mainly by breeding earlier
than inexperienced birds, as a result of retaining their mate from the
previous breeding season or by pairing earlier than inexperienced birds do.
The temporal advantage translates into a production advantage: experi-

enced pairs produce twice as many fledglings in a breeding season as do inexperienced pairs.

Survival under dry-season food stress is determined by a combination of behavioral and morphological factors, and by chance. In the first year of life the most important determinant is the ability to acquire skills at exploiting one or more of the principal dry-season foods, which are seeds in *Opuntia* (cactus) fruits, arthropods in rotting *Opuntia* pads, and arthropods beneath the bark of trees and shrubs. Different beak shapes are best suited to these tasks. Thus relatively long bills are suited to pecking and probing *Opuntia* fruits to extract the seeds and remove the surrounding fleshy arils. Relatively deep beaks are best suited for the tearing and ripping that is needed to expose hidden arthropods. Dry-season survival of adults is determined by the availability of foods which they are best able to exploit by virtue of their beak size and shape, and by the feeding skills they have acquired.

These relationships were evident in the first half of the study, but they became especially important in the aftermath of the exceedingly long wet season of 1983. During that wet season vines grew rampantly and smothered much of the cactus, which died. This had the effect of increasing the populations of arthropods sustained by rotting pads, but of almost eliminating fruit production. Under these changed circumstances, particularly the scarcity of cactus flowers, adults with long bills were at a strong selective disadvantage. There followed a period of selection in favor of birds with deep and wide beaks capable of extracting arthropods from rotting *Opuntia* pads, which by then had become hard and crusty, and from beneath the bark of trees. Young birds born in 1983 that had acquired dry-season feeding skills by the end of the year were subjected to the same selection pressure.

The analysis of size-selective mortality helps us to understand why the population is morphologically variable. Variation must be maintained by generating and depleting processes. Directional selection is a depleting process. It also has the potential of producing a directional evolutionary change in the population because beak and body size traits are highly heritable. However, the change in mean bill length that occurred as a result of selection in the early phase of the drought was counteracted later in the drought by selection in the opposite direction on bill depth, which is positively correlated with bill length. The overall result was no net change between the end of breeding in 1983 and the resumption of breeding in 1986.

The main generating processes are mutation, which we cannot study, and introgression of genes. Genes could enter the population through interbreeding with other species or with conspecific immigrants from Española or Gardner, the only other islands supporting *conirostris* populations. Immigration of *conirostris* was not detected in our study, but hybridization at low frequency (1–3 percent) does occur with the two other resident populations of ground finches, *magnirostris* and *difficilis*. Hybridization was observed. In addition, 14 birds were identified by their measurements as probable hybrids. Including probable hybrids in the sample of breeding birds had the effect of increasing the coefficient of variation by 20 percent, and increasing the variance by 35 percent. Hybrids reproduced successfully but did not live long and contributed less than nonhybrids to total breeding.

Thus variation is maintained by the opposing processes of introgression and selection. They were not in a state of balance during the decade. By the end of the study a significant reduction in the phenotypic variance of beak depth and beak width had taken place among females, and nonsignificant differences in the same direction were apparent among the males. Given the high heritability, it is possible that a reduction in genetic variation had occurred as well, as a result of selection.

Variation is enhanced by certain population structures such as partial subdivision. Our initial observations in 1978 suggested that the population was subdivided ecologically and possibly reproductively. Two types of song (*A* and *B*) are sung by males, but each individual male sings only one type. The group of *A* singers had significantly longer beaks than the group of *B* singers. Moreover, in the dry season the *A* singers fed on arils surrounding the seeds in fruits of *Opuntia* cactus which they obtained by probing and tearing, and the *B* singers fed on arthropods in rotting pads which were excavated by tearing and ripping the pads.

A possible reproductive subdivision along the lines of male song was hinted at by a nonrandom arrangement of territories of mated males in 1978; no two adjacent, mated males sang the same song type. In contrast the distribution of territories of males without mates was random with respect to song type of their neighbors, suggesting that female choice of mates may have been responsible for the difference. If females chose mates on the basis of their song, they may have paired preferentially with males that sang the same song type as their own fathers. All these features are of theoretical interest because, carried to its extreme, the divergence of two groups in a population could continue to the point at which they no longer interbred; in other words, to speciation in sympatry.

These initial signs of population subdivision subsequently disappeared. Sons sing the same, simple, song type as their fathers, copying it from them in early life and retaining it for life. A continuity of song characteristics among relatives is thereby maintained between generations. Song plays a major role in species and individual recognition, possibly also in kin recognition, in the context of maintaining a territory and attracting and stimulating a female. Females, however, do not pair preferentially with males singing a particular song type; paternal song type does not influence their choice of mate, and so there is no reproductive subdivision of the population along the lines of song through assortative mating. Instead females choose mates on the basis of plumage and courtship behavior, which are reliable indicators of age, experience, and condition. They gain a reproductive advantage by choosing to breed with experienced males, probably as a result of enhanced paternal care, and possibly as a result of superior genes from such males. A genetic factor in mate choice is suggested by several indications of extra-pair matings. Preferential mating with males of high genetic quality tends to deplete genetic variance. Genetic variation is maintained in the face of this process by introgression.

Given the random mating with respect to song, it is not surprising that the beak length difference between the *A* males and the *B* males disappeared, especially since beak length is highly heritable. Nor is it surprising that the feeding differences between the song groups and the pattern of nonrandom distribution of territories also disappeared. Their origin, however, remains an enigma. The differences between the song groups may have arisen under the dry conditions prevailing in 1977, the year before our study began. They did not reappear in the next drought (1985), but we understand the reason for this; the food supply changed drastically under the delayed influence of the extremely wet conditions in 1983.

Even though the population structure along the lines of song was ephemeral, the population is not homogeneous. The cactus distribution is somewhat patchy, and this imposes patchiness on finch distribution. Dispersal distances from natal to breeding sites are low in both sexes. Individuals are potentially long-lived and males stay on the same territory for life, thus most matings occur locally and there is an inertia in the spread of alleles across the island. Gene frequencies vary in space, as indicated by differences between areas in the frequencies of nestling beak colors (pink or yellow) among the offspring of the two song groups of males; nestling beak color is under fairly simple genetic control. As a result of this structure, local groups of finches show semi-independent fluctuations in frequencies of alleles that are thought to be neutral or nearly neutral.

Summarizing, we can draw the following conclusions with respect to our original questions. First, there is a large amount of genetic variation underlying the large phenotypic variation in beak traits. Introgression of genes from *magnirostris* and *difficilis* is largely responsible for the elevated level of variation. The variation is subject to occasional directional selection but not to strong stabilizing selection. The niche of the population is moderately broad, and individuals specialize to some extent according to their beak size and shape. The relative fitness of individuals in the dry season is determined by their particular food supply, so a change in the composition of the food supply alters relative fitnesses. The relative fitness of individuals in the wet season is usually not influenced by morphology. Overall, relative success or failure in surviving and breeding repeatedly is determined more by behavioral factors associated with experience than by morphological features. Hence, to more fully understand the reasons for success and failure of individuals with different morphological traits, we need to know much more about behavior associated with feeding and breeding: how it changes with age and experience, its relationship to morphology, and the degree to which it has a heritable basis.

Results of this study throw light on the past evolution of the population and the current place it occupies in the community of finches on the island. There has probably been ample time, environmental pressure, and additive genetic variance for an approximate alignment to be reached on Genovesa between the mean phenotype and an environmentally determined optimum. This conclusion follows from an estimate of the forces of selection necessary to transform the Española mean phenotype into the Genovesa mean phenotype, based on the genetic covariance of the main morphological features of the finches on Genovesa. The environmental optimum is determined by the availability of different types of food (the adaptive landscape) and the exploitation of part of the food supply by other ground finch species.

A comparison with the food niches of related populations and species elsewhere in the archipelago helps us to understand why *conirostris* on Genovesa is largely but not entirely a cactus feeder, and hence why it has a relatively pointed beak. In the absence of *scandens*, it exploits *Opuntia* products. In the absence of *Camarhynchus psittacula,* a tree finch, it exploits the arthropod fauna beneath the bark of trees and shrubs. And in the presence of *magnirostris,* it does not feed on large and hard seeds, whereas it does so in the absence of *magnirostris* on Española and Gardner.

Although not restricted to the cactus food niche, *conirostris* is so dependent upon it on Genovesa that it would possibly become extinct if *Opuntia* were ever eliminated, by a pathogen for example. It is an impressive testimony to feeding specialization that a single, key species of plant among a community of 60 terrestrial plant species is crucial to the persistence of a finch species on the island.

Implications for Conservation

Our point of departure for discussing the relevance of the study to conservation is the genetic structure of small populations. We begin by outlining the relevant theory of genetic variation in small populations, establish connections with the *conirostris* population, and then proceed to consider the factors which promote high levels of variation in small populations that are at risk of becoming extinct.

GENETIC VARIATION IN SMALL POPULATIONS

Genetic variation becomes progressively reduced through the process of random drift in small finite populations in the absence of mutation, migration, and selection. This occurs because the samples of zygotes in successive generations are small enough to produce random fluctuations in allele frequencies between generations. Through these fluctuations alleles are continuously lost or fixed by chance, heterozygotes decrease in frequency, homozygotes increase, and individuals become progressively more similar to each other genetically. Even in a randomly mating population this increasing similarity eventually leads to problems of inbreeding depression caused by the expression of deleterious recessives and by developmental instability associated with increased homozygosity (Falconer 1981; Charlesworth and Charlesworth 1987). Thus random drift is a process that both increases the similarity between individuals and decreases genetic variation.

The rate of loss of genetic variation can be estimated from the effective population size. The effective population size (N_e) is the number that would give rise to the same sample variance in alleles or the same rate of inbreeding as is observed in the actual population if mutation, migration, and selection did not occur (Falconer 1981). It is calculated from the number of breeding individuals in the population. It is generally lower than the number of breeders as a result of variation among pairs in the number of offspring that survive to breed (see Appendix IV).

Mutation, migration, and selection must be considered because they also influence the amount of genetic variation. Alleles are lost through most forms of selection and are gained through mutation and migration; migration is used here in the sense of interbreeding with members of another population. Lande and Barrowclough (1987) use a quantitative genetics model to show that mutation, in the absence of migration, maintains genetic variation at equilibrium under conditions of stabilizing or fluctuating directional selection when the effective population size is approximately 500 or more.

A very small amount of immigration is sufficient to increase the genetic variation and hence the effective population size. Falconer (1981), using information from Wright (1951), showed that one breeding immigrant every second generation will counteract the loss of variation by random drift. If frequent enough, gene exchange can lead to panmixia. Crow and Kimura (1970) suggest that two populations become panmictic when $m > 1/4N_e$, where m is the fraction of breeding individuals immigrating from another population. Wright's (1943) formula, $F_{st} = (1 - m)^2 / [2N_e - (2N_e - 1)$

$(1 - m)^2$], has frequently been used to estimate the amount of divergence between populations when the effective population size and immigration rate are known (Corbin 1987; Greenwood 1987; Koenig and Mumme 1987; Lande and Barrowclough 1987; Rockwell and Barrowclough 1987; Slatkin 1987).

This formula ignores selection. There is some controversy over the influence of migration on populations subject to selection (see Slatkin 1987 for a review). On the one hand, immigration is considered to prevent local adaptation from occurring by flooding a population with new alleles (Cooke 1987; Rockwell and Barrowclough 1987). On the other hand, it can promote adaptation by increasing the variation on which selection can act (Ehrlich and Raven 1969; Endler 1977; Wright 1980; Slatkin 1987). Haldane (1948) and Endler (1977) argued that the outcome depends on the strength of selection in relation to the amount of migration. However, where correlated polygenic traits are involved, the outcome is likely to be complex.

Using the above theory as a guide, we now consider the interactions of effective population size, mutation, migration, and selection in the *conirostris* population. Taking study area 1 to be a representative sample (of 10 percent) of the island occupied by breeding cactus finches, we calculate the actual breeding population size to be approximately 700 individuals, on average, varying from 380 (1978) to 1780 (1983). The effective population size, calculated from the actual population size and assuming random mating throughout the island (see below), is less than 200 individuals (Appendix IV). Mating is not random through the island, and neighborhood sizes are relatively small (Appendix IV), so the genetically effective population size is likely to be distinctly lower than we have estimated it to be. Whatever its exact value, it is much lower than the 500 calculated by Lande and Barrowclough (1987) as being necessary to maintain polygenic variation under fluctuating directional selection. Many generations of random mating in an isolated population of effective size less than 200 would lead to a depletion of much of the genetic variation. We have no reason to believe that the population has ever been an order of magnitude greater than it is at present, since cores of lake sediments (Chapter 2) indicate a long-term persistence of current climatic conditions. Yet there is substantial genetic variation underlying quantitative, morphological variation in the population. It seems, therefore, that migration must be the crucial factor responsible for the high level of variation (Chapter 9).

The principal sources of new alleles are *magnirostris* and *difficilis*, not conspecific or heterospecific populations on other islands. No *conirostris* from another island has been recorded on Genovesa, and the occasional *scandens* and *fortis* immigrants have not stayed to breed. Interbreeding with the sympatric congeners *magnirostris* and *difficilis* occurs at a frequency of about 1–3 percent (Chapter 9). The hybrids were calculated to have increased the coefficient of variation for beak depth by 20 percent, and the variance by 35 percent.

Thus the fraction of breeding individuals entering the *conirostris* population (*m*) is approximately 1/100, which is sufficient to counteract the loss of variation through genetic drift. It is also greater than $1/4 \, N_e$, the level considered by Crow and Kimura (1970) to produce panmixia in the absence of selection. We calculated the degree of divergence between populations of the three species, using Wright's (1943) formula. The result is $F_{st} = 0.24$; 24 percent of the alleles are estimated to be unique to *conirostris,* and 76 percent are estimated to be shared with *magnirostris* and *difficilis,* in the absence of selection.

Selection was not absent, however. Hybrids, with bills intermediate between *conirostris* and the other two species, were at a disadvantage, and none survived the drought. The ecological reason for the species remaining distinct and not merging into a single population is the discreteness of their niches (Table 11.2) and the strongly peaked nature of the adaptive landscape they occupy. Moreover, when changes in the composition of the niches occur through climatic change, at least two of the species (*conirostris* and *magnirostris*) are subject to natural selection. The effect of natural selection is to maintain or even enhance the differences between the species (Fig. 12.1). In contrast, genetically controlled traits that are not subject to natural selection tend to become uniform among species that hybridize. A possible example is the nestling beak color polymorphism. Frequencies of the two morphs are similar although not identical in *conirostris* and *magnirostris* (Grant et al. 1979).

In conclusion, migration (hybridization) is the crucial factor contributing to the high levels of variation in the Genovesa population of *conirostris.* It is sufficiently frequent to counteract the loss of genetic variation by random drift, but not so frequent as to prevent selection from favoring certain combinations of alleles that are important for adaptation. The degree of differentiation between the occasionally hybridizing species depends on the strengths and targets of selection and the genetic correlations between traits.

CONSERVATION

King (1985) found that of the 93 species and 83 subspecies of birds that have become extinct since the seventeenth century, 93 percent lived on islands. Island populations have been particularly vulnerable as a result of their small size and because they have evolved in isolation. As shown above, low effective population size causes loss of variation through random drift, increasing homozygosity, and genetic uniformity of individuals. This leads to inbreeding depression and the inability of populations to respond to ecological changes. Possible examples are included among the species listed in Table 1.4. The extinct Hawaiian kona grosbeak and greater koa finch (Pratt et al. 1987) and the near-extinct Azores bullfinch (Bannerman and Bannerman 1966; van Vegten 1968) have particularly low coefficients of variation for bill dimensions (Table 1.4; see also Fig. 1.3). At the time the museum specimens were obtained each of these species was restricted to small areas on a single island, probably largely as a result of the destruction of much of their habitat. They may have suffered a loss of variation and, as a consequence, a loss of the potential to change.

From the study of *conirostris* we have learned three facts relevant to the conservation of small populations. These are: (1) hybridization between

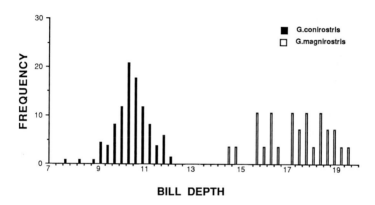

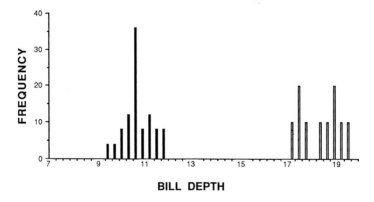

Figure 12.1 Frequency distributions of beak depths before and after the long dry period of 1984 and 1985; late 1983 above, and early 1986 below. The difference between *conirostris* and *magnirostris* mean beak depths (adult males and females combined) increased as a result of directional selection in *magnirostris*. The *conirostris* variance was reduced by 35 percent, but the mean did not change. The *magnirostris* variance was reduced by 64 percent.

closely related sympatric congeners provides added genetic variation that can facilitate adaptation to the environment by natural selection; (2) as a result, a full complement of associated species in a region aids in maintaining genetic variation in one species; and (3) the environment changes frequently, so populations are frequently subjected to directional natural selection. Evolutionary responses to these naturally occurring ecological changes are enhanced by large genetic variation.

Hybridization on Genovesa is by no means unique. Low levels of hybridization (1–4 percent) are found in the Galápagos in other populations of ground finch species; in *fortis*, *scandens*, *fuliginosa*, and *magnirostris* (Grant 1986a). Cade (1983) has emphasized that hybridization between closely related bird species should be considered a natural and regular phenomenon. Mayr and Short (1970) estimated that 10 percent of 516 nonmarine species of birds regularly hybridize. This number is certainly low, as long-term studies are required to detect hybridization at levels of 1 or 2 percent.

Our study is also not unique in suggesting that introgression of alleles provides genetic variation for adaptation, even when the hybrids are at a selective disadvantage. Lewontin and Birch (1966) demonstrated experimentally that gene flow between the fruit flies *Dacus neohumeralis* and *D. tryoni* provides the genetic variation for rapid adaptation to high temperatures, despite the hybrids being at a selective disadvantage. Gill (1980) suggested that gene flow from the golden-winged warbler population into the blue-winged warbler population has allowed the latter to increase its range through ecological and physiological adaptations.

Introgression from another population can either eliminate the genetic differences between populations or facilitate differentiation by providing additional variation on which selection can act. The outcome of introgression depends on the balance between the amount of hybridization, the strengths and targets of selection, and the genetic correlations between traits. Quantitative genetics models taking correlations between characters into account would be needed to tell us at which point this switch in outcomes occurs.

In our study the species exchanging alleles differed strongly in their ecological requirements, thus selection, acting on the increased population variation, helped to maintain the differences between the species. An opposite situation may exist at Bahía Academía on Santa Cruz, where the overlapping distributions of bill dimensions of *fortis* and *magnirostris* (Ford et al. 1973; Grant 1986a) could be the result of a small amount of hybridization and little or no selection. This environment has been made artificially benign close to human settlement by the provision of food and water.

Examples of the swamping of one species with alleles from another are found on the Mangere islands of the Chatham island group and on the Seychelles. The Chatham island yellow-crowned parakeet on the Mangere islands was reduced to six individuals in 1973 and underwent a change through extensive hybridization with a close relative, the red-crowned parakeet (Cade 1983; Taylor 1985). Likewise the Seychelles turtle dove was swamped with alleles from the introduced Madagascar turtle dove. However, as Cade (1983) points out, even in these two cases where the original phenotype is altered, the populations retain many of the original alleles, they are more genetically variable, and they are more capable of adapting to local ecological changes and surviving than were the original small populations.

The implication for conservation of the prevalence and effects of hybridization is that maintaining communities intact could be essential for the long-term persistence of their members. It is not enough to concentrate on one or two species of particular interest unless, by conserving them, all others are conserved as well, as might be the case with the conservation of top predators for example. The extinction of one species could eliminate an important supply of new genetic variation for another, thereby jeopardizing the long-term prospects of the other. Coadapted gene complexes for traits like genetic resistance to disease may be transferred from one species to another by hybridization.

Genetic interdependence is less well known than ecological interdependence of the sort exemplified by Terborgh's (1983) study of primates. He identified a few key species of plants that are necessary for populations of primates to survive through periods of food scarcity. If the key species were removed, extinction of populations of other organisms would follow. We identified *Opuntia helleri* as a key species for *conirostris*.

The third lesson to be learned from this study is that species may change within observable periods of time. Species are often considered to have evolved slowly in response to long-term environmental change and to be currently static, evolutionarily. Our study has revealed the dynamic nature of the environment and the change in selection pressures on heritable variation. During only 11 years of study there were two exceptionally wet years and two droughts. These fluctuations in climate altered the food supply and imposed strong selection favoring birds with bills of the appropriate shape to exploit the changing food supply. These naturally occurring events emphasize the need in conservation to maintain sufficient genetic variation in quantitative characters within a population for it to be able to respond to environmental fluctuations, be they natural or imposed by humans. Lande and Barrowclough (1987) give methods for continually monitoring the level of genetic variation in endangered species, and they suggest that every effort should be made to preserve variation in quantitative characters by either maintaining large populations or through breeding programs such as those used by Frankel and Soulé (1981) and Templeton and Read (1985).

In conclusion, if a species is to be managed in such a way that it will survive over the long term, it should not be considered a static entity but, like *conirostris,* a dynamic one constantly changing and capable of further change.

Appendixes

Appendix I. Scientific Names of Bird Species Referred to in the Text by Their Vernacular Names

Vernacular	Scientific
Large cactus finch	*Geospiza conirostris*
Cactus finch	*G. scandens*
Medium ground finch	*G. fortis*
Large ground finch	*G. magnirostris*
Small ground finch	*G. fuliginosa*
Sharp-beaked ground finch	*G. difficilis*
Warbler finch	*Certhidea olivacea*
Vegetarian finch	*Platyspiza crassirostris*
Large tree finch	*Camarhynchus psittacula*
Small tree finch	*C. parvulus*
Cocos finch	*Pinaroloxias inornata*
Song sparrow	*Melospiza melodia*
Swamp sparrow	*M. georgiana*
Fox sparrow	*Passerella iliaca*
Bridled sparrow	*Aimophila mysticalis*
Cinnamon-tailed sparrow	*A. sumichrasti*
Savannah sparrow	*Passerculus s. sandwichensis*
Ipswich sparrow	*P. sandwichensis princeps*
Seaside sparrow	*Ammospiza maritima*
Le Conte's sparrow	*A lecontei*
Sharp-tailed sparrow	*A. caudata*
White-crowned sparrow	*Zonotrichia leucophrys*
Dark-eyed junco	*Junco hyemalis*
Mexican junco	*J. phaeonotus*
Indigo bunting	*Passerina cyanea*
White-collared seedeater	*Sporophila torqueola*
Chestnut-sided seedeater	*S. telasco*
Parrot-billed seedeater	*S. (Neorhynchus) peruvianus*

Vernacular	Scientific
Crimson finch	*Rhodospingus cruentus*
Collared warbling finch	*Poospiza hispaniolensis*
Blue-black grassquit	*Volatinia jacarina*
Cardinal	*Cardinalis cardinalis*
Laysan finch	*Telespyza cantans*
Kona grosbeak	*Chloridops kona*
Greater koa finch	*Rhodacanthis palmeri*
O'u	*Psittarostra psittacea*
Palila	*Loxioides bailleui*
Small Tristan finch	*Nesospiza acunhae*
Gough Island finch	*Rowettia goughensis*
House finch	*Carpodacus mexicanus*
Scarlet rosefinch	*C. erythrinus*
Canary	*Serinus canaria*
Lesser goldfinch	*Spinus psaltria*
Gray-crowned rosy finch	*Leucosticte griseonucha*
Bullfinch	*Pyrrhula pyrrhula*
Blue chaffinch	*Fringilla teydea*
Chaffinch	*F. coelebs*
House sparrow	*Passer domesticus*
Eurasian tree sparrow	*P. montanus*
Zebra finch	*Taeniopygia guttata*
Red-winged blackbird	*Agelaius phoeniceus*
Eastern meadowlark	*Sturnella magna*
Western meadowlark	*S. neglecta*
Golden-winged warbler	*Vermivora chrysoptera*
Blue-winged warbler	*V. pinus*
Yellow warbler	*Dendroica petechia*
Marsh wren	*Cistothorus palustris*
Black-capped chickadee	*Parus atricapillus*
Great tit	*P. major*
Pied flycatcher	*Ficedula hypoleuca*
Collared flycatcher	*F. albicollis*
White-bearded manakin	*Manacus manacus*
Spotted antbird	*Hylophylax naevioides*
Bicolored antbird	*Gymnopithys bicolor*
Ocellated antbird	*Phaenostictus mcleanani*
Galápagos mockingbird	*Nesomimus parvulus*
Magpie	*Pica pica*
Scrub jay	*Aphelocoma coerulescens*
House martin	*Delichon urbica*
Meadow pipit	*Anthus pratensis*
Yellow-crowned parakeet	*Cyanoramphus auriceps*
Red-crowned parakeet	*C. novaezelandiae*

Vernacular	Scientific
Seychelles turtle dove	*Streptopelia picturata rostrata*
Madagascar turtle dove	*S. p. picturata*
Galápagos dove	*Zenaida galapagoensis*
European cuckoo	*Cuculus canorus*
Acorn woodpecker	*Melanerpes formicivorus*
Japanese quail	*Coturnix japonica*
Sparrowhawk	*Accipiter nisus*
Short-eared owl	*Asio flammeus*
Wedge-rumped storm petrel	*Oceanodroma tethys*
Band-rumped storm petrel	*O. castro*
Lava heron	*Butorides sundevalli*
Black-crowned night heron	*Nyctanassa violacea*
Masked booby	*Sula dactylatra*
Red-footed booby	*S. sula*

Appendix II. Annual Summaries of Reproductive Characteristics

Table A1 Mean Clutch Sizes

	Pairs		
Year	Inexperienced	Mixed	Experienced
1980	3.44 ± 0.18 8	3.59 ± 0.15 14	3.60 ± 0.11 11
1981	3.32 ± 0.15 19	3.79 ± 0.15 7	3.91 ± 0.10* 20
1982	3.44 ± 0.15 15	3.58 ± 0.22 12	3.45 ± 0.11 23
1983	3.91 ± 0.10 22	3.56 ± 0.20 20	4.16 ± 0.13 29
1986	3.74 ± 0.15 12	3.87 ± 0.12 14	3.72 ± 0.15 3
1987	4.25 ± 0.31 8	3.86 ± 0.13 14	4.02 ± 0.11 19
$\bar{x}$	3.68	3.71	3.81

Note: Either the male and female had bred before but not necessarily with each other (experienced pair), neither had bred before (inexperienced), or only one had previously bred (mixed). Sample size of pairs is shown in each case beneath the mean and standard error. A statistically significant difference between the means of experienced and inexperienced pairs in one year (t test) is indicated by * ($P < 0.05$).

Table A2 Mean Number of Clutches per Season, Excluding Replacement Clutches and Those Produced by Females That Changed Mates

	Pairs		
Year	Inexperienced	Mixed	Experienced
1980	1.7 ± 0.24 9	2.1 ± 0.23 8	2.7 ± 0.18** 7
1981	1.4 ± 0.11 19	1.9 ± 0.14 7	2.1 ± 0.09** 19
1982	1.7 ± 0.20 13	2.6 ± 0.30 7	2.6 ± 0.12** 23
1983	(2.2 ± 0.25) 27	(3.3 ± 0.53) 15	(4.1 ± 0.45) 17
1986	2.6 ± 0.29 9	2.7 ± 0.21 6	3.0 3
1987	1.7 3	4.0 2	4.0 ± 0.3** 12
$\bar{x}$	1.87	2.75	3.07

Note: Sample size is shown in each case beneath the mean and standard error. A statistically significant difference between the means of experienced and inexperienced pairs in a year (t test) is indicated by ** ($P < 0.01$). 1983 figures are placed in parentheses to reflect uncertainty in the number of clutches produced by many pairs in that year.

Table A3 Number of Nests That Failed Completely at Egg or Nestling Stage

	1980	1981	1982	1983	1986	1987	$\bar{x}$
Inexperienced pairs							
Failed	7	9	12	33	8	6	
Not failed	8	25	14	30	20	4	
% Failed	46.7	26.5	46.1	52.4	28.6	60.0	43.38
Mixed pairs							
Failed	3	5	8	34	5	15	
Not failed	21	11	17	23	23	20	
% Failed	14.3	31.2	32.0	59.6	21.7	42.9	33.62
Experienced pairs							
Failed	6	2	16	52	1	20	
Not failed	19	40	45	56	8	51	
% Failed	24.0	4.8	26.2	48.1	12.5	28.2	23.97

Table A4 Mean Number of Fledglings per Brood

	Pairs		
Year	Inexperienced	Mixed	Experienced
1980	2.22 ± 0.51 9	2.44 ± 0.27 14	2.43 ± 0.24 12
1981	2.15 ± 0.29 21	2.18 ± 0.45 7	3.14 ± 0.19 19
1982	1.39 ± 0.32 15	1.57 ± 0.31 13	2.05 ± 0.23 23
1983	1.54 ± 0.28 27	1.21 ± 0.23 22	1.84 ± 0.26 32
1986	2.33 ± 0.36 13	2.60 ± 0.28 15	2.53 ± 0.74 3
1987	1.34 ± 0.45 4	1.83 ± 0.31 14	2.43 ± 0.23 18
$\bar{x}$	1.83	1.97	2.40

Note: Sample size of pairs is shown in each case beneath the mean and standard error.

Table A5 Mean Annual Production of Fledglings per Pair in Those Years in Which It Could Be Accurately Estimated

	Pairs		
Year	Inexperienced	Mixed	Experienced
1980	3.0 ± 0.73 9	4.6 ± 1.00 8	6.3 ± 0.64** 7
1981	3.4 ± 0.54 19	4.4 ± 0.84 7	6.5 ± 0.38** 19
1982	2.3 ± 0.61 13	3.3 ± 0.89 11	5.6 ± 0.64** 23
1986	5.3 ± 1.08 10	5.5 ± 1.07 8	7.7 ± 2.19 3
1987	2.3 3	6.0 2	10.8 ± 0.96** 11
$\bar{x}$	3.27	4.77	7.38

Note: Birds who changed mates were excluded from the calculations. Sample size is shown in each case beneath the mean and standard error. A statistically significant difference between the means of experienced and inexperienced pairs in one year (t test) is indicated by ** ($P < 0.01$).

Appendix III. Demographic Characteristics of *A* Males and *B* Males

The question we address here is whether the two song groups of males differ demographically. They could differ because they have different intrinsic properties, or because they experience environmental factors differently. In the latter case, being relatively rare (like *A* males), for example, may confer a demographic advantage. If so, we may gain insights into why the frequencies of the two song groups were not equal or even more disparate than they were.

The two groups could differ in survival, recruitment, and reproduction, hence in relative success and rates of turnover.

Survival and Recruitment. Survival prior to breeding is estimated by comparing the proportions of banded fledglings of the two groups that survived to breed either in or outside the study areas. In the years 1978–86, 520 banded offspring fledged from the nests of song *A* males in areas 1 and 2, and 667 fledged from the nests of *B* males. Recruitment to the breeding population in the years 1980–87 was low from each group. Twenty-four (4.6 percent) of the offspring of *A* males and 26 (3.9 percent) of the offspring of *B* males survived to breed. The small difference in proportions can be attributed to chance ($\chi^2 = 0.13, P > 0.1$). We come to the same conclusion when considering just the male recruits. The numbers are 10 (1.9 percent) from the *A* males and 12 (1.8 percent) from the *B* males.

The two groups did not differ in the average age at first reproduction. After first reproduction, survival of the two groups was similar but not identical (Fig. A1). Relatively more *A* males disappeared after two years of age than *B* males, while after four years the difference was reversed. However, the frequency distributions of survival after first breeding do not differ (two-tailed Kolmogorov-Smirnov test, $P > 0.1$), and the small difference in central tendencies is not significant (median test, $\chi^2 = 0.29, P > 0.1$).

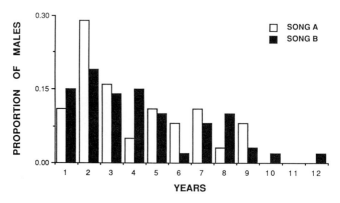

Figure A1 Number of years lived by *A* males and *B* males. Includes only birds breeding up to and during 1982. In 1988 two *A* males and seven *B* males were still alive. They were assumed to die in that year (in fact, four *B* males were alive in February 1989).

Table A6 Numerical Changes of *A* and *B* Males on Study Area 1

	Changes between Breeding Seasons					
	1978–80	1980–81	1981–82	1982–83	1983–86	1986–87
A males						
Surviving breeders	4	11	13	10	4	10
Loss of breeders	3	2	4	5	28	3
% loss	**42.9**	15.4	23.5	**33.3**	**87.5**	**23.1**
Gain of breeders	9	7	2	22	8	12
% gain	**69.2**	38.9	13.3	**68.8**	**66.7**	50.5
% turnover	60.0	29.0	18.8	**57.4**	**81.8**	**42.9**
New males	10	8	8	26	10	2
% mating success						
of new males	**90.0**	**75.0**	37.5	84.6	**80.6**	**100.0**
B males						
Surviving breeders	8	13	17	23	9	16
Loss of breeders	2	7	14	5	43	3
% loss	20.0	**35.0**	**45.2**	17.9	82.7	15.3
Gain of breeders	12	19	10	29	8	19
% gain	60.0	**59.4**	**37.0**	55.8	47.1	**54.3**
% turnover	**66.7**	**50.0**	**41.4**	42.5	73.9	40.7
New males	18	27	16	32	11	4
% mating success						
of new males	66.7	66.7	**68.8**	**90.6**	72.7	75.0

Note: Turnover is an index of gains and losses (see p. 302). In comparison of percentage values between *A* and *B* males, the larger value is shown in **boldface**.

Reproduction. On average *A* males and *B* males were approximately equally successful in attracting mates (Table A6). Type *A* males more often had the higher mating success, but the differences were generally small, and the greatest difference (in 1982) favored the *B* males.

From this starting point of rough equality, *A* males did better than the *B* males; they experienced higher breeding success than *B* males on area 1 in every year of the study (Table A7). They consistently fledged a larger number of offspring, and to a statistically significant extent in 1981 and 1982 (*t* tests, $P < 0.05$). The consistent difference did not arise from higher clutch sizes or larger numbers of clutches per season but from higher nesting success. Type *A* males suffered proportionately fewer total nest losses in each year (Table A7).

On average about half of the total nest failures can be attributed to predation by mockingbirds and owls (Table 4.5). A few of the remainder are caused by nests falling to the ground, especially in the high winds and heavy rain in El Niño years. Most of the remaining nests that fail are abandoned. We do not know if predators play a role in the abandonment. Restricting attention to just those nests known to be predated, but including all nests and not just those of pairs studied fully in a season, we find that *B* males consistently suffered relatively more complete nest

Table A7 Breeding Success of the Two Song Groups of Males on Study Area 1, for Mated Males Studied Fully in a Season

Year	Song Group	Males	Nests	% Nests Successful	Clutches $\bar{x} \pm$ SE	Fledglings $\bar{x} \pm$ SE
1978	A	7	19	94.7	3.0 ± 0.2	8.1 ± 0.7
	B	8	22	72.7	3.1 ± 0.4	7.6 ± 1.3
1980	A	13	26	80.7	2.0 ± 0.3	4.9 ± 0.5
	B	22	51	72.6	2.3 ± 0.2	4.6 ± 0.6
1981	A	17	34	91.2	2.0 ± 0.2	5.5 ± 0.6
	B	29	58	79.3	1.9 ± 0.1	4.0 ± 0.4
1982	A	17	27	81.5	2.5 ± 0.2	5.6 ± 0.9
	B	27	54	72.2	2.6 ± 0.1	4.3 ± 0.5
1986	A	11	25	84.0	2.3 ± 0.3	5.8 ± 0.7
	B	13	28	75.0	2.1 ± 0.2	5.1 ± 1.0
1987	A	11	33	78.8	3.0 ± 0.4	8.1 ± 1.1
	B	16	60	71.7	3.7 ± 0.2	7.7 ± 0.9

losses at the hands (and beaks) of predators in the breeding seasons. The consistent difference is not expected by chance (two-tailed sign test, $P < 0.02$), although the test is weakened by the lack of complete independence of the samples; some birds bred in two or more years. The consistent difference must be due at least partly to these experienced birds, because the independent samples of males of the two types breeding for the first time did not differ ($P > 0.1$).

Small sample sizes and small differences in nesting success in any one year make it difficult to probe the causes of the breeding success difference among the many factors that might be influential. Contributing factors include ages of the males, ages of their mates, past breeding experience, territory arrangement, temporal patterns of breeding within each season (Chapter 7), as well as the distribution and hunting behavior of predators.

We doubt if A males are intrinsically better than B males at avoiding nest predation, because if anything the loudness of their song makes them more conspicuous than B males to predators. Extrinsic factors are probably more important. Predators may hunt in a frequency-dependent manner, penalizing the more common B males proportionately more, although we have no idea why. An alternative cause lies in the breeding pattern. Experienced females tend to breed earlier than inexperienced ones, and to mate with males on territories with heterotypic neighbors in some years (Chapter 7). Type A males more often have heterotypic neighbors than B males, and so gain a frequency-dependent mating advantage. In short, the higher nesting success of A males may flow more from properties of their mates (age and experience) than from their own.

Relative Success. The consistently higher breeding success of A males should have led to a steady increase in the frequency of A males in the population, all else being equal. It did not: all other things were not equal. The frequency of *A* males declined from 1978 to 1980, remained constant for two years, increased in 1983, and declined once again between 1986 and 1987 (Table 6.4).

In the whole study period, therefore, the frequency increased only once. It increased following two years of significantly higher fledging success, ultimately as a result of that success. This is the only time that a change in frequency of *A* males can be linked to their previous breeding success.

For the rest of the data, there are three reasons why the frequency of *A* males declined when their breeding had been superior. The first is heavier mortality of the previous breeders, without compensating recruitment. This could arise entirely by chance; for example, if they happened to be older on average than the *B* males. In support of this possibility, losses of previous breeders were relatively higher among the *A* males at both times of decline in their frequencies, from 1978 to 1980, and again from 1986 to 1987 (Table A6).

The second possibility is heavier mortality among the *A* males from fledging to breeding. This explanation is not likely to be correct because direct evidence contradicts it. As discussed above, measures of recruitment of locally born birds show the two groups to survive equally well (or poorly). The figures of 1.9 percent for *A* males and 1.8 percent for *B* males need to be adjusted to remove those individuals who bred outside area 1, and to include those who established territories in area 1 but failed to get a mate. The adjusted figures are 1.1 percent for *A* males and 0.9 percent for B males.

The third possibility is that the frequency of *A* males declines as a result of a relatively greater influx of *B* males from the main source area of recruits, namely adjacent areas to the north and to the west. The greater influx, in turn, can be explained by a higher proportion of *B* males breeding in those areas, and by an amount which more than offsets a possibly higher per capita fledging success of *A* males. However, the decline in *A* male frequency from 1978 to 1980 coincided with a relatively low influx of *B* males. In the decline of 1986–87, the influx rate was marginally higher for *B* males than for *A* males.

Dispersal of potential recruits between areas should lead eventually to a stable homogeneous distribution of frequencies across the island, unless fluctuations in the fortunes of the two groups occur in different local areas out of phase with each other. Local fluctuations do occur, as can be seen in Table 6.4, but we do not know if they occur out of phase. The frequency of *A* males varied irregularly in area 1 from 48 to 31 percent.

Thus both stochastic and deterministic processes are likely to be responsible for changes in the frequencies of the two song groups, with events outside the main study areas contributing to events on it.

Turnover. As a further exploration of the changes in numbers of the two groups, we have calculated their rates of turnover on study area 1. Turnover is the product of gains and losses to the breeders. We have adapted a formula from Diamond (1969), who used extinction and immigration rates to quantify turnover of island avifaunas, to compare the two groups of males:

$$\frac{100 \ (L \ + \ G)}{N_t \ + \ N_{t+1}}$$

Thus turnover (T) of each group between one breeding season (t) and the next $(t + 1)$ is the sum of losses (L) and gains (G) expressed as a function of numbers in one breeding season (N_t) and numbers in the next (N_{t+1}).

Turnover rates were higher for B males than for A males in the first half of the study, and the reverse was true in the second half (Table A6). The switch came between 1982 and 1983, when the frequency of A males increased (Table 6.4) following two years of relatively high breeding success. The frequency of A males increased then because the relatively high recruitment gain more than compensated for the relatively high loss. In general the two components of turnover, gains and losses, contributed roughly equally to the turnover, in that the group with the higher turnover usually both lost and gained more individuals than the other.

The pattern of turnover may indicate that the two groups of males were influenced by different environmental factors or by the same factor to different degrees. Our simple exploration of this possibility finds no support for it. The numbers of the two groups do not fluctuate independently. In fact, there is a very strong positive correlation between them ($r = 0.98$, $P < 0.01$). Moreover, losses from each group are correlated, separately, with both numbers of A males and numbers of B males at time t ($P < 0.05$ in each case). In contrast to these strong correlations ($0.77 - 0.93$), gains in each group were not correlated with numbers of each group or with numbers of fledglings produced by each group in the preceding breeding season ($r = -0.23$ to 0.68, $P > 0.1$).

These results show two things. First, annual losses from each of the two groups vary in a density-dependent manner, whereas gains in each group vary in a density-independent manner (see also Chapters 3 and 4). Second, they provide no reason for believing that higher gains or losses are under different environmental control in the two groups of males.

Conclusion. Our overall conclusion is that A males and B males have similar but not identical demographies. The small differences that existed, at least during the study period on area 1, are not sufficient to account for the average frequencies of A and B males or for the annual variation in those frequencies. The most appealing hypothesis is that the rarer song group has a frequency-dependent mating and breeding advantage, which tends to produce a 1:1 ratio of the two groups, but that random events cause deviations with much stochastic fluctuation (Chapter 10) from that ratio.

Appendix IV. Effective Population Size and Neighborhood Size

In small finite populations in the absence of mutation, migration, and selection, genetic variance becomes progressively reduced through random drift (Wright 1969). This comes about through the random sampling of alleles from one generation to the next. The rate of loss can be estimated by calculating the genetically effective population size (see also Chapter 12). The effective population size (N_e) is the number that would give rise to the same sample variance in alleles (or the same rate of increase in genetic similarity among individuals) as is observed in the actual population, if mutation, migration, and selection did not occur (Falconer 1981).

For the *conirostris* population, the effective population size was calculated by using the formula for populations with nonoverlapping generations given by Lande and Barrowclough (1987):

$$N_e = (N\bar{k} - 1)/[\bar{k} + (\sigma_k^2/\bar{k}) - 1].$$

N is the actual population size, and $\bar{k}$ and σ_k^2 are the mean number of recruits per parent and the variance, respectively.

Our calculations yield 690.2 for the average population size, taken as the harmonic mean number of birds breeding on the island in the breeding seasons during 1978 to 1983. This period was selected to allow us to assemble a virtually complete register of the number of recruits per parent. To obtain the mean, we started with the number of breeding birds in study area 1 in each breeding season and multiplied by 10, because from aerial photographs we determined that area 1 is 10 percent of the island area occupied by cactus and by breeding *conirostris* (see also Chapter 2). The mean (0.3) and variance (0.53) of progeny production were calculated for the combined sample of breeding males and females, because the frequency distributions of lifetime production of recruits by the two sexes are almost identical.

The effective population size is estimated to be 193.2. It is much lower than the actual population sizes, which varied from 380 breeding individuals in 1978, to 1780 in 1983. The harmonic mean is more influenced by the low values than by the high ones. But the main factor responsible for the low effective population size is the large variance in the number of progeny produced per parent. It is almost twice as large as the mean because only 20.1 percent of the breeding population produce recruits.

Strictly we should use a model for populations with overlapping generations, but it is difficult to apply (Lande and Barrowclough 1987). It requires an adjustment to be made to the mean recruitment so that it equals 1.0, and it requires a concomitant adjustment to the variance. Estimations from the model for nonoverlapping generations should be approximately correct since the period of breeding they cover spans only six years, or two generations.

For the purpose of calculating the effective population size, we assumed that mating was random throughout the island, when in fact most matings occur locally. The nonrandomness comes about because individual dispersal distances from natal to breeding site are low in both sexes (Chapter 6). Under these circumstances there is an inevitable inertia to the spread of alleles across the island, reinforced by the slow turnover of breeders as a result of males retaining the same territory for life (Chapter 6) and of the lifespans of both sexes being long (Chapter 3).

Wright (1943) considered the effect of low dispersal on local genetic differentiation in an isolation-by-distance model, in which he treated a population as a series of interconnecting neighborhoods. Using Barrowclough's (1980) method, we calculated neighborhood size (N) from the formula $N = 4p\sigma^2$, where N is the number of individuals in a circle or radius 2σ, p is the population density, and σ^2 is the mean square dispersal distance. The components are $\sigma^2 = 0.39$ and $p = 54.3$/sq km from our data. This gives an estimate of the neighborhood size of 266.1 individuals, covering an area of 4.90 km^2 or 38.6 percent of the island area occupied by cactus (12.7 km^2).

If neighborhoods were discrete, then the effective population size would be much smaller than we have calculated it to be. By substituting 266.1 (neighorhood size) for 690.2 (harmonic mean total population size) in the first equation, we obtain an estimate for the effective population size of 73.9. But it must be emphasized that neighborhoods are interconnected and are not discrete units. Therefore the effective population size is larger than this, probably somewhere between 75 and 190 individuals. The main conclusion we draw is that gene frequencies in one area of the island could differ temporarily from those in another, giving rise to transient ecological and morphological subdivision. This arises from the nonrandom, local dispersal of potential breeders and is not the result of local heterogeneity in habitat features and selection pressures. In fact, the island is quite homogeneous.

Our data on the frequencies of nestling bill colors, which are considered to be under fairly simple genetic control and selectively neutral or nearly neutral (Chapter 8), are consistent with this view of local differentiation. Local differentiation of this sort may have contributed to the creation of the differences in average beak length between A males and B males (Chapter 10), and to differences in song subtype frequencies. Subtypes of song, which are not genetically determined but are transmitted culturally in fine detail from father to son (Chapter 6), show a patchy distribution across the island (Fig. A2).

The phenomenon of spatial differentiation in allele frequencies can locally increase the rate of inbreeding. Where dispersal is low and individuals are long-lived, many sibs that were not reared together are available as potential mates. Under these conditions there is a moderately high probability that the gametes from two parents will contain alleles identical by descent. If there is a cost to inbreeding in the elimination of progeny, then individuals that have the ability to recognize and avoid mating with close kin will be favored by selection. In spite of the high probability by chance of close relatives mating with each other (B. R. Grant 1984), we know of no incestuous matings and we recorded no female mated with a male of the same song subtype as her father. There is a possibility that females use detailed characteristics of song to avoid mating with close kin (Chapter 6), although our samples are too small to detect it. More detailed studies of island populations of birds have failed to demonstrate an avoidance of mating with close relatives (van Noordwijk and Scharloo 1981; van Tienderen and van Noordwijk 1988; Gibbs and

Grant 1989), although some evidence for inbreeding depression has been found (van Noordwijk and Scharloo 1981).

This discussion of the effects of small population size on its genetic structure has ignored mutation, migration, and selection. We discuss their interaction with random drift in Chapter 12.

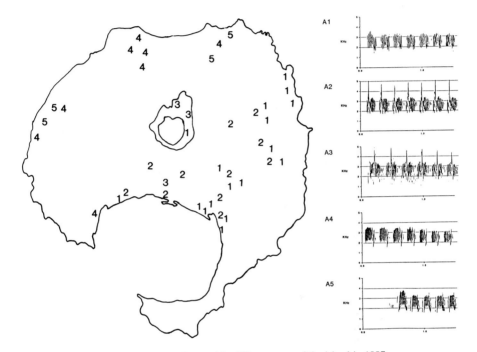

Figure A2 Frequencies of subtypes of song *A* in different parts of the island in 1987.

Appendix V. Nestling Beak Color Morphs Produced by *A* Males and *B* Males

Table A8 Frequencies of Nestling Beak Color Morphs among Offspring Produced by *A* Males and *B* Males on Areas 1 and 2

Year	Male Type	Males	Pairs	Nests	Offspring Pink	Yellow	Frequency of Yellows	Frequency of Males Producing Yellows
Area 1								
1978	A	7	7	19	38	21	0.36	1.00
	B	9	9	25	65	14	0.18	0.56
1980	A	13	14	22	43	26	0.38	0.77
	B	16	18	32	79	14	0.15	0.44
1981	A	17	18	30	67	26	0.28	0.71
	B	26	26	42	103	24	0.19	0.46
1982	A	15	15	31	74	22	0.23	0.80
	B	22	30	46	104	27	0.21	0.41
1983	A	32	36	58	100	47	0.32	0.81
	B	41	47	73	148	42	0.22	0.61
1984	A	—	—	—	—	—	—	—
	B	3	3	3	7	1	0.12	0.33
1986	A	12	12	23	51	16	0.24	0.58
	B	16	17	27	69	10	0.13	0.37
1987	A	12	15	31	68	34	0.33	0.67
	B	18	21	47	121	29	0.19	0.44
Area 2								
1980	A	2	2	5	11	3	0.21	0.50
	B	6	10	10	20	4	0.17	0.33
1981	A	3	3	3	9	1	0.10	0.33
	B	3	3	3	2	4	0.67	0.67
1982	A	2	2	4	9	1	0.10	0.50
	B	7	8	16	30	11	0.27	0.57
1983	A	7	7	10	28	13	0.32	0.29
	B	15	15	19	29	35	0.55	0.67
1986	A	4	4	7	18	4	0.18	0.75
	B	5	6	10	21	10	0.32	0.80
1987	A	5	5	13	38	6	0.14	0.40
	B	11	14	33	85	30	0.26	0.82

Note: The number of pairs often exceeds the number of males owing to polygamy and mate changes.

References

Abbott, I. 1978. The significance of morphological variation in the finch species on Gough, Inaccessible and Nightingale islands, South Atlantic Ocean. *Journal of Zoology, London* 184:119–25.

Abbott, I., L. K. Abbott, and P. R. Grant. 1977. Comparative ecology of Galápagos ground finches (*Geospiza* Gould): Evaluation of the importance of floristic diversity and interspecific competition. *Ecological Monographs* 47:151–84.

Alatalo, R. V., L. Gustafsson, and A. Lundberg. 1984. High frequency of cuckoldry in pied and collared flycatchers. *Oikos* 42:41–47.

Alatalo, R. V., and A. Lundberg. 1986. Heritability and selection on tarsus length in the pied flycatcher (*Ficedula hypoleuca*). *Evolution* 40:574–83.

Alatalo, R. V., A. Lundberg, and C. Glynn. 1986. Female pied flycatchers choose territory quality and not male characteristics. *Nature* 323:152–53.

Anderson, E. F., and D. L. Walkington. 1967. A study of some Neotropical *Opuntias*. Unpublished summary of MS in library of Charles Darwin Research Station, Galápagos.

Anderson, T. R. 1977. Reproductive responses of sparrows to a superabundant food supply. *Condor* 79:205–8.

Andersson, M. 1982a. Sexual selection, natural selection and quality advertisement. *Biological Journal of the Linnean Society* 17:375–93.

———. 1982b. Female choice selects for extreme tail length in a widowbird. *Nature* 299:818–20.

———. 1986. Evolution of condition-dependent sex ornaments and mating preference: Sexual selection based on viability differences. *Evolution* 40:804–16.

Arcese, P., and J. N. M. Smith. 1988. Effects of population density and supplemental food on reproduction in song sparrows. *Journal of Animal Ecology* 57:119–36.

Arnold, S. J. 1983a. Sexual selection: The interface of theory and empiricism. In *Mate choice*, ed. P. Bateson, 67–107. Cambridge: Cambridge University Press.

———. 1983b. Morphology, performance and fitness. *American Zoologist* 23:347–61.

Arnold, S. J., and M. J. Wade. 1984a. On the measurement of natural and sexual selection: Theory. *Evolution* 38:709–19.

————. 1984b. On the measurement of natural and sexual selection: Applications. *Evolution* 38:720–34.

Ashmole, N. P. 1963. The regulation of numbers of tropical oceanic birds. *Ibis* 103b:458–73.

Baker, M. C., and P. Marler. 1980. Behavioural adaptations that constrain the gene pool in vertebrates. In *Evolution of social behaviour: Hypotheses and empirical tests,* ed. H. Markl, 59–80. Dahlem Conference, 1980. Weinheim: Verlag Chemie.

Baker, M. C., L. R. Mewaldt, and R. M. Stewart. 1981. Demography of white-crowned sparrows (*Zonotrichia leucophrys nuttallii*). *Ecology* 62:636–44.

Banks, R. C. 1964. Geographic variation in the white-crowned sparrow *Zono trichia leucophrys*. *University of California Publications in Zoology* 70:1–123.

Bannerman, D. A., and W. M. Bannerman. 1966. *Birds of the Atlantic islands.* Vol. 3. *A history of the birds of the Azores.* Edinburgh: Oliver and Boyd.

Baptista, L. F., and L. Petrinovich. 1984. Social interaction, sensitive phases and the song template hypothesis in the white-crowned sparrow. *Animal Behaviour* 32:172–81.

Barrowclough, G. F. 1980. Gene flow, effective population sizes, and genetic variance components in birds. *Evolution* 34:789–98.

Bateson, P. 1982. Preferences for cousins in Japanese quail. *Nature* 295:236–37.

————. 1983. Optimal outbreeding. In *Mate choice,* ed. P. Bateson, 257–78. Cambridge: Cambridge University Press.

Bateson, W. 1902. Experiments with poultry. *Reports of the Evolution Commission of the Royal Society* 1:87–124.

Beebe, W. 1924. *Galápagos: World's end.* New York: Putnams.

————. 1926. *Arcturus adventure.* New York: Putnams.

Beecher, M. D. 1982. Signature systems and kin recognition. *American Zoologist* 23:477–90.

Beecher, M. D., and I. M. Beecher. 1979. Sociobiology of bank swallows: Reproductive strategy of the male. *Science* 205:1282–85.

Berndt, R., and H. Sternberg. 1963. Ist die Mortalitatsrate adulter *Ficedula hypoleuca* wirklich unabhängig von Lebensalter? *Proceedings of the 13th International Ornithological Congress* (Ithaca, 1962), 675–84.

Bjilsma, R. G. 1982. Breeding season, clutch size and breeding success in the bullfinch *Pyrrhula pyrrhula*. *Ardea* 70:25–30.

Black, J. 1974. *Galápagos: Archipiélago de Colón.* Quito: Imprenta Europa.

Boag, P. T. 1983. The heritability of external morphology in Darwin's ground finches (*Geospiza*) on Isla Daphne Major, Galápagos. *Evolution* 37:877–94.

Boag, P. T., and P. R. Grant. 1978. Heritability of external morphology in Darwin's Finches. *Nature* 274:793–94.

————. 1981. Intense natural selection in a population of Darwin's Finches (*Geospizinae*) in the Galápagos. *Science* 214:82–85.

————. 1984. Darwin's Finches (*Geospiza*) on Isla Daphne Major, Galápagos: Breeding and feeding ecology in a climatically variable environment. *Ecological Monographs* 54:463–89.

Boag, P. T., and A. van Noordwijk. 1987. Quantitative genetics. In *Avian genetics: A population and ecological approach*, ed. F. Cooke and P. A. Buckley, 45–78. New York: Academic Press.

Böhner, J. 1983. Song learning in the zebra finch (*Taeniopygia guttata*): Selectivity in the choice of tutor and accuracy of song copies. *Animal Behaviour* 31:231–37.

Borgia, G. 1985. Bower quality, number of decorations, and mating success of male satin bowerbirds (*Ptilinorhynchus violaceus*): An experimental analysis. *Animal Behaviour* 33:266–71.

Botkin, D. B., and R. S. Miller. 1974. Mortality rates and survival of birds. *American Naturalist* 108:181–92.

Bowman, R. I. 1961. Morphological differentiation and adaptation in the Galápagos finches. *University of California Publications in Zoology* 58:1–302.

———. 1979. Adaptive morphology of song dialects in Darwin's Finches. *Journal für Ornithologie* 120:353–89.

———. 1983. The evolution of song in Darwin's Finches. In *Patterns of evolution in Galápagos organisms*, ed. R. I. Bowman, M. Berson, and A. E. Leviton, 237–537. San Francisco: American Association for the Advancement of Science, Pacific Division.

Boyce, M. S. 1984. Restitution of *r*- and *K*-selection as a model of density dependent natural selection. *Annual Review of Ecology and Systematics* 15:427–47.

Bradbury, J. W., and M. Andersson, eds. 1987. *Sexual selection: Testing the alternatives*. New York: Wiley.

Brockway, B. F. 1969. Roles of budgerigar vocalizations in the integration of breeding behaviour. In *Bird vocalizations*, ed. R. A. Hinde, 131–58. Cambridge: Cambridge University Press.

Brooks, L. D. 1988. The evolution of recombination rates. In *The evolution of sex*, ed. R. E. Michod and B. R. Levin, 87–105. Sunderland, Mass.: Sinauer.

Broughey, M. J., and N. S. Thompson. 1981. Song variety in the brown thrasher (*Toxostoma rufum*). *Zeitschrift für Tierpsychologie* 56:47–58.

Brown, D. 1988. Components of lifetime reproductive success. In *Reproductive success*, ed. T. H. Clutton-Brock, 439–53. Chicago: University of Chicago Press.

Bryant, D. M. 1988. Lifetime reproductive success of house martins. In *Reproductive success*, ed. T. H. Clutton-Brock, 173–88. Chicago: University of Chicago Press.

Bryant, E. H., S. A. McCommas, and L. M. Combs. 1986. The effect of an experimental bottleneck upon quantitative genetic variation in the housefly. *Genetics* 114:1191–211.

Burke, T., and M. W. Bruford. 1987. DNA fingerprinting in birds. *Nature* 327:149–52.

Burley, N. 1981. Sex-ratio manipulation and selection for attractiveness. *Science* 211:721–22.

———. 1986. Sexual selection for aesthetic traits in species with biparental care. *American Naturalist* 127:415–45.

Burley, N., G. Krantzberg, and P. Radman. 1982. Influence of colour-banding on the conspecific preferences of zebra finches. *Animal Behaviour* 30:444–55.

Bush, G. L. 1975. Modes of animal speciation. *Annual Review of Ecology and Systematics* 6:339–64.

Cade, T. J. 1983. Hybridization and gene exchange among birds in relation to conservation. In *Genetics and conservation,* ed. C. M. Schonewald-Cox, S. M. Chambers, B. MacBryde, and W. L. Thomas, 288–309. Menlo Park, Calif.: Benjamin/Cummings.

Calder, W. A., III. 1984. *Size, function, and life history.* Cambridge, Mass.: Harvard University Press.

Cane, M. A. 1983. Oceanographic events during El Niño. *Science* 222:1189–95.

Cane, M. A., and S. E. Zebiak. 1985. A theory for El Niño and the southern oscillation. *Science* 228:1085–87.

Carson, H. L., and A. R. Templeton. 1984. Genetic revolutions in relation to speciation phenomena: The founding of new populations. *Annual Review of Ecology and Systematics* 15:97–131.

Catchpole, C. K. 1980. Sexual selection and the evolution of complex song among European warblers of the genus *Acrocephalus. Behaviour* 74:149–66.

Catchpole, C. K., J. Dittami, and B. Leisler. 1984. Differential responses to male song repertoires in female songbirds implanted with oestradiol. *Nature* 312:563–64.

Cate, C. ten, and G. Mug. 1984. The development of mate choice in zebra finch females. *Behaviour* 90:125–50.

Cavalli-Sforza, L. L., M. W. Feldman, K. H. Chen, and S. M. Dornbusch. 1982. Theory and observation in cultural transmission. *Science* 218:19–27.

Charlesworth, B. 1980. *Evolution in age-structured populations.* Cambridge: Cambridge University Press.

Charlesworth, D., and B. Charlesworth. 1987. Inbreeding depression and its evolutionary consequences. *Annual Review of Ecology and Systematics* 18:237–68.

Charnov, E. L., and W. M. Schaffer. 1973. Life history consequences of natural selection: Cole's result revisited. *American Naturalist* 107:791–93.

Cliff, A. D., and J. K. Ord. 1973. *Spatial autocorrelation.* London: Pion.

Clutton-Brock, T. H., ed. 1988. *Reproductive success.* Chicago: University of Chicago Press.

Cody, M. L. 1966. A general theory of clutch size. *Evolution* 20:174–84.

————. 1971. Ecological aspects of reproduction. In *Avian biology,* ed. D. S. Farner and J. R. King, 1:461–512. New York: Academic Press.

Colinvaux, P. A. 1972. Climate and the Galápagos Islands. *Nature* 240:17–20.

————. 1984. The Galápagos climate: Present and past. In *Galápagos,* ed. R. Perry, 55–69. Oxford: Pergamon Press.

Collias, N. E., E. C. Collias, C. H. Jacobs, C. R. Cox, and F. A. McAlary. 1986. Old age and breeding behavior in a tropical passerine bird *Ploceus cucullatus* under controlled conditions. *Auk* 103:408–19.

Connor, E. F., and D. S. Simberloff. 1978. Species number and compositional similarity of the Galápagos flora and avifauna. *Ecological Monographs* 48:219–48.

Cooke, F. 1978. Early learning and its effect on population structure: Studies of a wild population of snow geese. *Zeitschrift für Tierpsychologie* 46:344–58.

―――. 1987. Synthesis II—Moulding genetic variation. In *Avian genetics: A population and ecological approach,* ed. F. Cooke and P. A. Buckley, 355–59. New York: Academic Press.

Corbin, K. W. 1987. Geographic variation and speciation. In *Avian genetics: A population and ecological approach,* ed. F. Cooke and P. A. Buckley, 321–54. New York: Academic Press.

Coulson, J. C., and C. S. Thomas. 1983. Mate choice in the kittiwake gull. In *Mate choice,* ed. P. Bateson, 361–76. Cambridge: Cambridge University Press.

Cox, A. 1983. Ages of the Galápagos Islands. In *Patterns of evolution in Galápagos organisms,* ed. R. I. Bowman, M. Berson, and A. E. Leviton, 11–23. San Francisco: American Association for the Advancement of Science, Pacific Division.

Crow, J. F., and M. Kimura. 1970. *An introduction to population genetics theory.* New York: Harper and Row.

Crowell, K. L., and S. I. Rothstein. 1981. Clutch size and breeding strategies among Bermudan and North American passerines. *Ibis* 123:42–50.

Curio, E. 1965. Zur geographischen Variation des Feinderkennens einiger Darwinfinken (Geospizidae). *Zoologische Anzeiger,* suppl. 28:466–92.

―――. 1969. Funktionsweise und Stammesgeschichte des Flugfeinderkennens einiger Darwinfinken (Geospizidae). *Zeitschrift für Tierpsychologie* 26:394–487.

―――. 1983. Why do young birds reproduce less well? *Ibis* 125:400–404.

Curio, E., and P. Kramer. 1965. *Geospiza conirostris* auf Abingdon und Wenman entdeckt. *Journal für Ornithologie* 106:355–57.

Curry, R. L., and P. R. Grant. 1989. Demography of the cooperatively breeding Galápagos mockingbird, *Nesomimus parvulus,* in a climatically variable environment. *Journal of Animal Ecology* 58:441–63.

Darwin, C. R. 1871. *The descent of man, and selection in relation to sex.* London: John Murray.

Davies, N. B., and A. Lundberg. 1984. Food distribution and a variable mating system in the dunnock, *Prunella modularis. Journal of Animal Ecology* 53:895–912.

Davies, S. J. J. F. 1976. Environmental variables and the biology of Australian arid zone birds. *Proceedings of the 16th International Ornithological Congress* (Canberra, 1974), 481–88.

―――. 1977. The timing of breeding by the zebra finch *Taeniopygia castanotis* at Mileura, Western Australia. *Ibis* 119:369–72.

Dawson, E. Y. 1962. Cacti of the Galápagos Islands and coastal Ecuador. *Cactus and Succulent Journal* 34:67–74, 99–105.

―――. 1965. Further studies of *Opuntia* in the Galápagos archipelago. *Cactus and Succulent Journal* 37:135–48.

Deevey, E. S., Jr. 1947. Lifetables for natural populations of animals. *Quarterly Review of Biology* 22:283–314.

Dhondt, A. A. 1982. Heritability of blue tit tarsus length from normal and cross-fostered broods. *Evolution* 36:418–19.

———. 1985. Do old great tits forego breeding? *Auk* 102:870–72.

———. 1987. Reproduction and survival of polygynous and monogamous blue tit *Parus caeruleus. Ibis* 129:327–34.

Diamond, J. M. 1969. Avifaunal equilibria and species turnover rates on the Channel Islands of California. *Proceedings of the National Academy of Sciences USA* 64:57–63.

———. 1985. Population processes in island birds: Immigration, extinction and fluctuations. In *Conservation of island birds,* ed. P. J. Moors, 85–100. International Council for Bird Preservation, Technical Publications No. 3. Norwich, England: Paston Press.

———. 1988. Experimental study of bower decoration by the bowerbird *Amblyornis inornatus* using colored poker chips. *American Naturalist* 131:631–53.

Dickinson, H., and J. Antonovics. 1973. Theoretical considerations of sympatric divergence. *American Naturalist* 107:256–74.

Dooling, R. Y. 1982. Ontogeny of song recognition in birds. *American Zoologist* 22:571–80.

Downhower, J. F. 1976. Darwin's Finches and the evolution of sexual dimorphism in body size. *Nature* 263:558–63.

———. 1978. Observations on the nesting of the small ground finch *Geospiza fuliginosa* and the large cactus ground finch *G. conirostris* on Española, Galápagos. *Ibis* 120:340–46.

Downhower, J. F., and C. H. Racine. 1976. Darwin's Finches and *Croton scouleri:* An analysis of the consequences of seed predation. *Biotropica* 8:66–70.

Dunn, L. C. 1925. The genetic relation of some shank colors of the domestic fowl. *Anatomical Record* 31:343–44.

Dunning, J. B., Jr. 1984. Body weights of 686 species of North American birds. Western Bird Banding Association, Monograph No. 1. Cave Creek, AZ: Eldon Publishing.

Ehrlich, P. R. 1988. The loss of diversity: Causes and consequences. In *Biodiversity,* ed. E. O. Wilson, 21–27. Washington, D.C.: National Academy Press.

Ehrlich, P. R., and P. H. Raven. 1969. Differentiation of populations. *Science* 165:1228–31.

Ekman, J., and C. Askenmo. 1986. Reproductive cost, age-specific survival and a comparison of the reproductive strategy in two European tits (genus *Parus*). *Evolution* 40:159–68.

Emlen, J. M. 1970. Age specificity and ecological theory. *Ecology* 51:588–601.

Emlen, S. T. 1971. The role of song in individual recognition in the indigo bunting. *Zeitschrift für Tierpsychologie* 28:241–46.

———. 1972. An experimental analysis of the parameters of bird song eliciting species recognition. *Behaviour* 41:130–71.

Endler, J. A. 1977. *Geographic variation, speciation, and clines.* Princeton, N.J.: Princeton University Press.

———. 1986. *Natural selection in the wild.* Princeton, N.J.: Princeton University Press.

Eriksson, D., and L. Wallin. 1986. Male bird song attracts females—A field experiment. *Behavioral Ecology and Sociobiology* 19:297–99.

Falconer, D. S. 1981. *Introduction to quantitative genetics*. 2d ed. New York: Longman.

Falls, J. B., and L. G. P. d'Agincourt. 1981. A comparison of neighbour-stranger discrimination in eastern and western meadowlarks. *Canadian Journal of Zoology* 59:2380–85.

Falls, J. B., A. G. Horn, and T. E. Dickinson. 1988. How western meadowlarks classify their songs: Evidence from song matching. *Animal Behaviour* 36:579–85.

Felsenstein, J. 1981. Skepticism towards Santa Rosalia, or Why are there so few kinds of animals? *Evolution* 35:124–38.

Fisher, R. 1930. *The genetical theory of natural selection*. Oxford: Clarendon Press.

Fogden, M. P. L. 1972. The seasonality and population dynamics of equatorial forest birds in Sarawak. *Ibis* 114:307–43.

Ford, H. A., D. T. Parkin, and A. W. Ewing. 1973. Divergence and evolution in Darwin's Finches. *Biological Journal of the Linnean Society* 5:289–95.

Frankel, O. H., and M. E. Soulé. 1981. *Conservation and evolution*. Cambridge: Cambridge University Press.

Frankham, R. 1980. The founder effect and response to artificial selection in *Drosophila*. In *Proceedings of the International Symposium on Selection Experiments in Laboratory and Domestic Animals*, ed. A. Robertson. London: CAB Books.

Futuyma, D. J. 1983. Interspecific interactions and maintenance of genetic diversity. In *Genetics and conservation*, ed. C. M. Schonewald-Cox, S. M. Chambers, B. MacBryde, and W. L. Thomas, 364–73. Menlo Park, Calif.: Benjamin/Cummings.

———. 1986. *Evolution*. 2d ed. Sunderland, Mass.: Sinauer.

Futuyma, D. J., and G. C. Mayer. 1980. Non-allopatric speciation in animals. *Systematic Zoology* 29:254–71.

Gibbons, J. R. H. 1979. A model for sympatric speciation in *Megarhyssa* (*Hymenoptera: Ichneumonidae*): Competitive speciation. *American Naturalist* 114:719–41.

Gibbs, H. L. 1988. Heritability and selection on clutch size in Darwin's medium ground finches (*Geospiza fortis*). *Evolution* 42:750–62.

Gibbs, H. L., and P. R. Grant. 1987a. Adult survival in Darwin's ground finch (*Geospiza*) populations in a variable environment. *Journal of Animal Ecology* 56:797–813.

———. 1987b. Ecological consequences of an exceptionally strong El Niño event on Darwin's Finches. *Ecology* 68:1735–46.

———. 1987c. Oscillating selection on Darwin's Finches. *Nature* 327:511–13.

———. 1989. Inbreeding in Darwin's medium ground finches (*Geospiza fortis*). *Evolution* 43.

Gibbs, H. L., P. R. Grant, and J. Weiland. 1984. Breeding of Darwin's Finches at an unusually early age in an El Niño year. *Auk* 101:872–74.

Gibson, R. M., and J. W. Bradbury. 1985. Sexual selection in lekking sage grouse: Phenotypic correlates of male mating success. *Behavioral Ecology and Sociobiology* 18:117–23.

Gifford, E. W. 1919. Field notes on the land birds of the Galápagos Islands and of Cocos Island, Costa Rica. *Proceedings of the California Academy of Sciences,* ser. 4, 2:189–258.

Gill, F. B. 1980. Historical aspects of hybridization between blue-winged and golden-winged warblers. *Auk* 97:1–18.

Gimmelfarb, A. 1986. Additive variation maintained under stabilizing selection: A two-locus model of pleiotropy for two quantitative characters. *Genetics* 112:717–25.

Gladstone, D. E. 1979. Promiscuity in monogamous colonial birds. *American Naturalist* 114:545–57.

Goodman, D. 1972. The paleoecology of the Tower Island bird colony: A critical examination of the complexity-stability theory. Ph.D. diss., Ohio State University, Columbus.

————. 1982. Optimal life histories, optimal notation, and the value of reproductive value. *American Naturalist* 119:803–23.

Göransson, G., G. Högstedt, J. Kårlsson, H. Källander, and S. Ulfstrand. 1974. Sångens roll for revirhallandet hos naktergal *Luscinia luscinia*—Några experiment med play-back-teknik. *Vår Fågelvärld* 33:201–9.

Gosler, A. G. 1987. Pattern and process in the bill morphology of the great tit *Parus major. Ibis* 129:451–76.

Gottlander, K. 1987. Variation in the song rate of the male pied flycatcher *Ficedula hypoleuca*—Causes and consequences. *Animal Behaviour* 35:1037–45.

Gowaty, P., and A. A. Karlin. 1984. Multiple maternity and paternity in single broods of apparently monogamous eastern bluebirds (*Sialia sialis*). *Behavioral Ecology and Sociobiology* 15:91–95.

Graham, N. E., and W. B. White. 1988. The El Niño cycle: A natural oscillator of the Pacific Ocean–Atmospheric system. *Science* 240:1293–1302.

Grant, B. R. 1984. The significance of song variation in a population of Darwin's Finches. *Behaviour* 89:90–116.

————. 1985. Selection on bill characters in a population of Darwin's Finches: *Geospiza conirostris* on Isla Genovesa, Galápagos. *Evolution* 39:523–32.

Grant, B. R., and P. R. Grant. 1979. Darwin's Finches: Population variation and sympatric speciation. *Proceedings of the National Academy of Sciences USA* 76:2359–63.

————. 1981. Exploitation of *Opuntia* cactus by birds on the Galápagos. *Oecologia* (Berlin) 49:179–87.

————. 1982. Niche shifts and competition in Darwin's Finches: *Geospiza conirostris* and congeners. *Evolution* 36:637–57.

————. 1983. Fission and fusion in a population of Darwin's Finches: An example of the value of studying individuals in ecology. *Oikos* 41:530–47.

————. 1987. Mate choice in Darwin's Finches. *Biological Journal of the Linnean Society* 32:247–70.

————. 1989. Natural selection in a population of Darwin's Finches. *American Naturalist* 133:377–93.

Grant, P. R. 1965. A systematic study of the terrestrial birds of the Tres Marías Islands, Mexico. *Yale Peabody Museum of Natural History, Postilla* 90:1–106.

————. 1967. Bill length variability in birds of the Tres Marías Islands, Mexico. *Canadian Journal of Zoology* 45:805–15.

————. 1979. Evolution of the chaffinch, *Fringilla coelebs*, on the Atlantic Islands. *Biological Journal of the Linnean Society* 11:301–32.

————. 1981a. Patterns of growth in Darwin's Finches. *Proceedings of the Royal Society of London B* 212:403–32.

————. 1981b. Speciation and the adaptive radiation of Darwin's Finches. *American Scientist* 69:653–63.

————. 1981c. The feeding of Darwin's Finches on *Tribulus cistoides* (L.) seeds. *Animal Behaviour* 29:785–93.

————. 1982. Variation in the size and shape of Darwin's Finch eggs. *Auk* 99:15–23.

————. 1983a. The relative size of Darwin's Finch eggs. *Auk* 100:228–30.

————. 1983b. Inheritance of size and shape in a population of Darwin's Finches, *Geospiza conirostris*. *Proceedings of the Royal Society of London B* 220:219–36.

————. 1985. Climatic fluctuations on the Galápagos Islands and their influence on Darwin's Finches. In *Neotropical ornithology*, ed. P. A. Buckley, M. S. Foster, E. S. Morton, R. S. Ridgely, and F. G. Buckley, 471–83. American Ornithologists' Union Monograph, No. 36. Washington, D.C.: AOU.

————. 1986a. *Ecology and evolution of Darwin's Finches*. Princeton, N.J.: Princeton University Press.

————. 1986b. Interspecific competition in fluctuating environments. In *Community ecology*, ed. J. Diamond and T. J. Case, 173–91. New York: Harper and Row.

Grant, P. R., I. Abbott, D. Schluter, R. L. Curry, and L. K. Abbott. 1985. Variation in the size and shape of Darwin's Finches. *Biological Journal of the Linnean Society* 25:1–39.

Grant, P. R., and P. T. Boag. 1980. Rainfall on the Galápagos and the demography of Darwin's Finches. *Auk* 97:227–44.

Grant, P. R., and B. R. Grant. 1980. The breeding and feeding characteristics of Darwin's Finches on Isla Genovesa, Galápagos. *Ecological Monographs* 50:381–410.

————. 1985. Responses of Darwin's Finches to unusual rainfall. In *El Niño in the Galápagos Islands: The 1982–1983 event*, ed. G. Robinson and E. del Pino, 417–447. Quito: ISALPRO.

————. 1987. The extraordinary El Niño event of 1982–83: Effects on Darwin's Finches on Isla Genovesa, Galápagos. *Oikos* 49:55–66.

————. 1989. Sympatric speciation and Darwin's Finches. In *Speciation and its consequences*, ed. D. Otte and J. A. Endler, 433–57. Sunderland, Mass.: Sinauer.

Grant, P. R., B. R. Grant, J. N. M. Smith, I. Abbott, and L. K. Abbott. 1976. Darwin's Finches: Population variation and natural selection. *Proceedings of the National Academy of Sciences USA* 73:257–61.

Grant, P. R., and K. T. Grant. 1979. Breeding and feeding ecology of the Galápagos dove. *Condor* 81:397–403.

Grant, P. R., and N. Grant. 1979. Breeding and feeding of Galápagos mocking-birds, *Nesomimus parvulus*. *Auk* 96:723–36.

Grant, P. R., and T. D. Price. 1981. Population variation in continuously varying traits as an ecological genetics problem. *American Zoologist* 21:795–811.

Grant, P. R., D. Schluter, and P. T. Boag. 1979. A bill color polymorphism in young Darwin's Finches. *Auk* 96:800–802.

Grant, P. R., J. N. M. Smith, B. R. Grant, I. Abbott, and L. K. Abbott. 1975. Finch numbers, owl predation and plant dispersal on Isla Daphne Major, Galápagos. *Oecologia* (Berlin) 19:239–57.

Greenwood, P. J., and P. H. Harvey. 1982. The natal and breeding dispersal of birds. *Annual Review of Ecology and Systematics* 13:1–21.

Gustafsson, L. 1986. Lifetime reproductive success and heritability: Empirical support for Fisher's fundamental theorem. *American Naturalist* 128:761–64.

_____. 1989. Collared flycatcher. In *Lifetime reproductive success in birds*, ed. I. Newton. New York: Academic Press.

Hails, C. J., and D. M. Bryant. 1979. Reproductive energetics of a free-living bird. *Journal of Animal Ecology* 48:471–82.

Haldane, J. B. S. 1948. The theory of a cline. *Journal of Genetics* 48:277–84.

Hall, M. L., P. Ramon, and H. Yepes. 1983. Origin of Española Island and the age of terrestrial life on the Galápagos Islands. *Science* 221:545–47.

Halliday, T. R. 1983. The study of mate choice. In *Mate choice*, ed. P. Bateson, 257–78. Cambridge: Cambridge University Press.

Hamann, O. 1981. Plant communities of the Galápagos Islands. *Dansk Botanisk Archiv* 34:1–163.

Hamilton, W. D. 1966. The moulding of senescence by natural selection. *Journal of Theoretical Biology* 12:12–45.

_____. 1982. Pathogens as causes of genetic diversity in their host populations. In *Population biology of infectious disease agents*, ed. R. M. Anderson and R. M. May, 269–96. Berlin: Springer-Verlag.

Hamilton, W. D., and M. Zuk. 1982. Heritable true fitness and bright plumage in birds: A role for parasites? *Science* 218:384–87.

Harper, D. G. C. 1985. Pairing strategies and mate choice in female robins *Erithacus rubecula*. *Animal Behaviour* 33:862–75.

Harvey, P. H., P. J. Greenwood, and C. M. Perrins. 1979. Breeding area fidelity of great tits (*Parus major*). *Journal of Animal Ecology* 48:305–13.

Harvey, P. H., P. J. Greenwood, C. M. Perrins, and A. R. Martin. 1979. Breeding success of great tits *Parus major* in relation to age of male and female parent. *Ibis* 121:216–19.

Harvey, P. H., M. J. Stenning, and B. Campbell. 1985. Individual variation in seasonal breeding success of pied flycatchers (*Ficedula hypoleuca*). *Journal of Animal Ecology* 54:391–98.

_____. 1988. Factors influencing reproductive success in the pied flycatcher. In *Reproductive success*, ed. T. H. Clutton-Brock, 189–200. Chicago: University of Chicago Press.

Hendrickson, H. T. 1978. Sympatric speciation: Evidence? *Science* 200:345–46.

Higuchi, H. 1976. Comparative study on the breeding of mainland and island subspecies of the varied tit, *Parus varius*. *Tori* 25:11–20.

Hoeck, H. N. 1984. Introduced fauna. In *Galápagos*, ed. R. Perry, 233–45. Oxford: Pergamon.

Högstedt, G. 1980. Evolution of clutch size in birds: Adaptive variation in relation to territory quality. *Science* 210:1148–50.

———. 1981. Should there be a positive or negative correlation between survival of adults in a bird population and their clutch size? *American Naturalist* 118:568–71.

Holmes, W. G., and P. W. Sherman. 1982. The ontogeny of kin recognition in two species of ground squirrels. *American Zoologist* 22:491–517.

Horn, H. S., and D. I. Rubenstein. 1984. Behavioral adaptations and life history. In *Behavioural ecology*, ed. J. R. Krebs and N. B. Davis, 2d ed., 279–98. Sunderland, Mass.: Sinauer.

Hötker, H. 1988. Lifetime reproductive output of male and female meadow pipits *Anthus pratensis*. *Journal of Animal Ecology* 57:109–17.

Houvenaghel, G. T. 1984. Oceanographic setting of the Galápagos Islands. In *Galápagos*, ed. R. Perry, 43–54. Oxford: Pergamon Press.

Howard, R. D. 1974. The influence of sexual selection and interspecific competition on mockingbird song. *Evolution* 28:428–38.

———. 1979. Estimating reproductive success in natural populations. *American Naturalist* 114:221–31.

Hussell, D. J. T., and T. E. Quinney. 1987. Food abundance and clutch size of tree swallows *Tachycineta bicolor*. *Ibis* 129:243–58.

Hutt, F. B. 1949. *Genetics of the fowl*. New York: McGraw-Hill.

Immelmann, K. 1969. Song development in the zebra finch and other *estrildid* finches. In *Bird vocalizations*, ed. R. A. Hinde, 61–74. Cambridge: Cambridge University Press.

———. 1971. Ecological aspects of periodic reproduction. In *Avian biology*, ed. D. S. Farner and J. R. King, 1:341–89. New York: Academic Press.

———. 1975. Ecological significance of imprinting and early learning. *Annual Review of Ecology and Systematics* 6:15–37.

———. 1980. Genetical constraints on early learning: A perspective from sexual imprinting in birds. In *Theoretical advances in behavior genetics*, ed. J. R. Royce, 121–33. Amsterdam: Van Nijhoff.

Janetos, A. C. 1980. Strategies of female choice: A theoretical analysis. *Behavioral Ecology and Sociobiology* 7:107–12.

Jenkins, P. F. 1978. Cultural transmission of song patterns in dialect development in a free-living bird population. *Animal Behaviour* 26:50–78.

Johnson, R. E. 1977. Seasonal variation in the genus *Leucosticte* in North America. *Condor* 79:76–86.

Johnston, R. F. 1956. Population structure in salt marsh song sparrows. 2. Environment and annual cycle. *Condor* 58:22–44.

King, J. R. 1973. Energetics of reproduction in birds. In *Breeding biology of birds*, ed. D. S. Farner, 78–117. Washington, D.C.: National Academy of Sciences.

King, W. B. 1985. Island birds: Will the future repeat the past? In *Conservation of island birds,* ed. P. J. Moors, 3–15. International Council for Bird Preservation, Technical Publications No. 3. Norwich, England: Paston Press.

Kirkpatrick, M. 1982. Sexual selection and the evolution of female choice. *Evolution* 36:1–12.

———. 1986. The handicap mechanism of sexual selection does not work. *American Naturalist* 127:222–40.

Klimkiewicz, M. K., and A. G. Futcher. 1987. Longevity records of North American birds: Coerebinae through Estrildidae. *Journal of Field Ornithology* 58:318–33.

Klomp, H. 1980. Fluctuations and stability in great tit populations. *Ardea* 68:205–24.

Koenig, W. D., and R. L. Mumme. 1987. *Population ecology of the cooperatively breeding acorn woodpecker.* Princeton, N.J.: Princeton University Press.

Konishi, M. 1985. Birdsong: From behavior to neuron. *Annual Review of Neuroscience* 8:125–70.

Krebs, C. J. 1985. *Ecology: The experimental analysis of distribution and abundance.* New York: Harper and Row.

Krebs, J. R., R. Ashcroft, and M. I. Webber. 1978. Song repertoires and territory defence in the great tit. *Nature* 271:539–42.

Kroodsma, D. E. 1974. Song learning, dialects and dispersal in Bewick's wren. *Zeitschrift für Tierpsychologie* 35:352–80.

———. 1976. Reproductive development in a female songbird: Differential stimulation by quality of male song. *Science* 192:574–75.

———. 1978. Aspects of learning in the ontogeny of bird song: Where, from whom, when, how many, which and how accurately? In *Ontogeny of behavior,* ed. G. Burkhardt and M. Bekoff, 215–30. New York: Garland.

———. 1983. The ecology of vocal learning. *Bioscience* 33:165–71.

———. 1984. Songs of the alder flycatcher (*Empidonax alnorum*) and willow flycatcher (*Empidonax traillii*) are innate. *Auk* 101:13–24.

Kroodsma, D. E., and R. A. Canady. 1985. Differences in repertoire size, singing behavior, and associated neuroanatomy among marsh wren populations have a genetic basis. *Auk* 102:439–46.

Kroodsma, D. E., and L. D. Parker. 1977. Vocal virtuosity in the brown thrasher. *Auk* 97:783–85.

Kunkel, P. 1974. Mating systems of tropical birds: The effects of weakness or absence of external reproduction-timing factors, with special reference to prolonged pair bonds. *Zeitschrift für Tierpsychologie* 34:265–307.

Lack, D. 1945. The Galápagos finches (*Geospizinae*): A study in variation. *Occasional Papers of the California Academy of Sciences* 21:1–159.

———. 1947. *Darwin's Finches.* Cambridge: Cambridge University Press.

———. 1950. Breeding seasons in the Galápagos. *Ibis* 92:268–78.

———. 1954. *The natural regulation of animal numbers.* Oxford: Clarendon Press.

———. 1968. *Ecological adaptations for breeding in birds.* London: Methuen.

———. 1969. Subspecies and sympatry in Darwin's Finches. *Evolution* 23:252–63.

Lacy, R. C., and P. W. Sherman. 1983. Kin recognition by phenotype matching. *American Naturalist* 121:489–512.

Lambert, W. V., and C. W. Knox. 1927. Genetic studies in poultry. 2. The inheritance of skin colour. *Poultry Science* 7:24–30.

Lambrechts, M., and A. A. Dhondt. 1986. Male quality, reproduction and survival in the great tit (*Parus major*). *Behavioral Ecology and Sociobiology* 19:57–63.

———. 1988. Male quality and territory quality in the great tit *Parus major*. *Animal Behaviour* 36:596–601.

Lande, R. 1976. The maintenance of genetic variability by mutation in a polygenic character with linked loci. *Genetical Research* 26:221–35.

———. 1979. Quantitative genetic analysis of multivariate evolution, applied to brain:body size allometry. *Evolution* 33:402–16.

———. 1981. Models of speciation by sexual selection on polygenic traits. *Proceedings of the National Academy of Sciences USA* 78:3721–25.

———. 1988. Genetics and demography in biological conservation. *Science* 241:1455–60.

Lande, R., and S. J. Arnold. 1983. The measurement of selection on correlated characters. *Evolution* 37:1210–26.

Lande, R., and G. F. Barrowclough. 1987. Effective population size, genetic variation, and their use in population management. In *Viable populations for conservation,* ed. M. E. Soulé, 87–123. Cambridge: Cambridge University Press.

Lande, R., and M. Kirkpatrick. 1988. Ecological speciation by sexual selection. *Journal of Theoretical Biology* 133:85–98.

Lande, R., and D. W. Schemske. 1985. The evolution of self-fertilization and inbreeding depression in plants. I. Genetic models. *Evolution* 39:24–40.

Laskey, A. 1944. A mockingbird acquires his song repertory. *Auk* 61:211–19.

Lessells, C. M., and P. T. Boag. 1987. Unrepeatable repeatabilities: A common mistake. *Auk* 104:116–21.

Lewontin, R. C., and L. C. Birch. 1966. Hybridization as a source of variation for adaptation to new environments. *Evolution* 20:315–35.

Lints, F. A., and M. Bourgois. 1982. A test of the genetic revolution hypothesis of speciation. In *Advances in genetics, development and evolution of Drosophila,* ed. S. Lakovaara. New York: Plenum.

Loery, G., K. H. Pollock, J. D. Nichols, and J. E. Hines. 1987. Age-specificity of black-capped chickadee survival rates: Analysis of capture-recapture data. *Ecology* 68:1038–44.

Loeschcke, V., and F. B. Christiansen. 1984. Evolution and intraspecific exploitative competition. 2. A two-locus model for additive gene effects. *Theoretical Population Biology* 26:228–64.

Lott, D., and P. N. Brody. 1966. Support of ovulation in the ring dove by auditory and visual stimuli. *Journal of Comparative Physiology and Psychology* 62:311–13.

Lott, D., S. D. Scholz, and D. S. Lehrman. 1967. Exteroceptive stimulation of the reproductive system of the female ring dove (*Streptopelia risoria*) by the mate and by the colony milieu. *Animal Behaviour* 15:433–37.

Lyon, B. E., and R. D. Montgomerie. 1986. Delayed plumage maturation in passerine birds: Reliable signalling by subordinate males? *Evolution* 40:605–15.

MacArthur, R. H., and E. O. Wilson. 1967. *The theory of island biogeography.* Princeton, N.J.: Princeton University Press.

McCleery, R. H., and C. M. Perrins. 1988. Lifetime reproduction success of the great tit, *Parus major.* In *Reproductive success,* ed. T. H. Clutton-Brock, 136–53. Chicago: University of Chicago Press.

McGregor, P. K., and J. R. Krebs. 1982. Song types in a population of great tits (*Parus major*): Their distribution, abundance, and acquisition by individuals. *Behaviour* 72:126–52.

McGregor, P. K., J. R. Krebs, and C. M. Perrins. 1981. Song repertoires and lifetime reproductive success in the great tit (*Parus major*). *American Naturalist* 118:149–59.

McKinney, F., K. M. Cheng, and D. J. Bruggers. 1984. Sperm competition in apparently monogamous birds. In *Sperm competition and the evolution of animal mating systems,* ed. R. L. Smith, 523–45. Orlando: Academic Press.

Maclean, G. L. 1976. Arid zone ornithology in Africa and South America. In *Proceedings of the 16th International Ornithological Congress* (Canberra, 1974), 468–80.

Manly, B. F. J. 1985. *The statistics of natural selection.* London: Chapman and Hall.

Manning, J. T. 1985. Choosy females and correlates of male age. *Journal of Theoretical Biology* 116:349–54.

Marchant, S. 1958. The birds of the Santa Elena peninsula, S.W. Ecuador. *Ibis* 100:349–87.

_____. 1959. The breeding season in S.W. Ecuador. *Ibis* 101:139–52.

_____. 1960. The breeding of some S.W. Ecuadorean birds. *Ibis* 102:349–81, 584–99.

Marler, P. 1960. Bird songs and male selection. In *Animal sound and communication,* ed. W. E. Lanyon and W. Tavolga, 348–67. Washington, D.C.: American Institute of Behavioral Sciences.

Marler, P., and P. C. Mundinger. 1971. Vocal learning in birds. In *The ontogeny of vertebrate behavior,* ed. H. Moltz, 389–449. New York: Academic Press.

Marler, P., and S. S. Peters. 1977. Selective vocal learning in a sparrow. *Science* 198:519–21.

_____. 1981. Sparrows learn adult song and more from memory. *Science* 213:780–82.

_____. 1982. Structural changes in song ontogeny in the swamp sparrow, *Melospiza georgiana. Auk* 99:446–58.

_____. 1988a. Sensitive periods for song acquisition from tape recordings and live tutors in the swamp sparrow, *Melospiza georgiana. Ethology* 77:76–84.

_____. 1988b. The role of song phenology and syntax in vocal learning preferences in the song sparrow *Melospiza melodia. Ethology* 77:125–49.

Marler, P., and M. Tamura. 1964. Culturally transmitted patterns of vocal behavior in sparrows. *Science* 146:1483–86.

Maynard Smith, J. 1966. Sympatric speciation. *American Naturalist* 100:637–50.

———. 1979. The effects of normalizing and disruptive selection on genes for recombination. *Genetical Research* 33:121–28.

———. 1985. Mini review. Sexual selection, handicaps and true fitness. *Journal of Theoretical Biology* 115:1–8.

———. 1988. The evolution of recombination. In *The evolution of sex*, ed. R. E. Michod and B. R. Levin, 106–25. Sunderland, Mass.: Sinauer.

Maynard Smith, J., and R. Hoekstra. 1980. Polymorphism in a varied environment: How robust are the models? *Genetical Research* 35:45–57.

Mayr, E. 1963. *Animal species and evolution.* Cambridge, Mass.: Harvard University Press.

Mayr, E., and L. L. Short. 1970. Species taxa of North American birds, and their contribution to comparative systematics. Publication of the Nuttall Ornithological Club, No. 9., Cambridge, Mass.

Medawar, P. B. 1952. *An unsolved problem of biology.* London: H. K. Lewis.

Miller, A. H. 1941. Speciation in the avian genus *Junco. University of California Publications in Zoology* 44:173–434.

Millington, S. J., and P. R. Grant. 1983. Feeding ecology and territoriality of the cactus finch *Geospiza scandens* on Isla Daphne Major, Galápagos. *Oecologia* (Berlin) 58:76–83.

———. 1984. The breeding ecology of the cactus finch *Geospiza scandens* on Isla Daphne Major, Galápagos. *Ardea* 72:177–88.

Mitchell-Olds, T., and R. G. Shaw. 1987. Regression analysis of natural selection: Statistical inference and biological interpretation. *Evolution* 41:1149–61.

Mock, D. W. 1985. An introduction to the neglected mating system. In *Avian monogamy*, ed. P. A. Gowaty and D. W. Mock, 1–10. American Ornithologists' Union Monograph, No. 37. Washington, D.C.: AOU.

Møller, A. P. 1987a. Mate guarding in the swallow *Hirundo rustica:* An experimental study. *Behavioral Ecology and Sociobiology* 21:119–23.

———. 1987b. Behavioral aspects of sperm competition in swallows (*Hirundo rustica*). *Behaviour* 100:92–104.

———. 1988. Female choice selects for male sexual tail ornaments in the monogamous swallow. *Nature* 332:640–42.

Moreau, R. E. 1944. Clutch size: A comparative study, with special reference to African birds. *Ibis* 86:286–347.

Morton, E. S. 1975. Ecological sources of selection on avian sounds. *American Naturalist* 109:17–34.

Morton, M. L., M. E. Pereyra, and L. F. Baptista. 1985. Photoperiodically induced ovarian growth in the white-crowned sparrow (*Zonotrichia leucophrys gambelii*) and its augmentation by song. *Comparative Biochemistry and Physiology* 80:93–97.

Mousseau, T. A., and D. A. Roff. 1987. Natural selection and the heritability of fitness components. *Heredity* 59:181–97.

Murphy, G. I. 1968. Pattern in life history and the environment. *American Naturalist* 102:390–404.

Murray, B. G., Jr. 1969. A comparative study of the Le Conte's and sharp-tailed sparrows. *Auk* 86:199–231.

Nagata, H. 1986. Female choice in Middendorff's grasshopper-warbler (*Locustella ochotensis*). *Auk* 103:694–700.

Newton, I. 1972. *Finches*. London: Collins.

―――. 1988. Age and reproduction in the sparrowhawk. In *Reproductive success,* ed. T. H. Clutton-Brock, 201–19. Chicago: University of Chicago Press.

Nice, M. M. 1937. Studies in the life history of the song sparrow. Part 1. *Transactions of the Linnaean Society of New York* 4:1–247.

Nicolai, J. 1956. Zur Biologie und Ethologie des Gimpels (*Pyrrhula pyrrhula* L.). *Zeitschrift für Tierpsychologie* 13:93–132.

―――. 1959. Familientradition in der Gesangsentwicklung des Gimpels (*Pyrrhula pyrrhula* L.). *Journal für Ornithologie* 100:39–46.

Nol, E., and J. N. M. Smith. 1987. Effects of age and breeding experience on seasonal reproductive success in the song sparrow. *Journal of Animal Ecology* 56:301–13.

Nolan, V., Jr. 1978. The ecology and behavior of the prairie warbler *Dendroica discolor*. American Ornithologists' Union Monograph, No. 26. Washington, D.C.: AOU.

Norberg, U. M., and R. Å. Norberg. 1989. Ecomorphology of flight and tree-trunk climbing in birds. In *Proceedings of the 19th International Ornithological Congress* (Ottawa, 1986), in press.

Nottebohm, F. 1969. The "critical period" for song learning. *Ibis* 111:386–87.

―――. 1972. The origins of vocal learning. *American Naturalist* 106:116–40.

Nottebohm, F., and M. Nottebohm. 1971. Vocalizations and breeding behaviour of surgically deafened ring doves *Streptopelia risoria*. *Animal Behaviour* 19:313–28.

Nur, N. 1984. The consequences of brood size for breeding blue tits. 1. Adult survival, weight change and the cost of reproduction. *Journal of Animal Ecology* 53:479–96.

―――. 1988. The consequences of brood size for breeding blue tits. 3. Measuring the cost of reproduction: Survival, future fecundity, and differential dispersal. *Evolution* 42:351–62.

O'Donald, P. 1960. Assortive mating in a population in which two alleles are segregating. *Heredity* 15:389–96.

―――. 1980. *Genetic models of sexual selection*. Cambridge: Cambridge University Press.

―――. 1983a. *The Arctic skua: A study of the ecology and evolution of a seabird*. Cambridge: Cambridge University Press.

―――. 1983b. Sexual selection by female choice. In *Mate choice*, ed. P. Bateson, 53–66. Cambridge: Cambridge University Press.

Orr, R. T. 1945. A study of captive Galápagos finches of the genus *Geospiza*. *Condor* 47:177–201.

Payne, R. B. 1981. Song learning and social interaction in indigo buntings. *Animal Behaviour* 29:688–91.

―――. 1982. Ecological consequences of song matching: Breeding success and intraspecific song mimicry in indigo buntings. *Ecology* 63:401–11.

―――. 1989. Indigo bunting. In *Lifetime reproductive success in birds*, ed. I. Newton. New York: Academic Press.

Payne, R. B., W. L. Thompson, K. L. Fiala, and L. L. Sweany. 1981. Local song traditions in indigo buntings: Cultural transmission of behavior patterns across generations. *Behaviour* 77:199–201.

Perrins, C. M. 1979. *British tits.* London: Collins.

Perrins, C. M., and P. J. Jones. 1974. The inheritance of clutch size in the great tit (*Parus major* L.). *Condor* 76:225–29.

Perrins, C. M., and R. H. McCleery. 1985. The effect of age and pair bond on the breeding success of great tits *Parus major. Ibis* 127:306–15.

Peters, S. S., W. A. Searcy, and P. Marler. 1980. Species song discrimination in choice experiments with territorial male swamp and song sparrows. *Animal Behaviour* 28:393–404.

Philander, S. G. H. 1983. El Niño southern oscillation phenomena. *Nature* 302:295–301.

Pimm, S. L. 1979. Sympatric speciation: A simulation model. *Biological Journal of the Linnean Society* 11:131–39.

Pinowski, J. 1968. Fecundity, mortality, numbers and biomass dynamics of a population of the tree sparrow (*Passer m. montanus* L.). *Ekologia Polska,* ser. A, 16:1–58.

Power, D. 1983. Variability in island populations of the house finch. *Auk* 100:180–87.

Pratt, H. D., P. L. Bruner, and D. G. Berrett. 1987. *The birds of Hawaii and the tropical Pacific.* Princeton, N.J.: Princeton University Press.

Price, T. D. 1984a. Sexual selection on body size, territory and plumage variables in a population of Darwin's Finches. *Evolution* 38:327–41.

———. 1984b. The evolution of sexual size dimorphism in a population of Darwin's Finches. *American Naturalist* 123:500–518.

———. 1985. Reproductive responses to varying food supply in a population of Darwin's Finches: Clutch size, growth rates and hatching synchrony. *Oecologia* (Berlin) 66:411–16.

———. 1987. Diet variation in a population of Darwin's Finches. *Ecology* 68:1015–28.

Price, T. D., and P. T. Boag. 1987. Selection in natural populations of birds. In *Avian genetics: A population and ecological approach,* ed. F. Cooke and P. A. Buckley, 257–87. New York: Academic Press.

Price, T. D., and H. L. Gibbs. 1987. Brood division in Darwin's ground finches. *Animal Behaviour* 35:299–300.

Price, T. D., and P. R. Grant. 1984. Life history traits and natural selection for small body size in a population of Darwin's Finches. *Evolution* 38:483–94.

———. 1985. The evolution of ontogeny in Darwin's Finches: A quantitative genetic approach. *American Naturalist* 125:169–88.

Price, T. D., P. R. Grant, and P. T. Boag. 1984. Genetic changes in the morphological differentiation of Darwin's ground finches. In *Population biology and evolution,* ed. K. Wöhrmann and V. Loeschcke, 49–66. Berlin: Springer-Verlag.

Price, T. D., P. R. Grant, H. L. Gibbs, and P. T. Boag. 1984. Recurrent patterns of natural selection in a population of Darwin's Finches. *Nature* 309: 787–89.

Price, T. S., Millington, and P. Grant. 1983. Helping at the nest in Darwin's Finches as misdirected parental care. *Auk* 100:192–94.

Pulliam, H. R. 1985. Foraging efficiency, resource partitioning, and the coexistence of sparrow species. *Ecology* 66:1829–36.

Pyke, D. A., and J. N. Thompson. 1986. Statistical analysis of survival and removal rate experiments. *Ecology* 67:240–45.

Racine, C. H., and J. F. Downhower. 1974. Vegetative and reproductive strategies of *Opuntia* (*Cactaceae*) in the Galápagos Islands. *Biotropica* 6:175–86.

Rahn, H., C. V. Paganelli, and A. Ar. 1975. Relation of avian egg weight to body weight. *Auk* 92:750–65.

Ratcliffe, L. M. 1981. Species recognition in Darwin's ground finches (*Geospiza, Gould*). Ph.D. diss., McGill University, Montreal.

Ratcliffe, L. M., and P. R. Grant. 1983a. Species recognition in Darwin's Finches (*Geospiza, Gould*). 1. Discrimination by morphological cues. *Animal Behaviour* 31:1139–53.

———. 1983b. Species recognition in Darwin's Finches (*Geospiza, Gould*). 2. Geographic variation in mate preference. *Animal Behaviour* 31:1154–65.

———. 1985. Species recognition in Darwin's Finches (*Geospiza, Gould*). 3. Male responses to playback of different song types, dialects and heterospecific songs. *Animal Behaviour* 33:290–307.

Ratcliffe, L., R. F. Rockwell, and F. Cooke. 1988. Recruitment and maternal age in lesser snow geese *Chen caerulescens caerulescens*. *Journal of Animal Ecology* 57:553–63.

Reznick, D. 1985. Cost of reproduction: An evaluation of the empirical evidence. *Oikos* 44:257–67.

Rice, W. R. 1984. Disruptive selection on habitat preference and the evolution of reproductive isolation: A simulation study. *Evolution* 38:1251–60.

———. 1987. Speciation via habitat specialization: The evolution of reproductive isolation as a correlated character. *Evolutionary Ecology* 1:301–14.

Ricklefs, R. E. 1973. Fecundity, mortality and avian demography. In *Breeding biology of birds,* ed. D. S. Farner, 366–435. Washington, D.C.: National Academy of Sciences.

———. 1974. Energetics of reproduction in birds. In *Avian energetics,* ed. R. A. Paynter, 152–292. Publication of the Nuttall Ornithological Club, No. 15, Cambridge, Mass.

———. 1980. On geographical variation in clutch size among passerine birds: Ashmole's hypothesis. *Auk* 97:38–49.

———. 1983. Avian demography. *Current Ornithology* 1:1–32.

Rockwell, R. F., and G. F. Barrowclough. 1987. Gene flow and genetic structure of populations. In *Avian genetics: A population and ecological approach,* ed. F. Cooke and P. A. Buckley, 223–56. New York: Academic Press.

Rohwer, S. 1985. Dyed birds achieve higher social status than controls in Harris' sparrows. *Animal Behaviour* 33:1325–31.

Rohwer, S., and G. S. Butcher. 1988. Winter versus summer explanations of delayed plumage maturation in temperate passerine birds. *American Naturalist* 131:556–72.

Rohwer, S., and F. C. Rohwer. 1984. Status signalling in Harris' sparrows: Experimental deception achieved. *Animal Behaviour* 26:1012–22.

Roughgarden, J. 1972. Evolution of niche width. *American Naturalist* 106:683–718.

———. 1986. A comparison of food-limited and space-limited animal competition communities. In *Community ecology,* ed. J. Diamond and T. J. Case, 492–516. New York: Harper and Row.

Rowley, I. 1983. Re-mating in birds. In *Mate choice,* ed. P. Bateson, 331–60. Cambridge: Cambridge University Press.

Ruff, M. D., W. M. Reid, and J. K. Johnson. 1974. Lowered blood carotenoid levels in chickens infected with coccidea. *Poultry Science* 53:1801–9.

Schluter, D. 1984a. Feeding correlates of breeding and social organization in two Galápagos finches. *Auk* 101:59–68.

———. 1984b. Morphological and phylogenetic relations among the Darwin's Finches. *Evolution* 38:921–30.

Schluter, D., and P. R. Grant. 1982. The distribution of *Geospiza difficilis* in relation to *G. fuliginosa* in the Galápagos Islands: Tests of three hypotheses. *Evolution* 36:1213–26.

———. 1984a. Determinants of morphological patterns in communities of Darwin's Finches. *American Naturalist* 123:175–96.

———. 1984b. Ecological correlates of morphological evolution in a Darwin's Finch species. *Evolution* 38:856–69.

Schluter, D., and J. N. M. Smith. 1986. Genetic and phenotypic correlations in a natural population of song sparrows. *Biological Journal of the Linnean Society* 29:23–36.

Searcy, W. A. 1982. The evolutionary effects of mate selection. *Annual Review of Ecology and Systematics* 13:57–85.

Searcy, W. A., and M. Andersson. 1986. Sexual selection and the evolution of song. *Annual Review of Ecology and Systematics* 17:507–33.

Searcy, W. A., P. D. MacArthur, S. S. Peters, and P. Marler. 1981. Response of male song and swamp sparrows to neighbour, stranger and self song. *Behaviour* 77:152–63.

Searcy, W. A., and P. Marler. 1981. A test for responsiveness to song structure and programming in female sparrows. *Science* 213:926–28.

Searcy, W. A., P. Marler, and S. S. Peters. 1982. Species song discrimination in adult female song and swamp sparrows. *Animal Behaviour* 29:997–1003.

Searcy, W. A., M. H. Searcy, and P. Marler. 1982. The response of swamp sparrows to acoustically distinct song types. *Behaviour* 80:70–83.

Searcy, W. A., and K. Yasukawa. 1981. Does the "sexy son" hypothesis apply to mate choice in red-winged blackbirds? *American Naturalist* 117:343–48.

Seger, J. 1985. Intraspecific resource competition as a cause of sympatric speciation. In *Evolution: Essays in honour of John Maynard Smith,* ed. P. J. Greenwood, P. H. Harvey, and M. Slatkin, 43–53. Cambridge: Cambridge University Press.

Sherman, P. W., and M. L. Morton. 1988. Extra-pair fertilizations in mountain white-crowned sparrows. *Behavioral Ecology and Sociobiology* 22:413–20.

Sherry, T. W. 1986. Nest, eggs, and reproductive behavior of the Cocos flycatcher (*Nesotriccus ridgwayi*). *Condor* 88:531–32.

Sibley, C. G., J. E. Ahlquist, and B. L. Monroe, Jr. 1988. A classification of the living birds of the world based on DNA-DNA hybridization studies. *Auk* 105:409–23.

Simmons, M. J., and J. F. Crow. 1977. Mutations affecting fitness in *Drosophila* populations. *Annual Review of Genetics* 11:49–78.

Slatkin, M. 1987. Gene flow and the geographic structure of natural populations. *Science* 236:787–92.

Slud, P. 1967. The birds of Cocos Island (Costa Rica). *Bulletin of the American Museum of Natural History* 134:263–95.

Smith, D. G. 1976. An experimental analysis of the function of red-winged blackbirds song. *Behaviour* 56:136–56.

Smith, J. N. M. 1981. Does high fecundity reduce survival in song sparrows? *Evolution* 35:1142–48.

―――. 1988. Determinants of lifetime reproductive success in the song sparrow. In *Reproductive success,* ed. T. H. Clutton-Brock, 154–72. Chicago: University of Chicago Press.

Smith, J. N. M., P. Arcese, and D. Schluter. 1986. Song sparrows grow and shrink with age. *Auk* 103:210–12.

Smith, J. N. M., and A. A. Dhondt. 1980. Experimental confirmation of heritable morphological variation in a natural population of song sparrows. *Evolution* 34:1155–60.

Smith, J. N. M., P. R. Grant, B. R. Grant, I. Abbott, and L. K. Abbott. 1978. Seasonal variation in feeding habits of Darwin's ground finches. *Ecology* 59:1137–50.

Smith, J. N. M., Y. Yom-Tov, and R. Moses. 1982. Polygyny, male parental care and sex ratio in song sparrows: An experimental study. *Auk* 99:555–64.

Smith, J. N. M., and R. Zach. 1979. Heritability of some morphological characters in a song sparrow population. *Evolution* 33:460–67.

Snow, D. W. 1962. A field study of the black and white manakin, *Manacus manacus,* in Trinidad. *Zoologica* 47:65–104.

―――. 1966. Moult and the breeding cycle in Darwin's Finches. *Journal für Ornithologie* 107:283–91.

Snow, D. W., and A. Lill. 1974. Longevity records for some Neotropical land birds. *Condor* 76:262–67.

Soulé, M., and B. R. Stewart. 1970. The ''niche-variation'' hypothesis: A test and alternatives. *American Naturalist* 104:85–97.

Steadman, D. W. 1986. Holocene vertebrate fossils from Isla Floreana, Galápagos. Smithsonian Contributions to Zoology, No. 413.

Stearns, S. C. 1976. Life-history tactics: A review of the ideas. *Quarterly Review of Biology* 51:3–47.

Stjernberg, T. 1979. Breeding biology and population dynamics of the scarlet rosefinch *Carpodacus erythrinus. Acta Zoologica Fennica* 157:1–88.

Stobo, W. T., and I. A. McLaren. 1975. The Ipswich sparrow. *Proceedings of the Nova Scotian Institute of Science,* suppl., 27:1–105.

Stoddard, P. K., M. D. Beecher, M. S. Willis. 1988. Response of territorial male song sparrows to song types and variation. *Behavioral Ecology and Sociobiology* 22:125–30.

Svenson, H. K. 1946. Vegetation of the coast of Ecuador and Peru and its relation to the Galápagos archipelago. *American Journal of Botany* 33:394–426.

Swarth, H. S. 1929. A new bird family (Geospizidae) from the Galápagos Islands. *Proceedings of the California Academy of Sciences* 18:29–43.

Tauber, C. A., and M. J. Tauber. 1977a. Sympatric speciation based on allelic changes at three loci: Evidence from natural populations in two habitats. *Science* 197:1298–99.

———. 1977b. A genetic model for sympatric speciation through habitat diversification and seasonal isolation. *Nature* 268:702–5.

Tauber, C. A., M. J. Tauber, and J. R. Nechols. 1977. Two genes control seasonal isolation in sibling species. *Science* 197:592–93.

Taylor, R. H. 1985. Status, habits and conservation of *Cyanoramphus* parakeets in the New Zealand region. In *Conservation of island birds,* ed. P. J. Moors, 195–212. International Council for Bird Preservation, Technical Publications No. 3. Norwich, England: Paston Press.

Templeton, A. R. 1981. Mechanisms of speciation—A population genetic approach. *Annual Review of Ecology and Systematics* 12:23–48.

Templeton, A. R., and B. Read. 1985. The elimination of inbreeding depression in a captive herd of Speke's gazelle. In *Genetics and conservation,* ed. C. M. Schonewald-Cox, S. M. Chambers, B. MacBryde, and W. L. Thomas, 241–62. Menlo Park, Calif.: Benjamin/Cummings.

Terborgh, J. T. 1983. *Five New-World primates.* Princeton, N.J.: Princeton University Press.

———. 1986. Keystone plant resources in the tropical forest. In *Conservation biology,* ed. M. E. Soulé, 330–44. Sunderland, Mass.: Sinauer.

Thorpe, W. H., and E. W. North. 1965. The origin and significance of the power of vocal imitation with special reference to the antiphonal singing of birds. *Nature* 208:219–22.

Tompa, F. S. 1964. Factors determining the numbers of song sparrows, *Melospiza melodia* (Wilson), on Mandarte Island, B.C., Canada. *Acta Zoologica Fennica* 109:1–73.

Trivers, R. L. 1972. Parental investment and sexual selection. In *Sexual selection and the descent of man, 1871–1971,* ed. B. Campbell, 136–79. Chicago: Aldine.

Turelli, M. 1984. Heritable genetic variation via mutation-selection balance: Lerch's zeta meets the abdominal bristle. *Theoretical Population Biology* 25:138–93.

———. 1985. Effects of pleiotropy on predictions concerning mutation-selection balance for polygenic traits. *Genetics* 111:165–95.

———. 1988. Phenotypic evolution, constant covariances, and the maintenance of additive variance. *Evolution* 42:1342–47.

van Balen, J. H. 1980. Population fluctuations of the great tit and feeding conditions in winter. *Ardea* 68:143–64.

van Balen, J. H., A. J. van Noordwijk, and J. Visser. 1987. Lifetime reproductive success and recruitment in two great tit populations. *Ardea* 75:1–11.

van Noordwijk, A. J. 1984. Quantitative genetics in natural populations of birds, illustrated with examples from the great tit, *Parus major.* In *Population biology and evolution,* ed. K. Wöhrmann and V. Loeschcke, 67–79. Berlin: Springer-Verlag.

van Noordwijk, A. J., and G. de Jong. 1986. Acquisition and allocation of resources: Their influence on variation in life history tactics. *American Naturalist* 128:137–42.

van Noordwijk, A. J., and W. Scharloo. 1981. Inbreeding in an island population of the great tit. *Evolution* 35:674–88.

van Noordwijk, A. J., and J. H. van Balen. 1988. The great tit, *Parus major.* In *Reproductive success,* ed. T. H. Clutton-Brock, 119–35. Chicago: University of Chicago Press.

van Noordwijk, A. J., J. H. van Balen, and W. Scharloo. 1980. Heritability of ecologically important traits in the great tit. *Ardea* 68:193–203.

———. 1981. Genetic and environmental variation in the clutch size of the great tit (*Parus major*). *Netherlands Journal of Zoology* 31:342–72.

van Riper, C., III. 1980. Observations on the breeding of the palila *Psittarostra bailleui* of Hawaii. *Ibis* 122:462–75.

van Tienderen, P. H., and A. J. van Noordwijk. 1988. Dispersal, kinship and inbreeding in an island population of the great tit. *Journal of Evolutionary Biology* 2:117–37.

Van Valen, L. 1965. Morphological variation and the width of the ecological niche. *American Naturalist* 99:377–90.

———. 1978. The statistics of variation. *Evolutionary Theory* 4:33–43.

Van Valen, L., and P. R. Grant. 1970. Variation and niche width reexamined. *American Naturalist* 104:589–90.

van Vegten, J. A. 1968. The Azores bullfinch not extinct. *Ardea* 56:194.

Waser, M. S., and P. Marler. 1988. Song learning in canaries. *Journal of Comparative Physiology and Psychology* 91:1–7.

Watt, W. B., P. A. Carter, and K. Donohue. 1986. Female's choice of good genotypes as mates is promoted by an insect mating system. *Science* 233:1187–90.

Weathers, W. W. 1979. Climatic adaptation in avian standard metabolic rate. *Oecologia* (Berlin) 42:81–89.

Weathers, W. W., and C. van Riper III. 1982. Temperature regulation in two endangered Hawaiian honeycreepers: The palila (*Psittarostra bailleui*) and the Laysan finch (*Psittarostra cantans*). *Auk* 99:667–74.

Weeden, J. S., and J. B. Falls. 1959. Differential responses of male ovenbirds to recorded songs of neighboring and more distant individuals. *Auk* 76:343–51.

West-Eberhard, M. J. 1983. Sexual selection, social competition, and speciation. *Quarterly Review of Biology* 58:155–83.

Westneat, D. F. 1987a. Extra-pair copulations in a predominantly monogamous bird: Observations of behaviour. *Animal Behaviour* 35:865–76.

———. 1987b. Extra-pair copulations in a predominantly monogamous bird: Genetic evidence. *Animal Behaviour* 35:877–86.

Wetton, J. H., R. E. Carter, D. T. Parkin, and D. Walters. 1987. Demographic study of a wild house sparrow population by DNA fingerprinting. *Nature* 327:147–49.

Wickler, W., and U. Seibt. 1983. Monogamy: An ambiguous concept. In *Mate choice*, ed. P. Bateson, 33–52. Cambridge: Cambridge University Press.

Wiens, J. A. 1974. Climatic instability and the "ecological saturation" of bird communities in North American grasslands. *Condor* 76:385–400.

Wiggins, I. L., and D. M. Porter. 1971. *Flora of the Galápagos Islands.* Stanford, Calif.: Stanford University Press.

Williams, G. C. 1957. Pleiotropy, natural selection and the evolution of senescence. *Evolution* 11:398–411.

———. 1966. Natural selection, the costs of reproduction, and a refinement of Lack's principle. *American Naturalist* 100:687–90.

———. 1975. *Sex and evolution.* Princeton, N.J.: Princeton University Press.

Willis, E. O. 1974. Populations and local extinctions of birds on Barro Colorado Island, Panama. *Ecological Monographs* 44:153–69.

———. 1983. Longevities of some Panamanian forest birds with notes of low survivorship in old spotted antbirds (*Hylophylax naevioides*). *Journal of Field Ornithology* 54:413–14.

Wilson, D. S., and M. Turelli. 1986. Stable underdominance and the evolutionary invasion of empty niches. *American Naturalist* 127:835–50.

Wolf, L. L. 1977. Species relationships in the avian genus *Aimophila.* American Ornithologists' Union Monograph, No. 23. Washington, D.C.: AOU.

Woolfenden, G. E., and J. W. Fitzpatrick. 1984. *The Florida scrub jay: Demography of a cooperative-breeding bird.* Princeton, N.J.: Princeton University Press.

Wright, S. 1932. The roles of mutation, inbreeding, crossbreeding, and selection in evolution. *Proceedings of the 6th International Congress of Genetics* 1:356–66.

———. 1943. Isolation by distance. *Genetics* 28:114–38.

———. 1951. The genetical structure of populations. *Eugenics* 15:323–54.

———. 1969. *Evolution and the genetics of populations.* Vol. 1. Chicago: University of Chicago Press.

———. 1980. Genic and organismic selection. *Evolution* 34:825–43.

Wunderle, J. M., Jr. 1984. Mate switching and a seasonal increase in polygyny in the banaquit. *Behaviour* 88:123–44.

Yang, S.-Y., and J. L. Patton. 1981. Genic variability and differentiation in Galápagos finches. *Auk* 98:230–42.

Yasukawa, K. 1981. Song repertoires in the red-winged blackbird (*Agelaius phoeniceus*): A test of the Beau Geste hypothesis. *Animal Behaviour* 29:114–25.

Yasukawa, K., E. I. Bick, D. W. Wagman, and P. Marler. 1982. Playback and speaker-replacement experiments on song-based neighbour, stranger and self-discrimination in male red-winged blackbirds. *Behavioral Ecology and Sociobiology* 10:211–15.

Yasukawa, K., and W. A. Searcy. 1985. Song repertoires and density assessment in red-winged blackbirds: Further tests of the Beau Geste hypothesis. *Behavioral Ecology and Sociobiology* 16:171–75.

Zink, R. M. 1986. Patterns and evolutionary significance of geographic variation in the *schistacea* group of the fox sparrow. American Ornithologists' Union Monograph, No. 40. Washington, D.C.: AOU.

Author Index

Abbott, I., 6, 14, 24, 33, 51, 222, 247, 254, 265, 269, 277
Alatalo, R. V., 148, 155, 156, 159, 161, 182, 185
Anderson, E. F., 271
Anderson, T. R., 92
Andersson, M., 115, 148, 170
Antonovics, J., 236
Arcese, P., 92
Arnold, S. J., 148, 155, 208, 209, 227
Ashmole, N. P., 111
Askenmo, C., 101

Baker, M. C., 57, 65, 66, 92, 103, 104, 106, 141
Banks, R. C., 5, 6
Bannerman, D. A., 289
Bannerman, W. M., 289
Baptista, L. F., 122
Barrowclough, G. F., 279, 280, 286, 287, 291, 303, 304
Bateson, P., 142, 148
Bateson, W., 190
Beebe, W., 17, 34, 59
Beecher, I. M., 160
Beecher, M. D., 142, 160
Berndt, R., 67
Birch, L. C., 290
Bjilsma, R. G., 84
Black, J., 17
Boag, P. T., 12, 49, 58, 65, 66, 70, 73, 93, 107, 108, 113, 161, 167, 176, 180, 186, 203, 209, 224, 242, 274
Böhner, J., 118, 122
Borgia, G., 170
Botkin, D. B., 57, 63, 66, 67

Bourgois, M., 232
Bowman, R. I., 5, 7, 115, 117, 121, 138, 143, 201, 214, 217
Boyce, M. S., 110
Bradbury, J. W., 148, 170
Brockway, B. F., 115, 162
Brody, P. N., 162
Brooks, L. D., 197
Broughey, M. J., 115
Brown, D., 108, 208
Bruford, M. W., 160
Bryant, D. M., 46, 83, 108, 109
Bryant, E. H., 232
Burke, T., 160
Burley, N., 148
Bush, G. L., 236
Butcher, G. S., 171

Cade, T. J., 290, 291
Calder, W. A., III, 63, 67, 107
Canady, R. A., 139
Cane, M. A., 22, 23
Carson, H. L., 232
Catchpole, C. K., 115, 138, 148
Cate, C. ten, 148
Cavalli-Sforza, L. L., 118
Charlesworth, B., 142, 226, 286
Charlesworth, D., 142, 286
Charnov, E. L., 111
Christiansen, F. B., 236
Cliff, A. D., 136
Clutton-Brock, T. H., 108
Cody, M. L., 63, 107, 110
Colinvaux, P. A., 19, 38
Collias, N. E., 67
Connor, E. F., 269

Cooke, F., 152, 287
Corbin, K. W., 287
Coulson, J. C., 164, 168
Cox, A., 17, 254
Crow, J. F., 142, 286, 288
Crowell, K. L., 107
Curio, E., 59, 101, 108, 275
Curry, R. L., 61, 112

d'Agincourt, L. G. P., 141
Darwin, C. R., 148, 154
Davies, N. B., 148
Davies, S. J. J. F., 63, 70, 71
Dawson, E. Y., 271
Deevey, E. S., Jr., 66
de Jong, G., 101
Dhondt, A. A., 67, 139, 148, 150, 182, 185
Diamond, J. M., 170, 279, 301
Dickinson, H., 236
Dooling, R. Y., 117, 139
Downhower, J. F., 265, 270, 276
Dunn, L. C., 190
Dunning, J. B., Jr., 64

Ehrlich, P. R., 280, 287
Ekman, J., 101
Emlen, J. M., 67
Emlen, S. T., 115, 139
Endler, J. A., 209, 236, 287
Eriksson, D., 115

Falconer, D. S., 12, 93, 141, 176, 178, 183, 184, 188, 197, 207, 286, 303
Falls, J. B., 115, 117, 141
Felsenstein, J., 207, 236
Fisher, R., 148
Fitzpatrick, J. W., 67, 102, 103, 106
Fogden, M. P. L., 65
Ford, H. A., 290
Frankel, O. H., 292
Frankham, R., 232
Futcher, A. G., 63, 64
Futuyma, D. J., 235–37, 280

Gibbons, J. R. H., 236
Gibbs, H. L., 56–58, 65, 66, 73, 80, 84, 85, 87, 92, 96, 102, 108, 111, 230, 304
Gibson, R. M., 170
Gifford, E. W., 254
Gill, F. B., 290
Gimmelfarb, A., 11

Gladstone, D. E., 159, 163
Goodman, D., 38, 40, 105
Göransson, G., 128
Gosler, A. G., 176
Gottlander, K., 148
Gowaty, P., 160
Graham, N. E., 22
Grant, B. R., 3, 23–25, 31, 38, 47, 73, 75, 77, 107, 108, 111, 113, 117–19, 126, 129, 132, 133, 136, 140, 150, 151, 155, 156, 164, 166, 167, 172, 209, 210, 212, 214, 218, 219, 225, 229, 232, 235, 237, 240, 242–45, 258, 261, 265, 270–74, 276, 304
Grant, P. R., 1, 3–5, 6, 7, 10, 11, 14, 22, 23–25, 31, 37, 38, 49, 56–58, 61, 63, 65, 66, 70, 73, 75, 77, 81, 84, 85, 92, 102, 106–8, 111–13, 123, 126, 128, 129, 132, 133, 140, 143, 144, 148, 150, 151, 154, 156, 161, 164, 166, 167, 171, 172, 178–80, 182, 184, 186, 187, 190, 192, 193, 197, 203, 206, 209, 210, 212, 218, 219, 224, 225, 230, 235, 237, 238, 240, 242–45, 247, 248, 254, 258, 261, 265, 267, 269–77, 279, 288, 290, 304
Greenwood, P. J., 141
Gustafsson, L., 109, 182

Hails, C. J., 83
Haldane, J. B. S., 287
Hall, M. L., 254
Halliday, T. R., 158
Hamann, O., 24, 243
Hamilton, W. D., 67, 148, 158, 170
Harper, D. G. C., 148
Harvey, P. H., 92, 112, 141, 164
Hendrickson, H. T., 237
Higuchi, H., 107
Hoeck, H. N., 270
Hoekstra, R., 248
Högstedt, G., 101, 102
Holmes, W. G., 142
Horn, H. S., 101, 105, 111
Hötker, H., 101, 108, 109
Houvenaghel, G. T., 17, 19
Howard, R. D., 115, 138, 208
Hussell, D. J. T., 84, 92
Hutt, F. B., 190, 195

Immelmann, K., 70, 115, 118, 139, 148, 152

Janetos, A. C., 152
Jenkins, P. F., 115, 118, 122
Johnson, R. E., 6
Johnston, R. F., 65
Jones, P. J., 94

Karlin, A. A., 160
Kimura, M., 286, 288
King, J. R., 81
King, W. B., 288
Kirkpatrick, M., 148, 236
Klimkiewicz, M. K., 63, 64
Klomp, H., 111
Knox, C. W., 190
Koenig, W. D., 67, 102, 106, 287
Konishi, M., 117, 118, 138, 139
Kramer, P., 275
Krebs, C. J., 63, 102
Krebs, J. R., 115, 118, 128
Kroodsma, D. E., 115, 118, 122, 130, 138, 139, 162
Kunkel, P., 70

Lack, D., 4, 5, 63, 70, 107, 110, 111, 113, 140, 150, 203, 254, 265, 274
Lacy, R. C., 142
Lambert, W. V., 190
Lambrechts, M., 139, 148
Lande, 11, 131, 142, 148, 152, 155, 208, 209, 227, 229, 232, 236, 255, 279, 280, 286, 287, 291, 303
Laskey, A., 115, 138
Lessells, C. M., 93, 176
Lewontin, R. C., 290
Lill, A., 63–65, 112
Lints, F. A., 232
Loery, G., 67
Loeschcke, V., 236
Lott, D., 115, 162
Lundberg, A., 148, 182, 185
Lyon, B. E., 171

MacArthur, R. H., 110
McCleery, R. H., 92, 104–6, 108, 109, 164, 168
McGregor, P. K., 115, 118, 122, 138, 139, 148
McKinney, F., 171
McLaren, I. A., 5, 6, 92
Maclean, G. L., 63, 70
Manly, B. F. J., 209

Manning, J. T., 158
Marchant, S., 66, 70, 71, 106, 107
Marler, P., 115, 117, 118, 122, 123, 138, 139, 141
Maynard Smith, J., 148, 197, 207, 235, 236, 248
Mayr, E., 235, 290
Medawar, P. G., 67
Miller, A. H., 5, 6
Miller, R. S., 57, 63, 66, 67
Millington, S. J., 73, 84, 85, 92, 148, 154, 156
Mitchell-Olds, T., 209
Mock, D. W., 150, 171
Møller, A. P., 148, 160, 161
Montgomerie, R. D., 171
Moreau, R. E., 110, 113
Morton, E. S., 117, 138
Morton, M. L., 161, 162
Mousseau, T. A., 182
Mug, G., 148
Mumme, R. L., 67, 102, 106, 287
Mundinger, P. C., 115
Murphy, G. I., 110
Murray, B. G., Jr., 5

Nagata, H., 148
Newton, I., 81, 109
Nice, M. M., 65, 66
Nicolai, J., 115, 118, 122, 142, 171
Nol, E., 91, 92
Nolan, V., Jr., 150
Norberg, U. M., 150
Norberg, R. Å., 150
North, E. W., 115, 138
Nottebohm, F., 115, 118, 138, 162
Nottebohm, M., 162
Nur, N., 101

O'Donald, P., 131, 148, 152, 236
Ord, J. K., 136
Orr, R. T., 47, 170

Parker, L. D., 115
Patton, J. L., 203, 254, 260
Payne, R. B., 65, 109, 115, 118, 122, 139, 163
Perrins, C. M., 46, 92, 94, 104–6, 108, 109, 164, 168
Peters, S. S., 115, 117, 118, 123, 139
Petrinovich, L., 122

Philander, S. G. H., 22
Pimm, S. L., 236
Pinowski, J., 65
Porter, D. M., 17, 25, 269
Power, D., 6
Pratt, H. D., 289
Price, T. D., 70, 80, 85, 92, 123, 148, 151, 152, 154, 156, 158, 171, 176, 182, 184, 185, 209, 224, 227, 230, 255, 274, 276
Price, T. See Price, T. D.
Pulliam, H. R., 64
Pyke, D. A., 51

Quinney, T. E., 84, 92

Racine, C. H., 265, 270
Rahn, H., 107
Ratcliffe, L. M., 67, 115, 128, 140, 143, 240, 247
Raven, P. H., 287
Read, B., 292
Reznick, D., 101
Rice, W. R., 236
Ricklefs, R. E., 56, 57, 63, 83, 102, 103, 110, 111
Rockwell, R. F., 287
Roff, D. A., 182
Rohwer, F. C., 171
Rohwer, S., 171
Rothstein, S. I., 107
Roughgarden, J., 275
Rowley, I., 168
Rubenstein, D. I., 101, 105, 111
Ruff, M. D., 61, 190

Schaffer, W. M., 111
Scharloo, W., 304, 305
Schemske, D. W., 142
Schluter, D., 73, 177, 182, 231, 255, 258, 265, 267, 269, 270, 271, 275
Searcy, W. A., 115, 117, 138, 141, 145, 158, 170
Seger, J., 236
Seibt, U., 171
Shaw, R. G., 209
Sherman, P. W., 142, 161
Sherry, T. W., 113
Short, L. L., 290
Sibley, C. G., 7
Simberloff, D. S., 269

Simmons, M. J., 142
Slatkin, M., 287
Slud, P., 113
Smith, D. G., 144
Smith, J. N. M., 5, 51, 91, 92, 101, 102, 108, 150, 176, 177, 182, 185, 265, 271
Smith, J. Maynard. See Maynard Smith, J.
Snow, D. W., 47, 63–65, 70, 111–13
Soulé, M., 232, 292
Steadman, D. W., 253
Stearns, S. C., 101, 110
Sternberg, H., 67
Stjernberg, T., 84
Stobo, W. T., 5, 6, 92
Stoddard, P. K., 117
Svenson, H. K., 107
Swarth, H. S., 190

Tamura, M., 115, 138
Tauber, C. A., 236
Tauber, M. J., 236
Taylor, R. H., 291
Templeton, A. R., 232, 237, 292
Terborgh, J. T., 276, 291
Thomas, C. S., 164, 168
Thompson, J. N., 151
Thompson, N. S., 115
Thorpe, W. H., 115, 138
Tompa, F. S., 65
Trivers, R. L., 171
Turelli, M., 11, 236, 248, 255

van Balen, J. H., 46, 95, 101, 108, 109, 112
van Noordwijk, A. J., 12, 93, 95, 101, 108, 109, 182, 185, 189, 304, 305
van Riper, C., III, 65, 107
van Tienderen, P. H., 304
Van Valen, L., 6, 228, 276
van Vegten, J. A., 289

Wade, M. J., 208
Walkington, D. L., 271
Wallin, L., 115
Waser, M. S., 122
Watt, W. B., 170
Weathers, W. W., 65
Weeden, J. S., 115
West-Eberhard, M. J., 236
Westneat, D. F., 159, 161
Wetton, J. H., 160

White, W. B., 22
Wickler, W., 171
Wiens, J. A., 63
Wiggins, I. L., 17, 25, 269
Williams, G. C., 67, 101, 159
Willis, E. O., 63–65, 67, 112, 113
Wilson, D. S., 236, 248
Wilson, E. O., 110
Wolf, L. L., 5
Woolfenden, G. E., 67, 102, 103, 106

Wright, S., 232, 248, 286, 287, 288, 303, 304
Wunderle, J. M., Jr., 165

Yang, S. Y., 203, 254, 260
Yasukawa, K., 115, 128, 138, 145, 170

Zack, R., 5, 182
Zebiak, S. E., 22
Zink, R. M., 5
Zuk, M., 148, 158, 170

Subject Index

The main entry (with page numbers) for a species will be found under the name of the genus (e.g. *Asio flammeus*). A listing (without page numbers) is also given under the specific name (e.g. *flammeus, Asio*) and under the English vernacular name (short-eared owl) as a means of cross-reference to the main entry.

Abutilon depauperatum, 31, 32
Accipiter nisus (sparrowhawk), 295
Acorn woodpecker. See *Melanerpes*
Adaptation, adaptive, 271, 277, 290–91; adaptive landscape and peaks, 234, 265–69, 272–76, 278, 285; adaptive radiation, 7, 254
Age: at adult size, 178–81, 184–88; of community, 38–41; of islands, 17, 254; estimating males', 47–49; influence on mate choice, 153, 154, 158, 159, 170–73; and plumage, 87–92, 99, 102, 126, 151, 281, 284, 295–97
Age-specific survival. See Survival
Age-structure of population, 61–63, 68, 133
Agelaius phoeniceus (red-winged blackbird), 144, 294
Aimophila mysticalis (bridled sparrow), 5, 293
Aimophila sumichrasti (cinnamon-tailed sparrow), 5, 293
Alleles, allelic variation. See Variation
Allometry, 64
Allopatric speciation. See Speciation
Amaranthus sclerantoides, 25
Amaranthus squamulatus, 25
Ammospiza caudata (sharp-tailed sparrow), 5, 293

Ammospiza lecontei (Le Conte's sparrow), 5, 293
Ammospiza maritima (seaside sparrow), 5, 293
amplexicaulis, Chamaesyce
Ancestral populations, 254
Animal community on Isla Genovesa, 34, 37–40
Anthus pratensis (meadow pipit), 109, 294
Aphelocoma coerulescens (scrub jay), 106, 109, 294
Arcturus lake, 17, 19, 21, 38; sediment cores from, 40
Aristida repens, 25
Aristida subspicata, 25, 270
Arthropods: beneath bark; in *Opuntia* pads; on leaves, 38, 210, 212, 216, 220, 260, 261, 264, 271, 285; response to climatic fluctuations and rainfall, 37, 38, 40, 70–76, 220, 280–82; counts of, 14, 40; influence on survival and reproduction of finches, 70–76, 86, 98, 231, 281, 282, 285
Ashmole's hypothesis, 111
Asio flammeus (short-eared owl), 12, 37, 39, 59–61, 86, 107, 295
Assortative mating. See Mating, assortative
Asymptotic size, age when reached, 178–81, 184–88

atricapillus, Parus
auriceps, Cyanoramphus

bailleui, Loxioides
Banding, 12, 16
Band-rumped storm petrel. See *Oceanodroma*
Bark stripping. *See* Feeding behavior and habits
Basal metabolic rate, 64, 65
Beak. *See* Bill
Bet-hedging theory, 110–13
Bicolored antbird. See *Gymnopithys*
Bicolor, Gymnopithys
Bigamy, bigamous matings, 149, 150, 171, 172
Bill color polymorphism, 189–96, 239, 249–52, 306
Bill dimensions: correlations between, 176, 188, 189; difference between congeners, 198, 206; difference between Genovesa and Española *conirostris*, 258, 259, 272, 273, 277, 278; differences between song types, 237–48, 251, 252, 284; evolutionary response to selection, 196, 229, 230, 282; foraging efficiency as a function of, 214–18, 282; in relation to feeding skills, survival, mating and reproductive success, 14, 151–58, 169, 172, 214–18, 223–27, 238, 239, 265–73, 281–85; heritabilities of, 161, 180–89; hybrids, 198-200, 206, 207, 288, 289; measurements of, 12, 175, 176, 180, 206; mechanical properties of, 214–17; predicting mean phenotypes from food supply, 265–78, 285; selection on, 207–12, 218, 220–34, 282, 288, 289; variation, 1–10, 174–77, 228, 279
Biochemical analysis, electrophoresis, 254; DNA analysis, 254
Black-capped chickadee. See *Parus*
Black-crowned night-heron. See *Nyctanassa*
Blue-black grassquit. See *Volatinia*
Blue-chaffinch. See *Fringilla*
Blue-winged warbler. See *Vermivora*
Body size: correlations, phenotypic and genetic, between traits, 176, 188, 189; growth, 178–81, 184–88; heritabilities,

181–84; hybrids, 206, 207; in relation to longevity, 63–65; and mating and reproductive success, 151, 152, 169, 172, 220–23, 281, 282; measurements, 175, 176, 180, 206; selection on, 207–12, 220–23, 232; and songtypes, 237; variation, 174–77, 279
Boerhaavia caribaea, 25
Boerhaavia erecta, 270
Booby. See *Sula*
Borneo, 65
Borreria ericaefolia, 25
Bottlenecks. *See* Genetic characteristics
Brachycereus nesioticus, 25, 27
Breeding. *See* Reproduction
Bridled sparrow. See *Aimophila*
Brood. *See* Clutch
Brood division, 79, 80
Bullfinch. See *Pyrrhula*
Bursera graveolens, 24–27, 31–35, 41, 70, 72, 75, 222, 260, 264, 265, 269, 270
Butorides sundevalli (lava heron), 37, 61, 295

Cacabus miersii, 25, 31–33, 260, 265, 270
Cactus. See *Opuntia helleri; Brachycereus nesioticus*
Cactus, pad-ripping. See *Opuntia helleri*
Cactus finch. See *Geospiza*
Camarhynchus parvulus (small tree finch), 203, 293
Camarhynchus psittacula (large tree finch), 264, 265, 267, 272, 285, 293
Cambium, 264
canaria, Serinus
Canary. See *Serinus*
canorus, Cuculus
cantans, Telespyza
Capture of birds, 12, 16
Cardinal. See *Cardinalis*
Cardinalis cardinalis (cardinal), 6, 294
Cardueline finch, 7
Carotenoid pigments, 190
Carpodacus erythrinus (scarlet rosefinch), 84, 294
Carpodacus mexicanus (house finch), 6, 294
Cassia sp., 25

Caterpillars, 14, 28, 40, 86, 87; decline in numbers, 73–76, 281; influence on breeding, 70, 72–76, 79, 98, 99, 281; in nestling crops, 75; response to rain, 37, 40, 70–76, 86, 281. *See also* Arthropods
caudata, Ammospiza
Cenchrus platyacanthus, 25, 270
Cerambycid beetles, 38
Certhidea olivacea (warbler finch), 28, 38, 73, 260–62, 264, 265, 293
Chaffinch. See *Fringilla*
Chamaesyce abdita, 25
Chamaesyce amplexicaulis, 25, 27, 31, 75, 260
Chamaesyce nummularia, 25
Chamaesyce recurva, 25, 31, 33
Chamaesyce viminea, 24–27, 31
Chatham island, 291
Chestnut-sided seedeater. See *Sporophila*
Chloridops kona (kona finch), 6, 289, 294
chrysoptera, Vermivora
Cinnamon-tailed sparrow. See *Aimophila*
Cistothorus palustris (marsh wren), 139, 294
Climate, 19–24; day length, 63; droughts, 22, 69, 281; El Niño, 22–24, 69, 281; fluctuations in, 14, 15, 19, 22, 23, 63, 280, 281; long-term stability of, 38–41, 287; rainfall, 23, 24; seasons, 19, 63, 280; temperature, 22
Climatic fluctuations and age-structure of population, 68; and arthropod production, 37, 38, 72, 76, 98, 280, 281; and brood division, 80; and food supply of finches, 14, 15, 65, 68, 72, 98, 229, 231, 280–82, 284; and onset of breeding, 70–72, 87, 88, 98, 154, 162, 281; and plant phenology, 27, 31–38, 72, 280–82; and survival and reproduction of finches, 14, 15, 58, 59, 65, 66, 68, 73, 80, 81, 84–88, 98, 99, 113, 114, 174, 209–12, 281, 282
Clutch: energy commitment to, 81; failures, 83, 90, 296; initiation, 70–72, 79; intervals between and overlap, 77–79, 88, 92–94, 99, 167, 168, 281; number per season, 78, 81–86, 88, 90–94, 99, 167, 281, 296; renesting, conflict with feeding fledglings, 79, 80;

repeatability and heritability, 92–96; size of, 81–84, 88–94, 99, 107, 112, 113, 167, 281, 295; synchrony to reduce cuckoldry, 161–63
Coccidea parasites, 61, 190
Cocos finch. See *Pinaroloxias*
Cocos island, 1, 113
Coefficients of variation. *See* Variation, coefficients of
coelebs, Fringilla
coerulescens, Aphelocoma
Cohorts: survivorship of, 43, 49, 52–59, 68; survival and reproduction, 103–6, 114; influence on age-structure of population, 61–63, 68, 133
Colonization, 17
Collared seedeater. See *Sporophila*
Collared warbling finch. See *Poospiza*
Community: age of, 38–40; ecology, 260–65; intact, 291; on Genovesa, 24–27, 34–38, 41, 260, 261, 265–66, 269–72; on Genovesa and Española, 269–72; on Marchena, 261–65
Comparisons with other species: bigamy, 150; bill color polymorphism, 190, 192, 194–96; clutch initiation, heritability and size, 84, 85, 95, 96, 107; coefficients of variation in annual survival, 112; coefficients of variation in bill traits, 5–7, 289; demography, 63–68, 104–14; Genovesa, Española, Marchena, food supply, niche, diet and bill dimensions, 261–67, 269–73, 285; hybridization and selection, 290, 291, 230–32, kin recognition, 142; longevity, survival and reproductive success, 106, 108–14; mating and reproductive features, 102, 104–9, 148, 154, 160, 162, 170; measurement variation due to wear, 176; morphology, 206; predation, 107, 108; selection, 184; song characteristics, 115, 117, 118, 123, 128, 138, 139, 141, 143, 162; survivorship, 57, 63–68, 112; tropic and temperate zone characteristics, 63–66, 68
Competition, 265, 273–76
Congeneric species, 3–5; songs of, 143, 144; morphology, 206; reproductive characteristics, 107
Conirostris populations, 254

Conservation, 279–92
Copulations, extra-pair, 137, 146, 159–63, 172, 173
Cordia lutea, 24, 25, 27, 31–34, 38, 41, 72, 212, 231, 232, 260, 261, 264, 267, 270, 273, 275
Cores of lake sediments, 38, 40, 287
Correlations between traits: genetic, 188, 189, 208, 253, 290; genotype-environment, 184–86; phenotypic, 176, 177, 188, 208, 209, 217
Coturnix japonica (Japanese quail), 142, 295
Courtship, 150, 153, 154, 157–59, 170–73, 224
crassirostris, Platyspiza
Crimson finch. See Rhodospingus
Croton scouleri, 24–27, 30–34, 41, 75, 151, 264, 265, 268–70
cruentus, Rhodospingus
Cryptocarpus pyriformis, 27, 72, 260
Cuckoldry: and experience, 137, 150, 170, 172, 173; reducing the risk through synchronous breeding, 161–63, 173
Cuculus canorus (European cuckoo), 115
Cultural inheritance. See Song learning
Cuscuta acuta, 25
Cuscuta gymnocarpa, 25
cyanea, Passerina
Cyanoramphus auriceps (yellow-crowned parakeet), 291, 294
Cyanoramphus novaezelandiae (red-crowned parakeet), 291, 294
Cyperus spp., 25, 31–33, 270
cystoides, Tribulus

dactylatra, Sula
Dacus neohumeralis, 290
Dacus tryoni, 290
Daphne, Isla. See Islands
Dark-eyed junco. See Junco
Darwin, Isla. See Islands
Darwin's Finches, closest relatives of, 1, 106
Day length, 70, 281
Delichon urbica (house martin), 108, 109, 294
Demographic analysis, 102–6
Demography: characteristics of song A and song B males, 298–302; comparison with other species, 63–68, 104–14;

reproduction, 69–99, 100–114; survival, 42–68, 100–114
Dendroica petechia (yellow warbler), 34, 39, 294
Density. See Population
depauperatum, Abutilon
Desmanthus virgatus, 270
Desertion of mate. See Mate changes
Desmodium canum, 25
Desmodium decumbens, 25
Desmodium procumbens, 25
Diets, 2, 210, 212–14, 218–23, 231, 232, 248, 260, 261, 264, 271, 282; comparison with finches on Marchena, 261–65; comparison with finches on Española, 270–73; differences between song types, 238, 239; key species in, 275, 276, 285; of nestlings, 75. See also Feeding behavior and habits; Food supply
difficilis, Geospiza
Disease, 59–61, 158, 190, 218, 280, 291
Distichlis spicata, 25
Dispersal: between islands, 201–3, 233, 279; by females who changed mates, 165; from natal to breeding site, 141–43, 146, 147, 251, 252, 284
DNA analyses, 254
domesticus, Passer
Dominance: hierarchies in relation to plumage and territory defense, 127, 128, 170, 171
Droughts. See Climate

Eastern meadowlark. See Sturnella
Ecology, ecological. See Community
Ecological isolation of song groups, 237–39, 241, 242
Ecuador, continental, 63, 66, 106–8, 110
Ecuadorian neorhynchus. See Neorhynchus
edmonstonei; Sesuvium
Eggs, energy devoted to, 81; incubation, 77, 99, 107, 113; laying, timing of, 70-73, 76, 87, 93–95, 98, 99; number of, 77, 81, 82, 85, 88–100, 107, 113, 295, 296; predation and failure, 83, 107, 113, 296. See also Clutch; Reproductive success
Effective population size. See Genetic
Eleusine indica, 25
El Junco Lake, 38

El Niño (Southern Oscillation). *See*
  Climate
Emberizine finches, 57, 63–65, 106
Endocrinological state, 71
Environmental change. *See* Climate
Eragrostis spp., 25, 29, 31–33, 270
erecta, Boerhaavia
Erythrina velutina, 24, 25
erythrinus, Carpodacus
Española, Isla. *See* Islands
Eurasian tree sparrows. *See Passer*
European cuckoo. See *Cuculus*
Evolvulus simplex, 25
Evolution, evolutionary: competition and
  replacement, 273–78; history, 253–60,
  285; longterm change, 274; short term
  change, 291; response to selection, 188,
  189, 196, 229, 230, 253, 255–60,
  289–91; transformation from one species
  to another, 255–60. *See also* Natural
  selection; Variation
Experience: and copulations and cuckoldry,
  137, 158–62; influence on food finding
  and foraging ability, 210–12, 282; and
  mate choice, 149, 151–59, 170–73,
  284; and reproductive success, 87–92,
  99, 102, 126, 137, 151, 158, 171, 174,
  281, 285, 295–97; and survival, 102,
  174; and territory retention and defense,
  126–28, 158–60
Extinctions, 280, 288

Family effect, 97, 246, 251, 252
fasciculatum, Panicum
Feeding, parental: of fledglings, 78–80,
  121, 122; of nestlings, 75–79
Feeding behavior and habits, 14; bark
  stripping, 2, 210–12, 214, 216, 221,
  222, 260, 261, 264, 265, 271, 272, 282;
  cactus pad ripping, 2, 151, 210–13,
  218, 221, 222, 238, 239, 260, 261, 264,
  271, 272, 282; caterpillar gleaning, 2,
  222, 260, 261, 264, 271; of
  magnirostris, 232, 260, 261, 272; nectar
  and pollen, 2, 73, 151, 212, 215, 222,
  260, 261, 264, 276; *Opuntia* arils, 2,
  151, 212, 222, 260, 261, 271; *Opuntia*
  fruits, 151, 238, 239; seeds, 2, 210–14,
  217, 222, 260, 261, 264, 270, 271, 285;
  in dry seasons and droughts, 2, 212–14,
  222, 238, 239, 260, 264, 270, 271,

212–14, 222, 223; in wet season, 2,
  222
Feeding differences: between Española and
  Genovesa, 270, 271, 273, 285; between
  Genovesa and Marchena, 261, 264, 265,
  269–73, 285; between seasons and
  years, 2, 210, 220; between sympatric
  congeners, 260, 261
Feeding skills: learning of, 122, 210–12,
  282; a function of bill dimensions, 212,
  214–18, 221–23, 238, 239, 271–73,
  282; influence on reproduction, 92, 285;
  influence on survival, 212, 222
Fernandina, Isla. *See* Islands
Fertile period (female): and cuckoldry,
  159, 172, 173; and intrusions into
  territories, 160, 172, 173; and mate
  guarding, 158, 159, 160, 172, 173; and
  song, 129, 130, 162; and synchronous
  breeding, 162, 163
Ficedula albicollis (collared flycatcher),
  109, 294
Ficedula hypoleuca (pied flycatcher), 67,
  112, 294
Fire ants, 7
Fish, 63
Fitness, 226, 227; absolute, 209;
  components of, 207, 208, 225; relative,
  209. *See also* Reproductive success
flammeus, Asio
Fledglings, feeding of, 79, 80, 121, 122;
  predation, 61; time spent in natal
  territory, 121, 122
Fledgling production, 82, 83, 85, 108,
  185, 297; lifetime, 98, 108, 109;
  influenced by age and experience, 90,
  91, 297; influenced by longevity, 100,
  101, 108; influenced by mate changes,
  167–69; influenced by rainfall and
  clutch size, 85, 86, 108; number per
  brood, 297
Floreana. *See* Islands
Florida Scrub Jay. *See Aphelocoma*
Flowers. *See* Plants, phenology
Food, food supply, 14, 27–34, and bill
  dimensions, 269–74, 278; on Cocos
  Island, 113; in droughts and dry season,
  212–14, 218, 220–23, 231, 260–62; in
  El Niño, 218, 220–23, 284; in
  Española, Marchena and Genovesa, 14,
  269–73; of *magnirostris*, 274;

Food (*continued*)
measurement of, 14; in nestling crops,
75; population density and bill shape,
265–73; on Santa Cruz, 290; seasonal
changes in, 210–18; influence on
survival and reproductive success, 58,
59, 65, 66, 68, 70–76, 77–80, 84, 92,
281, 282, 284, 285; and termination of
breeding, 73–76
Foraging. *See* Feeding behavior
formicivorus, Melanerpes
fortis, Geospiza
Fossils, 235
Fox Sparrows. See *Passerella*
Frequency-dependent mating advantage of
song types, 145. *See also* Song
Fringillinae; fringilline, 7, 64
Fringilla coelebs (chaffinch), 6, 294
Fringilla teydea (blue chaffinch), 6, 294
Fruits, production of. *See* Plants,
phenology
fuliginosa, Geospiza

Galápagos archipelago, 3; physical
environment, 17
Gardener, Isla. See Islands
Galápagos dove. See *Zenaida*
Galápagos mockingbird. See *Nesomimus*
Garúa, 19. *See also* Climate
Genetics; genetic characteristics: autosomal
inheritance, 190; bottlenecks, 232, 233;
correlations among traits, 188, 189;
diversity and mate changes, 170; genetic
drift, 232, 248, 303; effective population
size and neighborhood size, 142, 143,
146, 285–88, 303, 304; expression,
time-dependent, 194, 195; genetic
expression during ontogeny, 186–88;
genotype-environment interactions,
184–86; genotype-environment
interactions, 184–86; introgression of
alleles (*see* Hybrids; Heritability);
interdependence, 291; mutation, 197,
230; phylogenetic relationship, 254;
separation of species, 288; quality and
mate choice, 158, 159, 170–73, 284;
subdivision of population, 223, 239,
243–52, 284, 303, 304; transformation
of one species to another, 255–60,
273–78, 285

Genetic variation, 10–11, 16, 284–86,
292; continuously varying traits,
177–84, 195, 196; discontinuously
varying traits (bill color polymorphism),
189–96, 284; variance-covariance
matrix, 229, 230, 255, 277
Genealogical tree, 97
Genovesa, Isla. See Isla Genovesa
Geology. See Islands; Isla Genovesa,
physical features of
georgiana, Melospiza
Geospiza conirostris (large cactus finch),
Española, 3–5, 254, 255, 269, 271–78
Geospiza difficilis (sharp-beaked ground
finch): coefficients of variation, 3–5;
evolution and competition, 275; diet, 28,
260–62, 265; hybridization,
hybridization-selection balance, 198,
230, 231, 233, 234, 283, 287; prediction
of phenotype, 267, 269; reproductive
characteristics, 73, 107; songs, 138–40
Geospiza fortis (medium ground finch): bill
color polymorphism, 192, 194, 196;
coefficients of variation, 3–5, 7;
evolution and competition, 273, 274;
feeding characteristics, 264;
hybridization, 230, 233, 287, 290;
longevity, 63–68; maternal effects, 182;
misimprinting, 123; plumage, age and
experience, 154; reproductive
characteristics, 73, 80, 84, 85, 87, 88,
96, 106, 107; selection, 184, 224;
survival, 112
Geospiza magnirostris (large ground finch):
coefficients of variation, 3–7; evolution
and competition, 272–76; feeding
characteristics, 213, 260, 261, 264, 265;
hybridization, hybridization-selection
balance, 230, 231, 233, 234, 283, 285,
287, 290; predation, parasites, 59, 61;
prediction of phenotype, 267, 269;
reproductive characteristics, 73, 107;
songs, 138, 139
Geospiza scandens (cactus finch):
ancestor-descendant relationship, 254;
bill color polymorphism, 192, 194, 196;
coefficients of variation, 3–5; evolution,
competition, 273–75, 285; feeding
characteristics, 264–66; hybridization,
hybridization-selection balance, 230,
233, 287, 290; immigrants, 203;

longevity, survival, 56, 63–68, 112; misimprinting, 123; plumage, age and experience, 154; reproductive characteristics, 73, 80, 84, 85, 87; song, 143
Geospiza fuliginosa (small ground finch), 3–5, 73, 264, 269, 275, 276, 290, 293
glabra, Vallesia
Goat destruction, 256, 257, 270, 277, 278
Golden-winged warbler. See *Vermivora*
Gossypium, 34
Gough Island finch. See *Rowettia goughensis*
Grasses. See *Eragrostis* spp.
Gray-crowned rosy finch. See *Leucosticte*
Great tit. See *Parus*
Greater koa finch. See *Rhodacanthis*
graveolens, Bursera
Ground finches. See *Geospiza*
Growth curves, and growth of young, 178–81, 184–88
Guadalupe, 6
guttata, Taeniopygia
Gymnopithys bicolor (bicolored antbird), 65, 111, 112, 294

habeliana, Ipomoea
Habitat, 7, 13, 17–27, 30–34, 263; fragmentation, 279. *See also* Community; Plants
Hatching, 77, 81–83. *See also* Reproductive success
Hawaiian archipelago, 65
Hawaiian honeycreepers, 6, 7, 289
Heliotropium angiospermum, 25, 31–33, 270
Heliotropium curassavicum, 25, 270
Heritability, 12, 17, 161, 170, 172, 178–87, 195–97, 239, 283, 285; analysis, 178, 181–87; for detecting paternity, 161, 170; of clutch size, 92–96, 99
hispaniolensis, Poospiza
House martin. See *Delichon*
House sparrows. See *Passer*
howelli, Portulaca
Hybrids; hybridization: with immigrants, 201–3, 274, 283; effects on genetic and phenotypic variation, 198–200, 203–7, 228–31, 254, 286, 287, 289–91;

relative fitness of, 201; with residents, 197–201, 276, 282, 283; selection balance, 10–11, 16, 223, 228–31, 234, 278, 283, 284, 286–89, 290, 291; songs of, 200
hyemalis, Junco
Hylophylax naevioides (spotted antbird), 65, 111, 112, 294
hypoleuca, Ficedula

iliaca, Passerella, 5, 293
Immigration; immigrants, 201–3, 205. *See also* Hybrids
Imprinting: on phenotype, 152; on father's song, 118–23, 145–47; misimprinting on song, 123–25
Inbreeding: avoidance, 141–43, 147, 304; depression, 141, 142, 286–89, 304, 305
Incubation, 77, 107, 150
Indigo bunting. See *Passerina*
Individual recognition, 115, 128–31, 138, 139, 141, 143, 144, 146, 147
Inheritance. *See* Heritability
Insect larvae; pupae. *See* Arthropods
Interbreeding. *See* Hybridization
Introduction of animals and plants, 7, 10, 11, 16, 17, 24, 279
Introgression of alleles. *See* Hybrids
Ipomoea habeliana, 25, 31, 32, 34–36, 218, 260, 270
Ipomoea linearifolia, 25, 32, 33, 270
Ipswich sparrow. See *Passerculus*
Islands; Isla: Baltra, 3, 17; Champion, 3, 273, 274; Daphne, 3, 7, 14, 63, 67, 70, 87, 89, 143, 224; Darwin, 3, 7, 10, 275; Española, 3, 144, 235, 254, 256, 269–73, 276–78, 282, 285; Fernandina, 3, 7; Floreana, 3, 253, 274; Gardener, 3, 144, 235, 254, 256, 276–78, 282, 285; Genovesa (*see* Isla Genovesa); Marchena, 3, 17, 254, 261, 263–67, 272; Pinta, 3, 17, 73, 254, 272; Santiago, 3, 17; San Christóbal, 3, 38, 70; Santa Cruz, 3, 70, 143, 253, 272; Wolf, 3, 7, 275
Isla Genovesa, 3; age of, 17; animals on, 34–38; area of, 17; community, 38–41; elevation of, 17; physical features of, 17–21; vegetation on, 24–34
Isolation, 279

jacarina, Volatinia
Japanese quail. See *Corturnix*
japonica, Coturnix
juliflora, Prosopis
Junco hyemalis (Dark-eyed junco), 5, 6, 293
Junco phaeonotus (Mexican junco), 5, 293

Key species of plants, 275, 276, 285, 291
Kin-recognition system, 115, 138, 141–43, 146, 147, 283
kona, Chloridops

Lantana peduncularis, 25–28, 31–33, 270
Large ground finch. See *Geospiza*
Large tree finch. See *Camarhynchus*
Lava. *See* Isla Genovesa, physical features of
Lava heron. See *Butorides*
Laysan finch. See *Telespyza*
Learning: from experience, 126, 153, 154, 158, 159, 170, 172; of foraging techniques, 185, 210–12, 282; mate guarding skills, 158, 159; of song, 118–25
leucophrys, Zonotrichia
lecontei, Ammospiza
Le Conte's sparrow. See *Ammospiza*
Lesser goldfinch. See *Spinus*
Leucosticte griseonucha (gray-crowned rosy finch), 6, 294
Lifespan. *See* Longevity
Lifetime reproductive success. *See* Reproductive success
linearifolia, Ipomoea
Local settlement patterns. *See* Dispersal
Longevity, 51, 63–65, 284; and age-specific survival, 66–67; and body size, 64, 65; and latitudinal gradients, 110; and metabolism, 65; and song structure and song types, 143, 251, 298; survival and reproductive success, 63–65, 97–114, 280
Loxioides bailleu (Palila), 6, 294
lutea, Cordia
$L_x m_x$ curves (products of survival and reproduction in relation to age), 103–6

Madagascar turtle dove. See *Streptopelia*
Madeira, 6

magna, Sturnella
magnirostris, Geospiza
Magpie. See *Pica*
Manacus manacus (white-bearded manakin), 65, 111, 112, 294
Mangere islands, 291
mangle, Rhizophora
Mangrove. See *Rhizophora*
Marchena, Isla. *See* Islands
maritima, Ammospiza
Marsh wren. See *Cistothorus*
Masked booby. See *Sula*
Mate changes, 163–69, 172; causes of, 164, 165, 170; consequences for females, 167–69, 170, 172; consequences for males, 168, 169, 172; importance of location and age, 165, 166; influence on bill color morph ratio changes in nestlings, 193, 194; number of, 164; reproductive cost and benefits of, 167–69, 172; and song type, 132, 166
Mate choice; mating success, 15, 148–73; and bill and body dimensions, 151, 152, 172, 223, 224, 245; influence of preceding conditions, 223, 224, 226; the choosing sex, 150, 151; cues used in, 148, 151–59, 166, 169–73; experience and age, 148, 149, 151–59, 165, 166, 169–73, 245; female quality, 171, fitness consequences of, 156–58; genetic quality and relatedness, 143, 147, 148, 158, 159, 170, 172, 173; and onset of breeding, 154, 155; and parental care, 158, 170, 172, 173; and plumage, 148, 151–59, 166, 169–73; uncoupling of plumage and experience, 153–56, 166, 170; selection analyses on, 156, 157; song, 129, 131–38, 146–49, 172; song type and territory pattern, 132–38, 240–42, 245; territory size and quality, 148, 149, 151, 155, 156, 166
Mate desertion. *See* Mate changes
Mate guarding, 150; and experience, 158–60, 173; and extra-pair copulations, 160, 161, 173; and intrusions, 160, 170, 173; and female-fertile period, 158–60, 173; and plumage, 160, 173
Mating, assortative or random, 184, 236, 240, 241, 252, 284
Mating success, 15, 148–73

Mating system. *See* Monogamy; Bigamy

Maternal effects, 182. *See also* Parental care

mcleanani, Phaenostictus

Meadowlarks. See *Sturnella*

Meadow pipit. See *Anthus*

Measuring; measurements, 12, 13, 16; repeatability of, 175–77, 183, 243, 244

Medium ground finch. See *Geospiza*

Melanerpes formicivorus (Acorn woodpecker), 106, 295

melodia, Melospiza

Melospiza georgiana (swamp sparrow), 117, 139, 141, 293

Melospiza melodia (song sparrow), 5, 64, 65, 102, 108, 109, 117, 139, 141, 293

Mendelian inheritance, of bill color, 190–95

Mentzelia aspera, 25, 34

Merremia aegyptica, 25, 34

Metabolism, 64, 65

Mexican junco. See *Junco*

mexicanus, Carpodacus

miersii, Cacabus

Migration. *See* Hybrids; Immigration

minima, Rynchosia

Misimprinting of songs, 123–25

Mockingbird. See *Nesomimus*

Mollugo flavescens, 25

Molt, 47–49

Monogamy, 149, 150, 154, 171, 172

montanus, Passer

Morph ratios. *See* Bill color polymorphism

Morphology; morphological: characteristics of males and females, 174–77; characteristics of songtypes, 237–39; comparison with congeners, 198, 199, 206; distance and difference between species, 247, 255–59; variation (*see* Variation). *See also* Bill dimensions; Body Size

Mortality. *See* Survival

Mutation. *See* Genetic characteristics

mysticallis, Aimophila

naevioides, Hylophylax

Natural selection, 10–12, 15, 16, 197, 207–34; analyses of, 156, 157, 208, 209, 212, 221, 232; and bill color, 195;

and bill and body dimensions, 207, 209–12, 218, 220–23, 225, 226, 231–34, 282, 291; and components of fitness, 207–9; correlated traits, 155, 188, 189, 230, 282, 290, 291; directional, 210, 223, 227–30, 233, 234, 275, 284; disruptive, 207, 210, 223, 229, 244, 245, 252; during dry seasons and droughts, 209–12, 218, 220–23, 232, 234, 282; evolutionary response to, 196, 229, 230, 253, 282; episodes influenced by preceding conditions, 59, 223, 224, 226; in the first year, 184, 210–12, 282; fluctuating directional, 230; forces of, 255–58, 260, 278, 285, 290; on hybrids, 223–89; and hybridization balance, 10, 11, 16, 223, 228–31, 234, 278, 282–84, 286–91; and *magnirostris*, 231, 232, 288, 289, maintaining difference between species, 288, 289; and sexual selection, 223, 224, 227; stabilizing, 223, 227–30, 233, 276; and sympatric speciation, 235–48; and the transformation of one species to another, 253–55, 258–60, 273–78, 285; influence on variation, 10, 12, 207, 230–32, 277, 282, 283, 290, 291

Nectar: *Bursera*, 264; *Cordia*, 212, 260, 261; *Opuntia*, 73, 212, 260, 261; *Waltheria*, 28, 212, 231, 260–62, 264, 276

neglecta, Sturnella

Neighborhood size, 223–32, 251, 252, 284, 303–5

Nesospiza acunhae (Small Tristan finch), 6, 294

Nepotism, 141

Neptuna plena, 270

Nesomimus parvulus (Galápagos mockingbird), 12, 34, 37–39, 61, 73, 86, 107, 109, 112, 294

Nests: domed, 59, 60, 150, 158, 170; display, 150; nest failure, 59, 60, 83, 86, 90, 150, 296; sites, 29

Nesting success, 81–86, 98, 99; correlation between clutch size and onset of breeding, 93; for five geospizines, 107; influenced by experience, 87–92, 158, 159, 295, 296; individual variation, 92–98. *See also* Reproductive success

Nestlings: bill color polymorphism (*see* Bill color polymorphism); food demand and interbrood interval, 79; feeding, diet and starvation, 75–79, 86, 150; growth of, 178, 179, 185; measurements of, 12; number of, 82; nestling period, 77, 107; predation, 59–61, 83, 296; production of, 82, 83; comparison with other species, 107

nesioticus, Brachycereus

Niches: dry season, 222, 247, 248, 284, 285; comparison between related species and populations, 230, 260, 261, 266; on Marchena, 261–65; on Española and Gardner, 269–73, 276, 277; on Champion, 273; and prediction of mean phenotype, 265–69; comparison between song types, 237–39; evolution and competition, 273–76; multiple, 229

nisus, Accipiter

novaezelandiae, Cyanoramphus

Nutritional state: determined by bill shape, 223–24

Nyctanassa violacea (black-crowned night heron), 37, 61, 295

Ocean influence on climate, 22

Oceanodroma tethys (wedge-rumped storm petrel), 30, 37, 295

Oceanodroma castro (band-rumped storm petrel), 30, 37, 295

olivacea, Certhidea

Ontogeny, genetic variation during, 186–89

Opuntia helleri, 14, 24–34, 36–38, 41, 66, 128, 212, 213, 215, 260, 261, 264, 265, 269–72, 277, 282, 285, 291; arils, 75, 212–13, 282; destruction of, 34, 60, 256–57, 282, 285; distribution of, 27, 135, 151; flowers, 14, 74, 75, 151, 215, 226, 234, 238, 239, 260, 212, 261; flowers, decline of, 218–20, 234, 282; foraging on, 215, 260, 261; fruits, 14, 212, 214, 220; fruits, decline of, 218–20, 234, 282; nectar and pollen, 73, 75, 212, 219; pad ripping, 210–13, 220–22, 282, 260; stigma snipping, 75, 219, 221; seeds, 75, 212–14, 217, 231, 272, 282; and territories, 135, 151

O'u. See *Psittarostra psittacea*

Owl pellets, 59

Ocellated antbird. See *Phaenostictus mcleanani*

Paleoecological analysis, 38–41

Palila. See *Loxioides*

palmeri, Rhodacanthis

palustris, Cistothorus

Panama, 65

Panicum fasciculatum, 270

Parakeet, red-crowned. See *Cyanoramphus*

Parakeet, yellow-crowned. See *Cyanoramphus*

Parasites, 61, 158, 218, 285, 290

Parental care, 83, 91, 121, 122, 150, 151, 158, 159, 170–73, 224, 284; feeding of young, 77–80, 122

Parent-offspring resemblance. *See* Heritability

Parrot-billed seedeater. See *Sporophila*

Parus atricapillus (black-capped chickadee), 67, 294

Parus major (great tit), 95, 96, 105, 106, 108, 109, 112, 139, 294

parvulus, Camarhynchus

Passer montanus (Eurasian tree sparrow), 65, 294

Passer domesticus (house sparrow), 65, 294

Passerculus sandwichensis princeps (Ipswich sparrow), 6, 293

Passerculus sandwichensis sandwichensis (savannah sparrow), 5, 293

Passerella iliaca (fox sparrow), 5, 293

Passerina cyanea (indigo bunting), 65, 109, 293

Paternity, detection of, 160, 161, 163. *See also* Copulations, extra-pair

pedunculata, Lantana

petechia, Dendroica

phaeonotus, Junco

Phaenostictus mcleanani (ocellated antbird), 65, 112, 294

Phenotypes; phenotypic: correlations (*see under* Correlations); environmental determined optimum of, 265–78, 285; preference in mate choice, 151, 152, 172, 223, 224; variation (*see under* Variation)

phoeniceus, Agelaius

Physiological condition, 71

Pica pica (magpie), 102, 294

picturata, Streptopelia
Pied flycatcher. See *Ficedula*
Pinaroloxia inornata (Cocos finch), 1, 113, 293
Pink morph. *See* Bill color polymorphism
Pinta, Isla. *See* Islands
pinus, Vermivora
Plants; vegetation; habitats: age of community, 38–40; community on Genovesa, 24–35, 38–40, 270–71; on Española, 270–71; introduced, 7, 24; key species, 275, 276, 285, 291; on Marchena, 261, 263, 265; items in nestling crops, 75; phenology, 27–34; pollen profiles from Acturus Lake, 40; quadrats, 14, 32, 33, 270; response to fluctuations in rainfall, 27, 31, 32, 34–37, 41, 280–82
platyacanthus, Cenchrus
Platyspiza crassirostris (vegetarian finch), 264, 265, 268, 293
Playback experiments. *See* Song
Plumage: rate of acquisition of adult plumage, 46–49; categories in relation to age, 46–49; of congeners, 140; cost and benefits of delay in acquiring adult plumage, 170, 171; and cuckoldry, 37, 170; and extra-pair copulations, 159–61, 170–73; and experience, the influence on mate choice, 137, 151–59, 170–73, 284; and mate guarding, response to intruders and territorial defense, 127, 128, 137, 160, 170–73
Pollen: as food, 2, 73, 151, 212, 215, 222, 260, 261, 264, 276; spores in lake cores, 38–41
Polygamy, 149, 150, 171, 172
Polymorphic traits. *See* Bill color polymorphism of nestlings
Polypodium insularum, 25
Poospiza hispaniolensis (Collared warbling finch), 294
Population; population size; density: age structure, 61–63, 68, 133; ancestral, 254; influenced by bill shape, food supply and preceding conditions, 58, 59, 68, 223, 267, 269; effective (*see under* Genetics); fluctuations in numbers, 58, 59, 68, 223, 227, 228, 231, 232, 303; recruitment into (*see also* Recruits), 86, 87, 96–98, 108, 109, 280; small

populations, variation in, 232, 233, 286–88
Population subdivision, 235–52, 283, 284, 303, 305; ecological, 235–39, 252, 283; genetic, 239, 248–52; morphological, 237–39; spatial heterogeneity in allele frequency, 232, 251, 252, 284, 303–5; song types, 242–46
portulacastrum, Trianthema
Portulaca howelli, 25, 270
Pox, avian, 61
pratensis, Anthus
Predators; predation, 37, 83, 107, 108, 113, 185; comparison with other species, 107, 108
princeps, Passerculus sandwichensis
Prosopis juliflora, 269–71, 277
psaltria, Spinus
psittacea, Psittarostra
psittacula, Camarhynchus
Psittarostra psittacea (O'u), 6, 294
pyriformis, Cryptocarpus
Pyrrhula pyrrhula (Bullfinch), 6, 84, 118, 142, 289, 294

Quail, Japanese. See *Coturnix*
Quadrats. *See* Plants

Rainfall. *See* Climate
R- and k-selection theory, 110–13
Recruits; recruitment into breeding population, 43, 69, 86, 87, 96–98, 108, 109, 280; influenced by experience and age of parents, 91; maximum per female, 109; correlated with number of fledglings, 227; and longevity, 100, 101, 108, 109; of song A, song B males, 298; individual variation in, 92–99, 109
Red-crowned parakeet. See *Cyanoramphus*
Red-footed booby. See *Sula*
Red-winged blackbird. See *Agelaius*
Rhizophora mangle, 24, 25, 27
Rhynchosia minima, 25, 264
Repeatability: analysis of measurements, 175–77, 183, 243, 244; of intervals between clutches and clutch sizes, 93–96
Repertoires, song. *See* Song
Reproduction; reproductive: influence of age and experience, 87–92, 99, 102, 126, 151, 154–59, 295–97; amount of,

Reproduction (*continued*)
80–89; annual variation in, 69–99, 295–97; and climatic fluctuations and food supply, 69–80, 281; cost of, 101, 102; costs and benefits of mate changes, 167–69; demography, 69–99; effort, 83; feeding of fledglings, 79, 80; onset of, 70–73, 87, 88, 93–94, 151, 154–55; phenology, 69; rate of, 69, 76–80, 88; recruitment, 69, 86, 87, 96, 98; retention of territories and onset, 126; subdivision of population, 239–42; and survival, 100–114; synchrony to reduce cuckoldry, 161–63; timing of, 70–76; termination of, 73–76

Reproductive success: influenced by age and experience, 89–92, 151, 171, 174, 281, 284, 285, 295–97; and bill dimensions, 174, 225–27, 281, 285; comparison between species, 104–11; influenced by clutch size and onset of breeding, 93; life-time, 96–99, 226, 227; longevity and survival, 97, 100–114, 174, 209, 227, 280; longevity and latitudinal variation, 110; maximum number of recruits, 109; percent producing recruits, 109; monogamy and bigamy, 105; in relation to potential, 83; of song types, 299–302; tropical and temperate characteristics of, 281; value increases with age, 103–6; individual variation in, 92–99, 280

Retention of territories. *See* Territories
Rhodacanthis palmeri (greater Koa finch), 6, 289, 294
Rhodospingus cruentus (crimson finch), 107, 294
Rhynchosia minima, 265
rostrata picturata, Streptopelia
Rowettia goughensis (Gough Island finch), 6, 294

salviifolia, Sida
sandwichensis sandwichensis, Passerculus
San Christóbal, Isla. *See* Islands
Santa Cruz, Isla. *See* Islands
Santa Elena peninsula, 66, 107
Santiago, Isla. *See* Islands
Savannah sparrow. See *Passerculus*
scandens, Geospiza
Scarlet rosefinch. See *Carpodacus*

Scientific and vernacular names of birds, 293–95
Scrub jay, Florida. See *Aphelocoma scouleri, Croton*
Seaside sparrow. See *Ammospiza*
Seasons. *See* Climate
Seeds: foraging on (*see* Feeding behavior); size-hardness values, 33, 222, 270, 271; production, 27–34, 66. *See also* Plants
Selection. *See* Natural selection
Senility, 67, 89
Sensitive period for song learning. *See* Song learning
Serinus canaria (canary), 138, 162, 294
Sesuvium edmonstonei, 25, 27
Sex ratio, 149–52
Sexual selection; sexually selected traits, 12, 16, 156–58, 169–73, 223–24, 227. *See also* Mate choice
Sexual dimorphism, 204
Seychelles turtle dove. See *Streptopelia*
Sharp-beaked ground finch. See *Geospiza*
Sharp-tailed sparrow. See *Ammospiza*
Short-eared owl. See *Asio*
Sida salviifolia, 25, 31–33, 270
Small ground finch. See *Geospiza*
Small tree finch. See *Camarhynchus*
Small Tristan finch. See *Nesospiza acunhae*
Snakes, 107
Songs; song types: attenuation of, 138, 146; attraction and stimulation of females, 115, 125, 129–38, 145, 146, 283, 284; bilingual birds, 122–25; bill dimensions and bill color polymorphism, 237–48, 306; characteristics, 115–18; of congeners, 140, 141, 143, 144, 202; demographic characteristics of, 298–308; discrimination between, 128–31, 138, 139, 141, 143, 145–47; of hybrids and immigrants, 200–203; and feeding young, 122; frequency of and frequency-dependent advantage, 132, 243, 299–300; learning (transmission) of, 118–25, 142, 146, 147, 251; in mate choice and mating success, 131–38, 148, 158, 169, 172, 240, 241; misimprinting, 115, 123–25; origin and maintenance of, 143–46; playback experiments, 128, 129, 130, 131, 140, 146, 159; and population subdivision, 283, 306; rate changes

during breeding cycle, 129, 130, 154; repertoires, 115, 139, 141, 145, 147; reproductive success, 299, 300; spatial distribution of, 251, 252, 304, 305; species, individual and kin recognition of, 115, 138–43, 146, 147, 283; and synchronous breeding, 162, 163; and territory defense, 115, 125–29, 132–38, 145, 146, 283; and territory pattern, 122, 132–38, 146, 239–42, 244

Song sparrow. See *Melospiza*

Southern oscillation. *See* Climate, El Niño

Speciation, 279; allopatric, 235; evolutionary transformation, 255–60, 278, 285; sympatric, 15, 235–48, 252

Sphingid, 37

Spiders, 37

Spinus psaltria (lesser goldfinch), 6, 294

Sparrowhawk. See *Accipiter*

Sporobolus virginicus, 25

Sporophila (Neorhynchus) peruvianus (parrot-billed seedeater), 107, 293

Sporophila telasco (chestnut-sided seedeater), 107, 293

Sporophila torqueola (collared seedeater), 63, 107, 293

Spotted antbird. See *Hylophylax*

Starvation, 58, 68, 86, 174, 218

Stigma snipping. See *Opuntia*

Streptopelia picturata picturata (Madagascar turtle dove), 291, 295

Streptopelia picturata rostrata (Seychelles turtle dove), 291, 295

Study area and animal, choice of, 7, 16, 19, 44, 279

Study duration, 7, 14

Sturnella magna (Eastern meadowlark), 117, 141, 294

Sturnella neglecta (Western meadowlark), 117, 141, 294

Stylosanthes sympodiales, 25

subspicata, Aristida

Sula dactylatra (masked booby), 37, 295

Sula sula (red-footed booby), 37, 295

sumichrasti, Aimophila

sundevalli, Butorides

Survival; survivorship: determined by antecedent conditions, 59, 223; age-specific, 43–54, 61–63, 66–68; annual variation in, 65, 66, 68, 112, 174; influenced by bill and body

dimensions, 64, 65, 282; influence of brood number and clutch size, 87; of cohorts, 51–56, 68; comparison with other species, 57, 63–68; curve, 49, 56–58; demography of, 42–68; in dry seasons and droughts, 209–12, 218, 220–23, 282; influenced by climate, 57–61, 65, 66, 87, 287; influence of experience, 102; in first year, 185, 210–12; food supply and reproductive success, 42, 58, 59, 65, 66, 68, 69, 100–114, 174, 209, 280, 282; genetic quality, 158; influenced by learning, 210, 282; and longevity, 63–68; patterns of, 42–68; predation and disease, 59–61; and reproduction in relation to age ($1_x m_x$), 104; and recruitment of songtypes, 298; sex-specific, 54–56; tropical and temperate characteristics of, 281; variation in adult, 51, 110–13, 174, 220–23, 230; variation in juvenile, 110–13, 138, 174; coefficients of variations for, 66

Swamp sparrow. See *Melospiza*

Taeniopygia guttata (Zebra finch), 118, 139, 148, 294

Telespyza cantans (Laysan finch), 6, 65, 294

Territory; territories; territorial: changes, 126; establishment, 125; attraction of female to, 126–27, 283; defense of, 127–29, 154; importance of familiarity with, 127, 151, 165–66; intrusions into and response, 127, 128, 137, 158–60, 170, 172, 173; role in male choice, 148, 149, 151, 154–56, 165, 166, 169, 172; occupancy of, 126, 127; overlap, 139; pattern of, 134–37, 239–42, 244, 284; retention of, 126, 137, 141, 143, 146, 151, 165, 166, 251, 284; size and quality of, 151; in relation to song, 115, 122, 125–38, 146, 147

Termites, 38, 216, 261

teydea, Fringilla

Tiquilia fusca, 25

Topography. *See* Isla Genovesa, physical features of

torqueola, Sporophila

Trees, 24–27

Tree finches. See *Camarhynchus*

Tres Marías, 6
Trianthema portulacastrum, 25, 270
Trichoneura lindleyana, 25, 32, 33
Tribulus cistoides, 25, 270, 271
Trinidad, 65
Tropics; tropical, 65–66, 70, 281

urbica, Delichon

Vallesia glabra, 270
Variation; variance: in bill and body traits, 2–7, 174–96, 228–34, 279, 284, 289; in climate (*see* Climate); in clutch size, 83–85, 89; continuous, 177–84, 195, 196; discontinuous, bill color polymorphism, 189–96, 284; response to environmental changes, 280; in fledgling production, 86; in food supply (*see* Food supply); genetic, 10, 11, 16, 232, 235, 274, 279, 280, 284, 285, 292; genetic variance-covariance matrix, 229, 230, 255, 277; hybridization, selection and maintenance of, 1–12, 15, 16, 197–200, 203–7, 228–34, 276, 278, 282, 283, 288–91; phenotypic, 10, 11, 174–96, 206, 228, 232, 233, 253; influence of population size on, 232, 233, 285–92, 303–5; comparison between species and populations, 3–6, 66, 84, 230, 276–78, 289; in survival and reproduction success, 51–59, 61, 68, 92–99, 103–6, 109, 110, 112, 114; in territory quality, 171; due to wear, 176
Variation, coefficients of: for bill and body traits, 2–7, 230, 289; for clutch size, 84; for survival, 66
Vegetarian finch. See *Platyspiza*
Vegetation. *See* Plants

velutina, Erythrina
Vermivora chrysoptera (golden-winged warbler), 290, 294
Vermivora pinus (blue-winged warbler), 290, 294
Vernacular and scientific names of birds, 293–95
violacea, Nyctanassa
virgatus, Desmanthus
Volatinia jacarina (blue-black grassquit), 107, 294
V_x curves, 103–6; bimodality in, 106

Waltheria ovata, 25–28, 31, 32, 72, 212, 260, 261, 264, 265
Warbler finch. See *Certhidea*
Wedge-rumped storm petrel. See *Oceanodroma*
Western meadowlark. See *Sturnella*
White-bearded manakin. See *Manacus*
White-collared seedeater. See *Sporophila*
White-crowned sparrow. See *Zonotrichia*
Wolf, Isla. *See* Islands

Xanthophyll, 190

Yellow-crowned parakeet. See *Cyanoramphus*
Yellow pigment; carotenoids; xanthophyll, 190
Yellow warbler. See *Dendroica*

Zebra finch. See *Taeniopygia*
Zenaida galapagoensis (Galápagos dove), 34, 38, 73, 162, 295
Zonotrichia leucophrys (white-crowned sparrow), 5, 6, 57, 65, 66, 104, 106, 162, 293